The FOUR-COLOR THEOREM

and Basic GRAPH THEORY

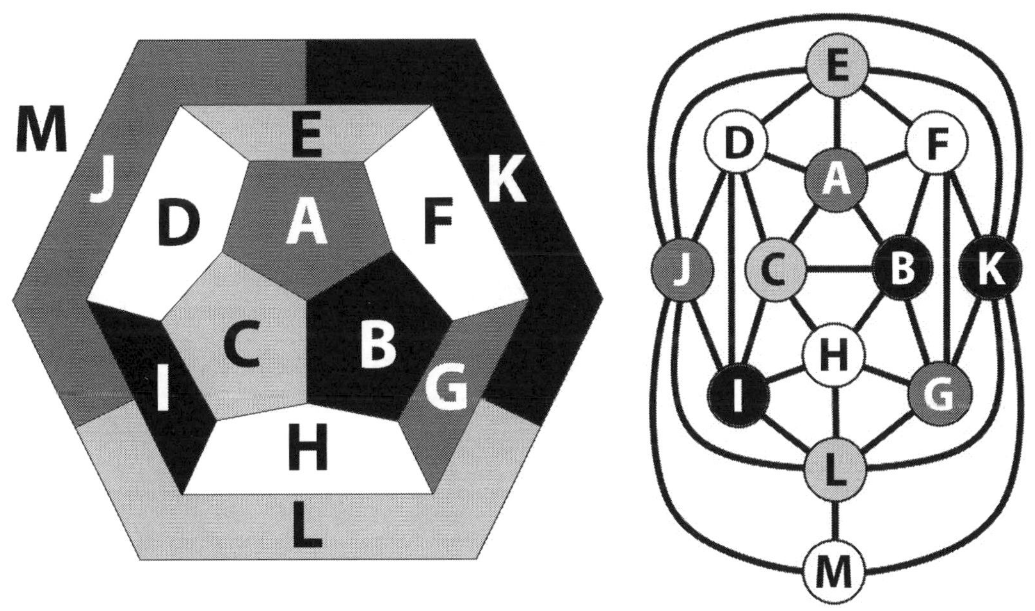

Chris McMullen, Ph.D.

The Four-Color Theorem and Basic Graph Theory

Chris McMullen, Ph.D.

www.improveyourmathfluency.com

www.monkeyphysicsblog.wordpress.com

www.chrismcmullen.com

Zishka Publishing

ISBN: 978-1-941691-09-0

Mathematics > Four-Color Theorem

Mathematics > Graph Theory

Contents

Introduction

This book will take you on a tour of the four-color theorem and related concepts from graph theory. Numerous illustrations are provided to help you visualize important ideas. Concepts are explained in clear, simple terms. No prior knowledge of graph theory is assumed.

Following is a sample of what you will find in this book:

- what the four-color theorem is
- a **novel explanation** for why the four-color theorem holds (Chapter 27)
- the reason for working with **graphs** instead of maps
- what **triangulation** is and the reason behind it
- visual examples of **Kempe chains** and Kempe's attempted proof
- the **three-edges theorem**: a **simplified approach** to the four-color theorem
- cool concepts like "**quadrilateral switching**" and "**vertex splitting**"
- the distinction between planar graphs and nonplanar graphs
- how to determine if a graph is a maximal planar graph or not
- **Euler's formula** and its relation to maximal planar graphs
- explanations of **Kuratowski's theorem** and **Wagner's theorem**
- complete graphs and complete bipartite graphs
- a survey of a few named graphs such as the Fritsch and Errera graphs
- how some maximal planar graphs can be trivially colored
- a **simple algorithm** for four-coloring a maximal planar graph (Chapters 22 and 24)
- counting how many ways a graph can be colored using no more than four colors
- comparing the coloring of a graph to a logic puzzle
- **Hamiltonian cycles** and polygon forms of maximal planar graphs
- what a **separating triangle** is and how to use it

May you enjoy this tour of the four-color theorem and basic graph theory.

1 Maps vs. Graphs

In a geography class, a **map** shows relationships between various regions, such as:

- the border surrounding each region

- the size and shape of each region

- which regions share borders with one another

- where one region is located relative to other regions

Notice that we're using the term **region**. The regions could be countries on a continent, but they could be states or provinces of a nation or they could be counties that make up a state. The term region allows for generic use. Of the features mentioned above, the only one that we will be concerned with in this book is "which regions share borders with one another."

In mathematics, especially as it relates to the four-color theorem (which we'll introduce in Chapter 2), maps and regions have slightly different meanings than they do in geography. For one, we won't allow a region to consist of two disjointed areas. For example, the United States wouldn't meet our definition of a region because Alaska and Hawaii are separated from the other 48 states. Another difference between math and geography is that we will require the regions of a map to be contiguous; there can't be gaps between the regions like lakes. A map doesn't need to show *real* places; we will imagine different ways maps can be drawn.

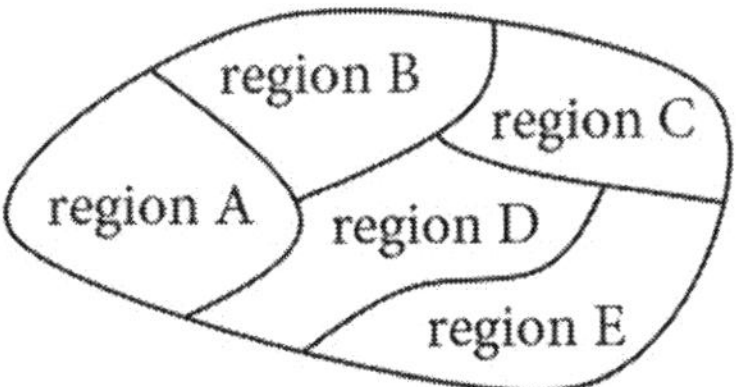

For a map, we will use the term **edge** to refer to any line or curve that separates one region from another region or any line or curve that separates an exterior region from the region outside of the map. Each edge begins at one vertex and ends at another vertex. Every line or curve on a map must adhere to this definition of an edge.

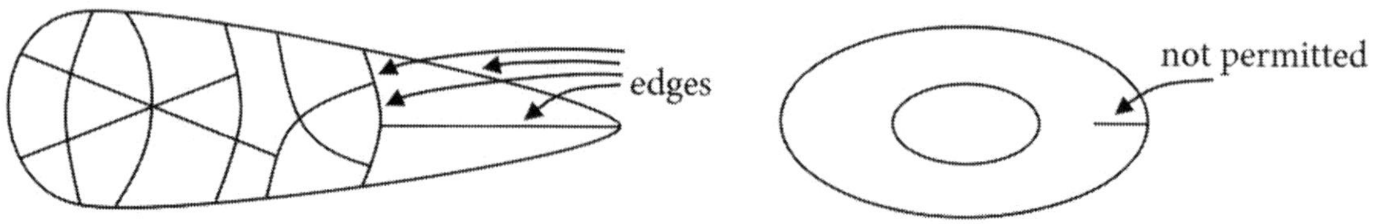

For a map, we will use the term **vertex** to refer to a point where three or more edges intersect. In plural form these points are called **vertices**. This definition is for a *map*. (When we learn about graphs, we will see that a vertex has a different definition for a *graph*, although it will still be a point where edges intersect.)

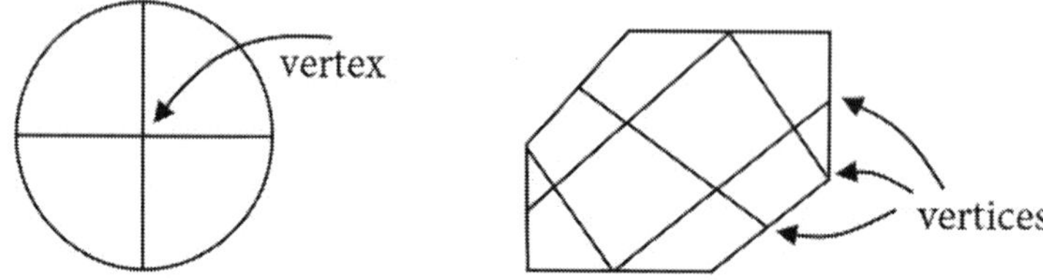

We will require each region of a map as well as the border that surrounds all of the regions to be a simple closed figure. A **closed** figure divides the plane into two distinct areas: the area inside the figure and the area outside the figure. In contrast, an **open** figure does not. By **simple**, we mean that the border of a *single* region doesn't cross itself like a figure eight; however, two *different* regions may join to form a figure eight. Some common examples of simple closed figures include circles, polygons, and ellipses, but the regions don't need to be common shapes; they just need to be simple closed figures.

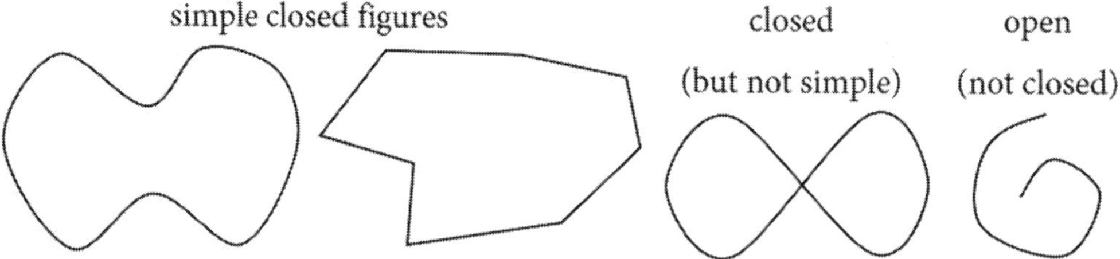

The only information that a map contains which is relevant to the four-color theorem (to be introduced in Chapter 2) is which regions share borders with which other regions. As far as this is concerned, it is simpler to redraw a *map* in the form of a **graph**. The distinction between a graph and a map is that a graph *only* shows which regions are neighbors; it doesn't show the size or shape of regions, relative locations, or other information that we won't need.

Any *map* can be represented by a **graph** as follows:

- For each region of the map, on the graph we will draw a small circle and place the corresponding label inside of it.
- For each pair of regions on the map that share an edge, on the graph we will connect these corresponding circles with a line or curve.

The example below shows a map on the left and its corresponding graph on the right.

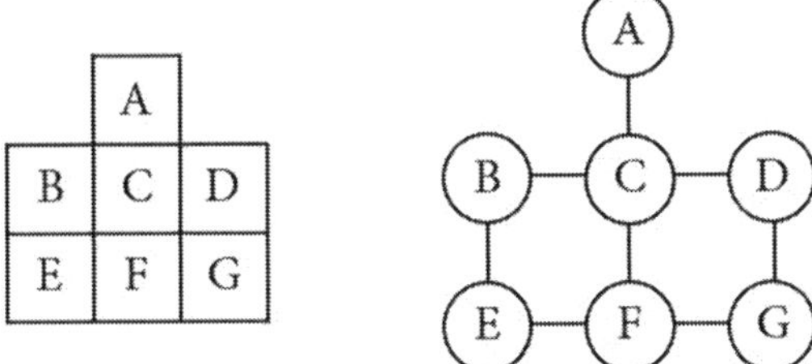

Notice how a graph corresponds to a map:

- On a map, each region is a face (a simple closed figure surrounded by edges). On a graph, each region is a **vertex** (a small circle where edges intersect).
- On a map, adjacent regions share an edge. On a graph, the edges indicate which pairs of regions share edges on the map. Any line or curve that connects two vertices on a graph is considered to be an **edge**.
- On a map, three or more edges intersect at a point called a vertex. On a graph, three or more edges surround a **face**.

The faces of a map become the vertices of a graph, while the vertices of a map become the faces of a graph. Since the roles of faces and vertices are swapped, the graph is sort of a dual representation of the map. (The solution to Problem 1 in Chapter 4 elaborates on this.)

The graph above makes it easy to see which regions of the map share edges:

- Region A shares an edge only with region C.
- Region B shares edges with regions C and E.
- Region C shares edges with regions A, B, D, and F.
- Region D shares edges with regions C and G.

- Region E shares edges with regions B and F.

- Region F shares edges with regions C, E, and G.

- Region G shares edges with regions D and F.

Note that a *vertex* on a *graph* (a small circle where edges intersect) corresponds to a *region* on a *map*. Also note that an edge on a graph represents two regions that share an edge on a map. It is essential to keep these two points in mind when analyzing a graph.

Why bother with graphs? Why not just work exclusively with maps? Here is why:

- A graph makes it visually easy to identify what really matters for our purposes; we can quickly tell which regions share edges with which other regions.

- A variety of different maps sometimes correspond to the exact same graph. Working with graphs instead of maps, we don't need to deal with as many types of diagrams.

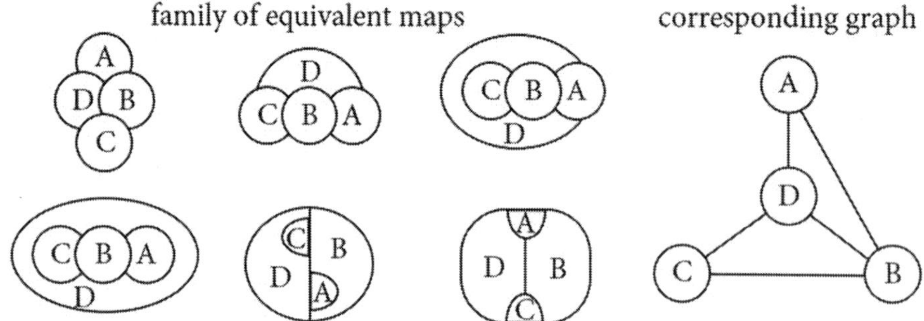

Note that the different maps shown above correspond to a *single* graph. In each case:

- Region A shares edges with regions B and D.

- Region B shares edges with regions A, C, and D.

- Region C shares edges with regions B and D.

- Region D shares edges with regions A, B, and C.

Since one graph can be equivalent to an entire family of maps with a variety of seemingly different structures, this makes it simpler to analyze graphs than to analyze maps. A graph contains all of the information that we need from a map as far as the four-color theorem is concerned; we just need to know which regions share edges with which other regions.

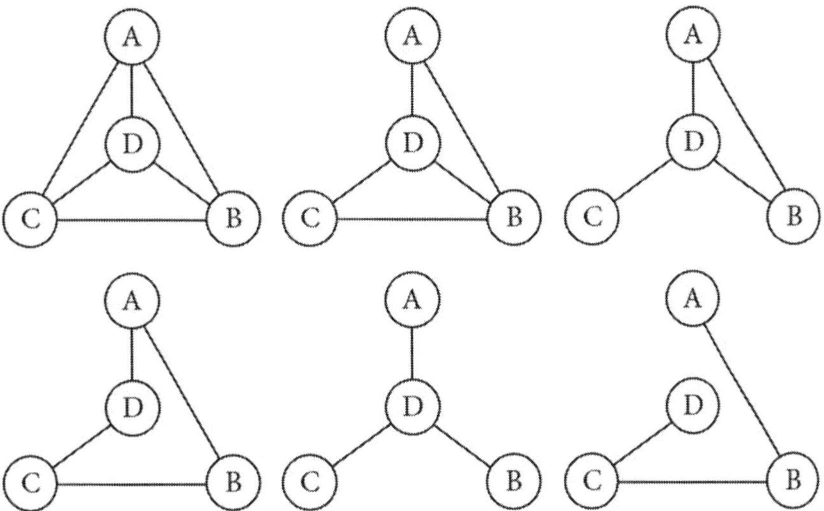

The diagram above shows six structurally different types of ways to connect four regions in a graph:

- Connect each region to all three other regions (top left).
- Connect two regions to all three other regions, but connect the other two regions to two other regions (top middle). If you remove *any* of the six edges from the top left graph, the result would be structurally equivalent to the top middle graph.
- Connect one region to all three other regions, connect two regions to two others, and connect one region only to one other region (top right). If you remove any pair of edges from the top left graph, the result would be structurally equivalent to the top right graph.
- Connect the four regions in a loop such that each region connects to two others (bottom left).
- Connect one region to all three other regions, but don't add any other connections (bottom middle). Regardless of which region connects to the other three regions, the result is structurally equivalent. (These last two graphs are called *trees*.)
- Connect the four regions in a chain that doesn't form a loop (bottom right). Regardless of which region you start with and end with, the result is structurally equivalent.

A graph simplifies the structure of a map without losing important information.

Note that it doesn't matter how the regions are organized, whether the edges are straight or curved, or how long the edges are when drawing or examining a graph. All that matters is which regions are connected to which other regions. All of the graphs below are effectively equivalent in the sense that the edges connect the same pairs of regions; in each case, region D connects to regions A, B, and C, without any other connections. The term **isomorphic** is used to describe two (or more) graphs where the same pairs of regions are connected.

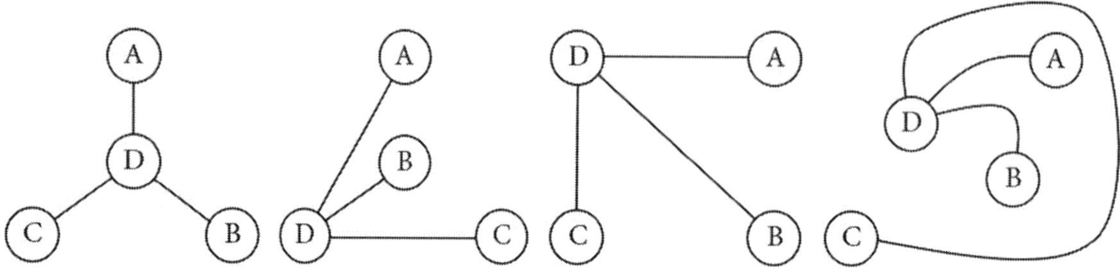

Although any map can be represented by a graph, not every graph can be represented by a map. It's possible to draw graphs that *can't* be represented by drawing a map in the plane. So when we study graphs, if we wish to ensure that the graphs actually correspond to maps that are possible to draw, we need to be able to determine whether or not a given graph can be represented by a map. One way to determine this has to do with the *crossing* of edges.

If it is possible to redraw a graph in the plane such that no edges cross (except at the vertices), then it is possible to draw a map in the plane that corresponds to the graph. If a crossing of edges in a graph can't be avoided, then it isn't possible to draw a corresponding map in the plane. The diagram below shows a case where the crossing of edges can be avoided.

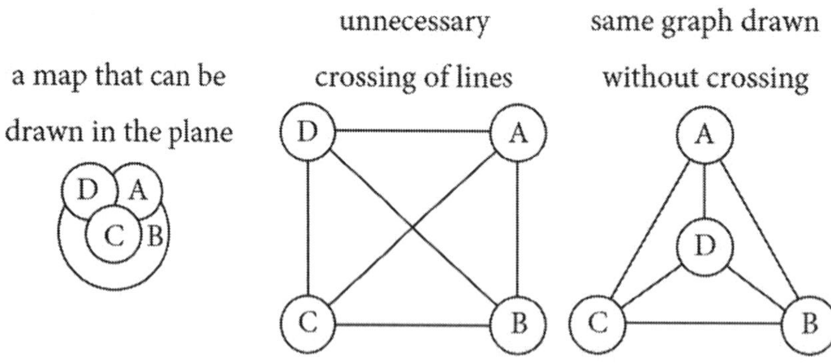

The previous example shows a map and corresponding graph of four regions where each region shares an edge with all three other regions. Although the edges cross in the graph in the middle, the graph to the right (which is isomorphic to the graph in the middle) shows that in this case the crossing of edges is avoidable.

It isn't possible to draw a map of five regions where each region shares an edge with all four other regions. You can draw such a graph, but it can only be done by crossing lines. In this case it isn't possible to draw a corresponding map in the plane.

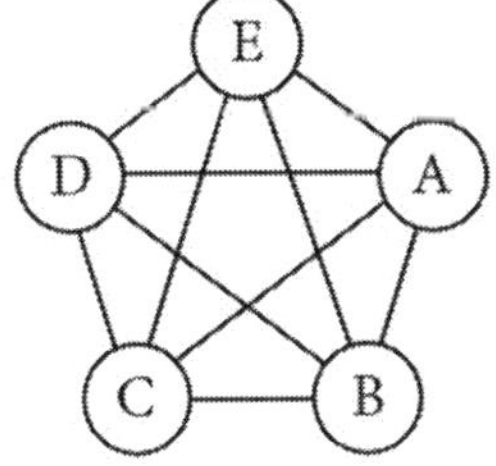

this graph can't be redrawn
without at least one crossing

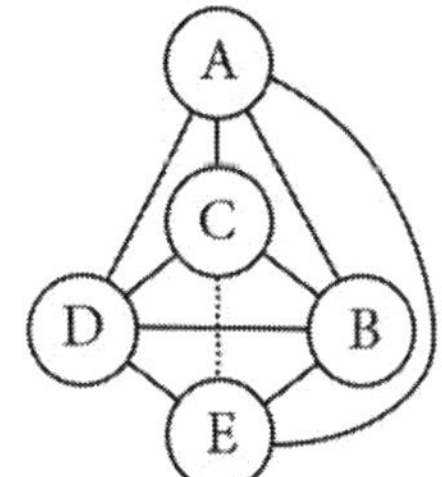

C and E can only be
joined by crossing lines

It is important to distinguish between avoidable and unavoidable crossings:

- A crossing is **<u>avoidable</u>** if the graph can be redrawn (by connecting the same pairs of vertices with edges, such that the new graph is isomorphic to the original graph) without any crossings. If all crossings are avoidable, the graph can be mapped in the plane.
- A crossing is **<u>unavoidable</u>** if the graph can't be redrawn without at least one crossing. If the crossing of edges can't be avoided, the graph can't be mapped in the plane.

We will use the term **<u>planar graph</u>** to refer to a graph where it is possible to draw the graph without crossings. We will use the abbreviation PG for a planar graph. PG's are graphs that can be mapped in the plane.

Beware that a PG may be drawn with an avoidable crossing. If you see a crossing in a graph, you must determine whether or not the crossing can be avoided to tell if it is a PG or not.

The **degree** of a vertex (which represents a region) equals the number of lines (or curves) connecting to the vertex. For example, a vertex of degree 3 has three lines connecting to it. Since each line (or curve) of a graph indicates two regions that share an edge on a map, the degree of a vertex tells you how many regions a given region shares an edge with. For example, a region of degree 3 shares an edge with three other regions.

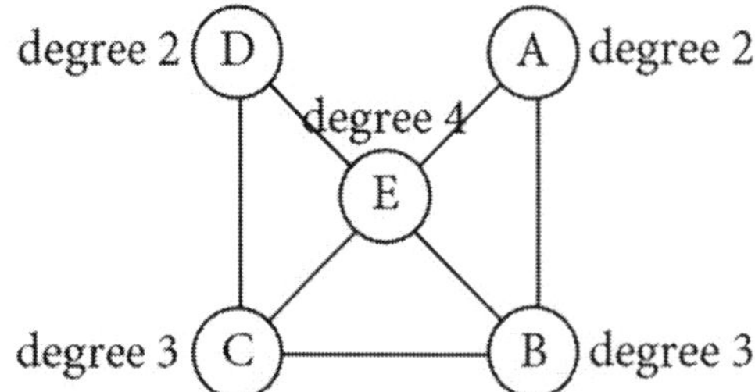

There are many different types of graphs. We will primarily be interested in the following types of graphs:

- Since PG's can be mapped in the plane and since the four-color theorem relates to coloring maps, we will be mainly concerned with PG's.

- We will only consider graphs that are undirected. (In contrast, a directed graph includes arrows indicating that edges only go one-way.)

- We will not draw any double edges. For example, if region A shares two different edges with region B in a map, we will only draw one edge to connect these regions in the corresponding graph. Two regions either connect or they don't; we won't care if they connect more than once in the map.

- We will not allow an edge that forms a loop joining a region to itself. That is, an edge can't leave one region and return to the same region. (Note that an edge begins at one region and ends at another region. We're saying that a single edge isn't permitted to connect the same region to itself.)

- We will not need to consider a graph that is disconnected. (If a graph is disconnected, this means that it may be separated into two or more separate graphs. We will treat them as individual connected graphs.)

Note that the surrounding unbounded area of any map may be considered as a region of its own. In the example below, region E represents the surrounding unbounded area.

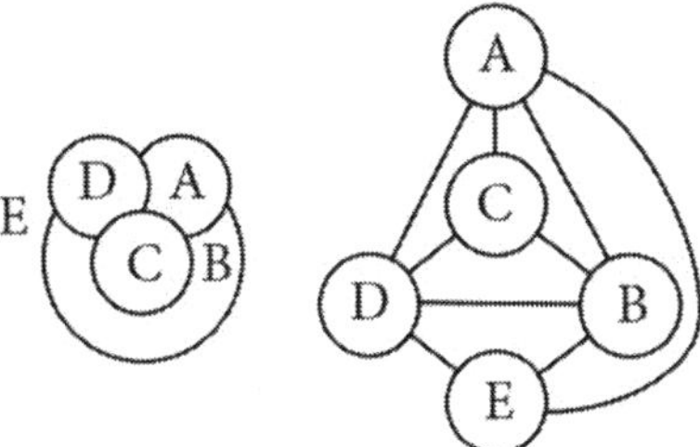

Looking at the map above, it's obvious that region E is the surrounding unbounded area. If you looked at the graph without seeing the map at all, how would you know that region A isn't the surrounding unbounded area? You wouldn't, as the picture below shows.

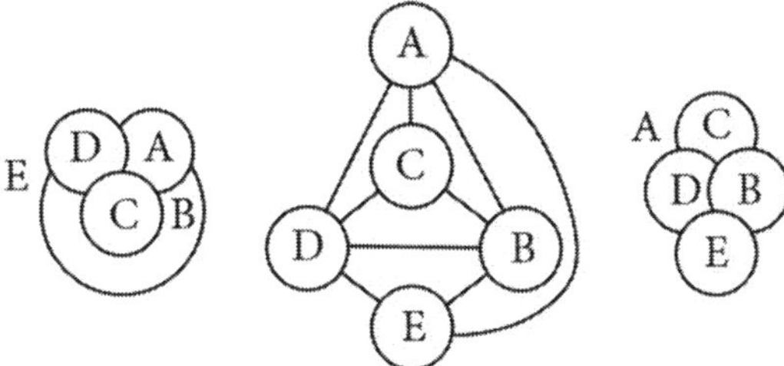

The map on the left interprets the graph with region E as the surrounding unbounded area, while the map on the right interprets the graph with region A as the surrounding unbounded area. The two maps are equivalent and are equivalent to the graph.

Let's go a step further. We can interpret the graph with *any* of its regions as corresponding to the surrounding unbounded area. The following map assigns the surrounding unbounded area to region C, which doesn't even appear to be on the "outside" of the graph.

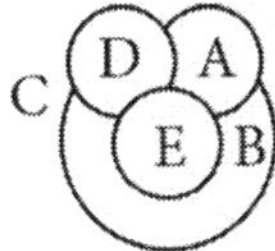

It's easy to verify that all of these maps are equivalent by looking at the "sharing":

- A connects to B, C, D, and E in each map (and the graph).
- B connects to A, C, D, and E in each map (and the graph).

- C connects to A, B, and D (but not E) in each map (and the graph).
- D connects to A, B, C, and E in each map (and the graph).
- E connects to A, B, and D (but not C) in each map (and the graph).

An interesting feature of a graph is that it can easily be inverted. In the two graphs below, we simply redrew the left graph to make D look like it is "outside" instead of "inside."

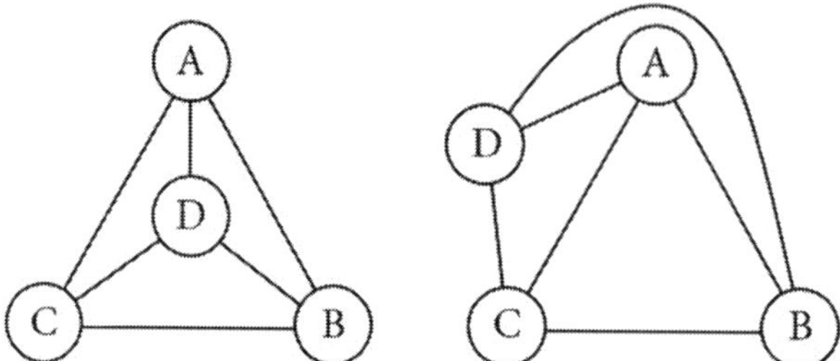

The two graphs shown previously are **<u>isomorphic</u>** in the sense that their edges connect the same pairs of vertices: AB, AC, AD, BC, BD, and CD. (Note that AB and BA are the same edge, for example; the order of the letters doesn't matter in an undirected graph.)

Any graph can similarly be inverted to move any point from "outside" to "inside" (or vice-versa) provided that all of the sharing between regions is the same before and after. For a graph, the concepts of "inside" and "outside" are not so obvious, and these notions can easily be interchanged by inverting the diagram. We will see additional examples of how to invert what appear to be inside and outside regions in Chapter 14, for example.

Style note: Some authors prefer to draw all graphs using only straight edges. In this book, many of our graphs will include one or more *curved* edges. This is purely a stylistic choice; it doesn't affect the concepts or results discussed in this text. In some cases, a curved edge may offer a little more space between edges, or it may bring to light a certain symmetry in a graph; but often, it was purely a stylistic choice. If you're accustomed to viewing graphs with only straight edges, note that there can sometimes be value in seeing things from a new perspective. Named graphs, like the Fritsch and Errera graphs of Chapter 8, can also be found online in case you may wish to see how they appear with only straight edges.

Chapter 1 Exercises

1. Draw and label a graph corresponding to the map below.

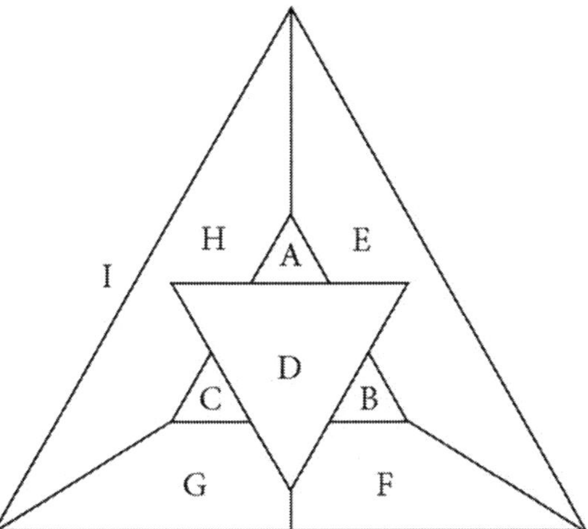

2. Draw and label a graph corresponding to the map below.

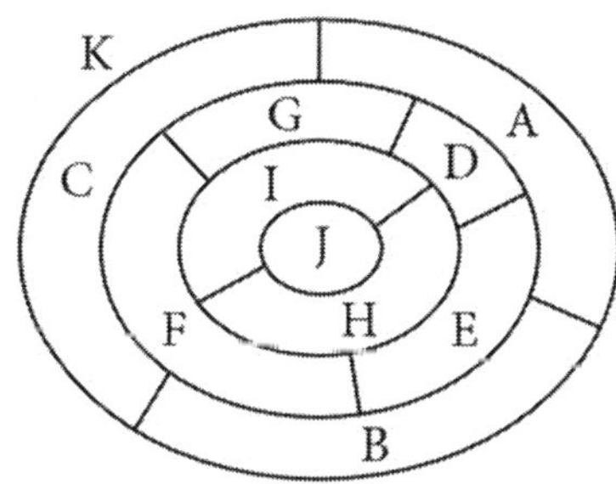

3. Draw and label at least two maps that look significantly different and which correspond to the graph below. Also determine the degree of each vertex.

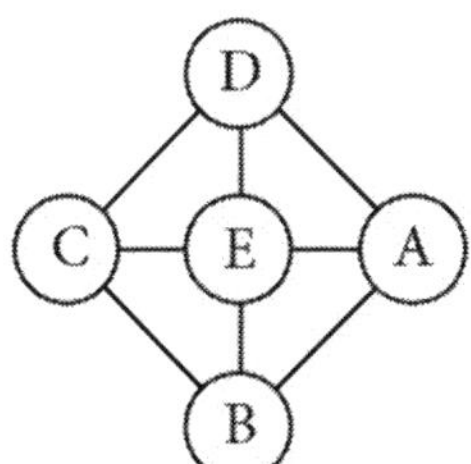

2 The Four-Color Theorem

According to the **four-color theorem**, the vertices of any PG may be colored using no more than four different colors (such as red, blue, green, and yellow) such that [Ref. 1]:

- No two vertices connected by an edge have the same color. (An edge connects two vertices and may be straight, curved, or bent.)
- Every vertex is colored. A single vertex can only be colored using a single color; a multi-colored vertex isn't allowed.
- The number of vertices is finite. The graph isn't an infinitely repeated design like a tessellation or fractal. (This particular assumption may not be necessary, but provides a simple starting point with which to approach the four-color theorem.)
- The graph is drawn in the plane or on the surface of a sphere (but other surfaces like a torus are not allowed). Chapter 14 illustrates the concept of the sphere.
- The graph is undirected (the edges don't have arrows). The graph isn't disconnected. No edge connects a region to itself (this is called a loop). There are no double edges.

Recall from Chapter 1 that PG stands for "planar graph." A PG is a graph that can be drawn in the plane without any crossings. PG's are special because they can be mapped in the plane.

Note that there are additional requirements for maps. For example, the regions of a map must be contiguous (there can't be any gaps or lakes between regions). Two regions of a map may be the same color if they meet only at a vertex (and not an edge). For a map, we can't allow regions to be disjointed (like the United States, for which Alaska and Hawaii are separated from the other 48 states).

As discussed in Chapter 1, a single graph may correspond to a multitude of different maps, which makes it simpler to analyze a graph. For this reason, this book will focus primarily on graphs from this point forward.

Coloring a graph is no different from coloring a map. Below, we colored both a map (lower figures) and its corresponding graph (upper figures). The numbers 1-4 represent four different colors (such as red, blue, green, and yellow). On the graph, no two regions connected by an edge have the same color. On the map, no two regions that share a border have the same color.

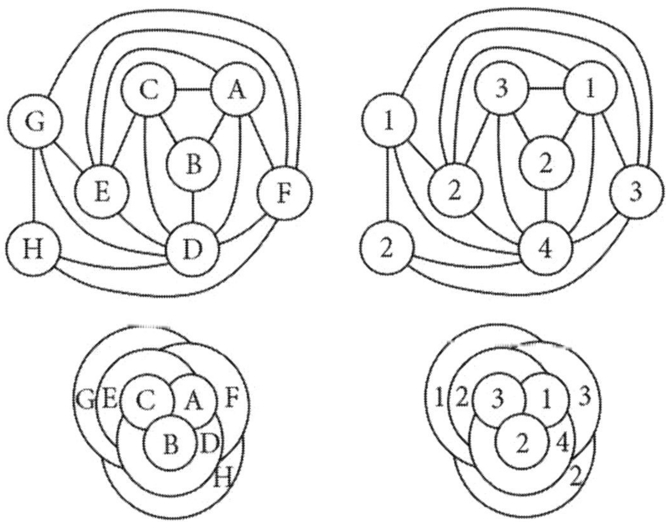

The map and graph on the left side are labeled with letters to help you see how the regions of the map and graph correspond. The numbers on the right side show how they are colored.

We encourage you to attempt to color the graph below using no more than four colors.

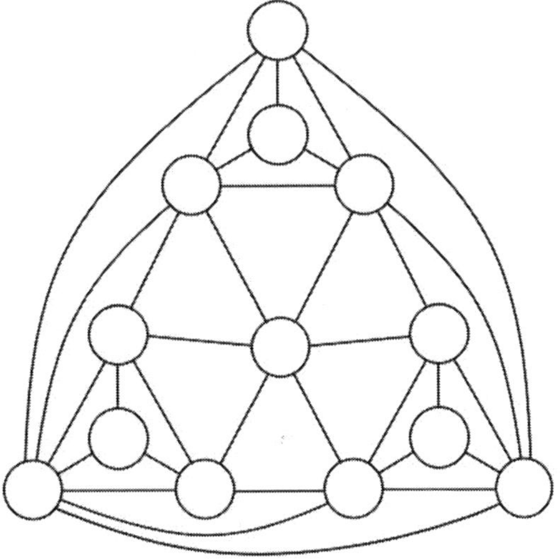

Note that it isn't necessary to actually use colors. You can simply write the numbers 1, 2, 3, and 4 in each region to represent four different colors (like red, green, blue, and yellow).

You can find the solution to the previous puzzle on the following page. Note that there is more than one possible solution. For example, you could get another solution by swapping all of the 1's and 3's, then you could get another solution by swapping all of the 2's and 3's of that graph, etc. (Yet there are additional solutions possible besides color swapping.)

After you attempt to color the graph (or any map or graph), check your answers carefully. It is really easy to make a mistake. The following approach can help you check your attempt to color a graph so that it satisfies the four-color theorem:

- Find all of the regions of the same color.
- Check these regions one pair at a time.
- For each pair of regions that are the same color, verify that no lines or curves connect these two regions.
- Repeat these steps for each of the remaining colors.

Once a math lover spends enough time drawing a variety of graphs (and perhaps maps), attempting to color the graphs, and attempting to disprove the four-color theorem, it often seems intuitive that a simple and convincing proof of the four-color theorem should exist. (You are highly encouraged to try these things; they are great ways to learn more about the four-color theorem.) However, once enough time is spent attempting to prove the four-color theorem by hand, a variety of challenges tend to become apparent:

- Even for a fairly small number of vertices like 20, there a great many ways to draw graphs that are structurally different, and it isn't easy to think of every possible way to draw the graph.
- As the number of vertices increases, the number of ways to draw graphs that differ in structure grows tremendously.

- Once a graph is drawn, it isn't always easy to tell at a glance whether or not it is a PG. That is, if there are crossings, it takes some effort to determine whether or not the crossings are avoidable. Recall that we are defining PG to include any graph that *can* be drawn in the plane without crossings (even if the graph happens to be drawn in a form that has avoidable crossings). Chapter 6 will discuss how to determine this.
- It is difficult to formulate a proof that covers every conceivable scenario.

Following is one possible solution to the previous puzzle:

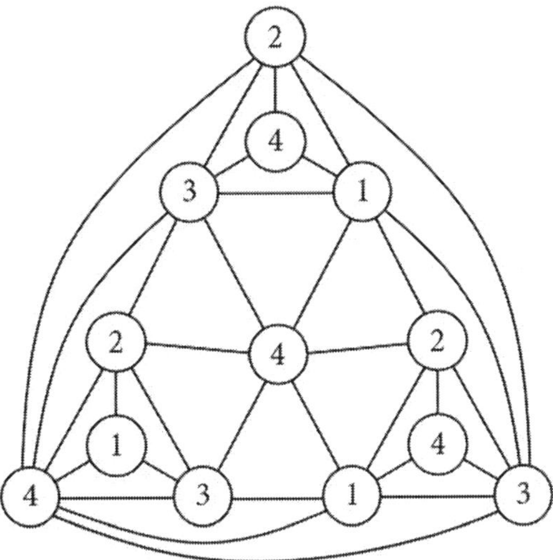

Check the solution:
- Find the three regions colored 1. Verify that no edge joins two of these 1's together.
- Find the three regions colored 2. Verify that no edge joins two of these 2's together.
- Find the three regions colored 3. Verify that no edge joins two of these 3's together.
- Find the four regions colored 4. Verify that no edge joins two of these 4's together.

As of the publication of this book, the only known proof of the four-color theorem involves computer calculations; the four-color theorem has yet to be proven by *hand* [Ref. 13]. We'll explore a variety of ways, including a few novel ideas, for approaching the problem of how one might go about proving the four-color theorem by hand throughout this book.

Chapter 2 Exercises

1. Color each region of the map below so that the coloring satisfies the four-color theorem. Treat the unbounded surrounding area (that is, the area outside of the map) as one of the regions.

2. Color each vertex of the graph below so that the coloring satisfies the four-color theorem.

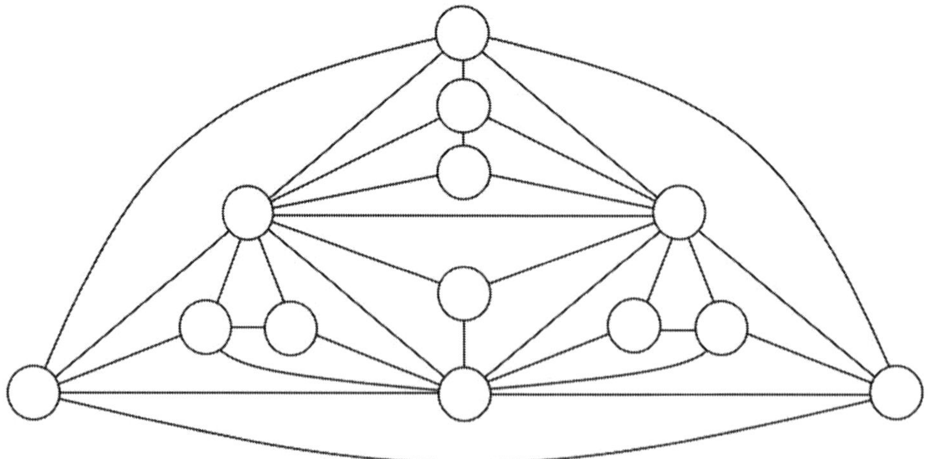

3. Color each vertex of the graph below so that the coloring satisfies the four-color theorem.

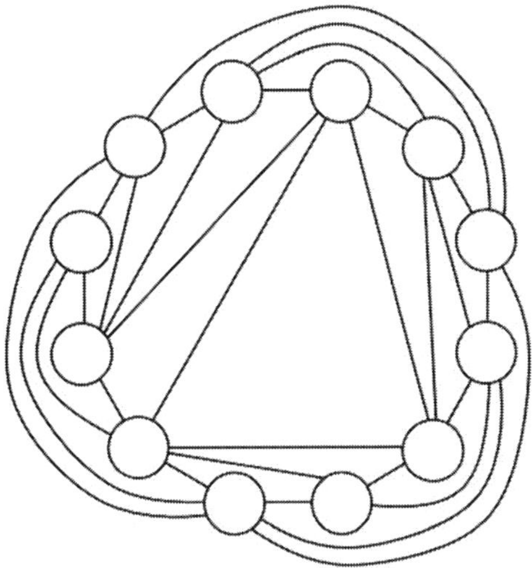

4. Color each vertex of the graph below so that the coloring satisfies the four-color theorem.

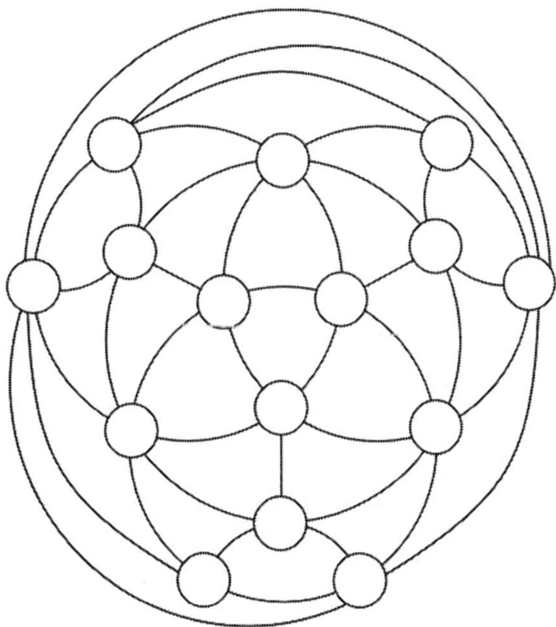

3 Triangulation

A graph is **<u>triangulated</u>** if every face is surrounded by three edges (which may be lines or curves), including the "face" that represents the infinite area "outside" of the graph.

- The left graph below isn't triangulated because ACDE is a quadrilateral (four-sided).
- The right graph below isn't triangulated because the infinite area "outside" of the graph has five sides (B, C, D, E, and F) instead of three.
- The center graph below is triangulated because every face, including the infinite area outside, has three sides. Its faces are ABC, ACE, AFE, ABF, BCD, CDE, DEF and BDF. Note that BDF is the infinite area outside (but recall from the end of Chapter 1 that any graph can be inverted to make any face correspond to the infinite area outside of the graph).

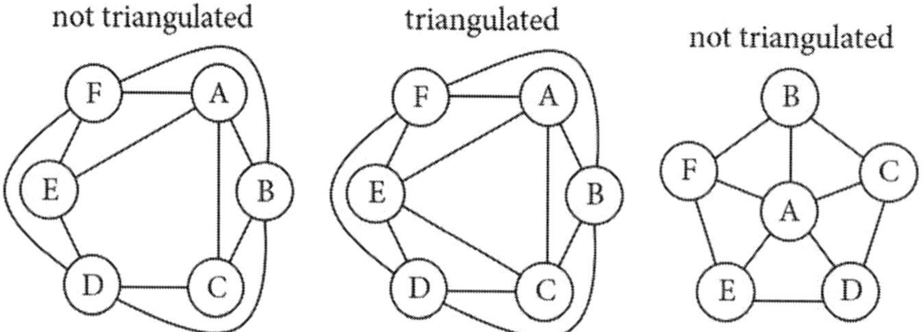

Any graph that isn't already triangulated can become triangulated by adding one or more edges to the existing graph. Consider the example below.

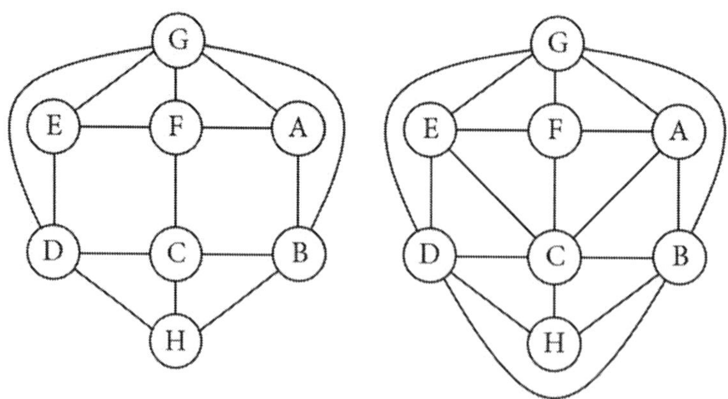

In the previous diagram, the left graph isn't triangulated because ABCF and CDEF each have four sides and because the infinite area outside BGDH also has four sides. If we add edges AC, CE, and BD (this one is curved), every face will be a triangle (including the infinite area outside, which is now BDG).

We will use the term **maximal planar graph** for any PG that has been triangulated in this sense (including the infinite area outside), and we will abbreviate this MPG. An alternative name that is also common is "triangulated graph." Since every face of a MPG is triangular (in a loose sense of the word, since any of its three edges may be curved), it may seem like triangulated graph would be the better choice. However, since the term triangulated graph is sometimes used with other meanings in mind, the term MPG is common in order to help avoid possible confusion. (We are abbreviating MPG since we will use this term frequently.)

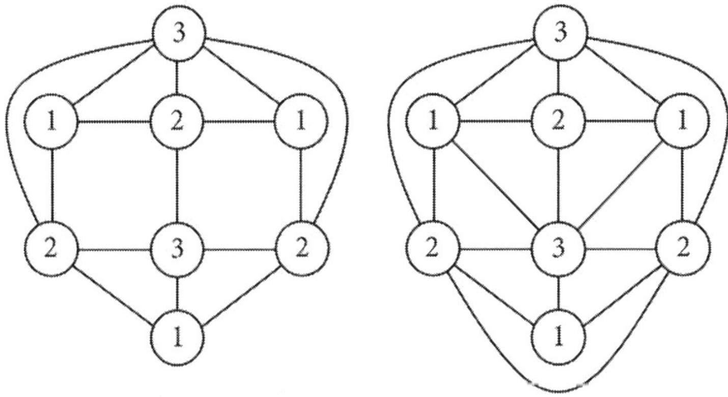

The following property makes it very useful to triangulate graphs to turn PG's into MPG's. If a graph is colored in such a way that it satisfies the four-color theorem, the same coloring will still satisfy the four-color theorem if one or more edges are removed from the graph. You can see that in the example above. We first colored the MPG on the right. (It turns out that this MPG can be colored using just three colors, but that is unimportant.) We obtained the graph on the left (which is a PG, not a MPG) by removing three edges from the MPG. You can see that the coloring from the MPG on the right still works for the PG on the left after removing the edges from the graph.

Recall from Chapter 1 that a PG is a graph that can be drawn in the plane without crossings. In contrast, a MPG is a special type of PG in that it is fully triangulated, including the infinite area outside.

Any MPG that is colored in such a way as to satisfy the four-color theorem will still satisfy the four-color theorem if any of its edges are removed from the graph. This important classic property of triangulation is the reason that most attempts to prove the four-color theorem only consider MPG's. If you can prove that the four-color theorem holds are all MPG's, you will have proven that it holds for all PG's.

MPG's are thus central to the four-color theorem. Note that a single MPG may be drawn more than one way, as illustrated below.

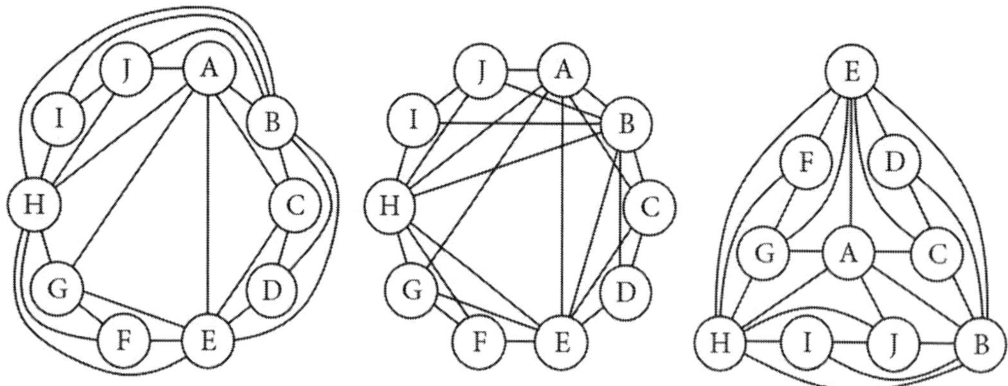

The three MPG's shown above are all isomorphic. The form of the graph in the middle has crossings, but the forms of the graph on the left and right show that these crossings are avoidable. (Recall from Chapter 1 that graphs are isomorphic if they are structurally equivalent in terms of edge sharing.)

The following diagrams show that two MPG's can have the same number of vertices, yet be structurally different. The MPG on the left has two vertices of degree 5, two vertices of degree 4, and two vertices of degree 3, whereas the MPG on the right has six vertices of degree 4.

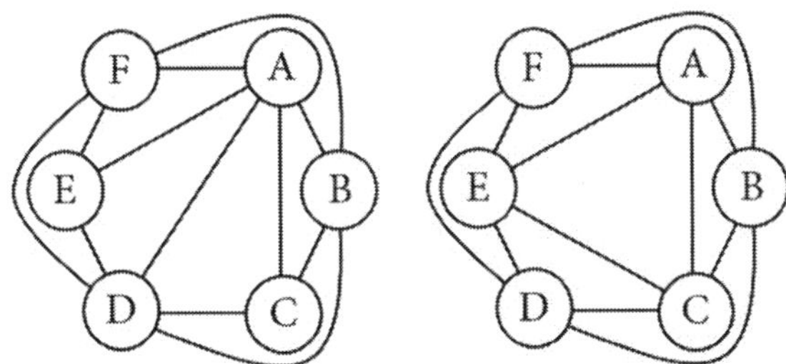

As the number of vertices increases, the number of MPG's with structural differences increases severalfold. For example, with eight vertices, one graph can have two vertices of degree 7, two vertices of degree 3, and four vertices of degree 4, a graph can have four vertices of degree 5 and four vertices of degree 4, and there are many other graphs between these two extremes. Three of the many possibilities are shown below.

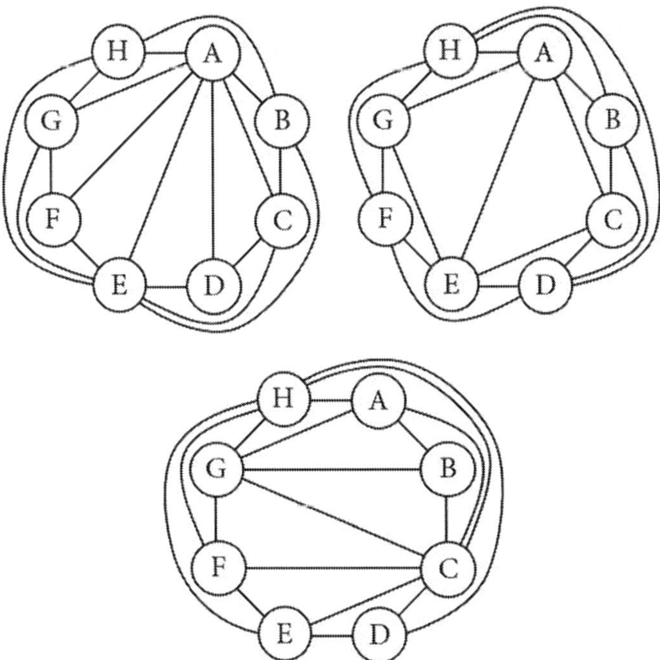

Chapter 3 Exercises

1. For each graph below:

- Indicate whether the graph is a PG or MPG.

- If the graph is a PG, add edges to the graph to turn it into a MPG.

- Once it is a MPG, identify the face that corresponds to the infinite area outside.

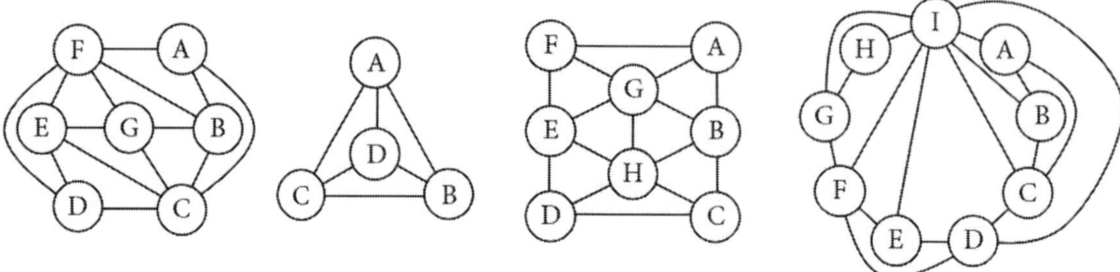

2. Add edges to triangulate the left graph below. Once the graph is triangulated, color the graph so that the coloring satisfies the four-color theorem. Color the right graph below the same way. If the added edges are removed from the MPG, will the coloring still satisfy the four-color theorem? Does the original PG have fewer, the same, or more ways to color it (so that it satisfies the four-color theorem) compared to the MPG?

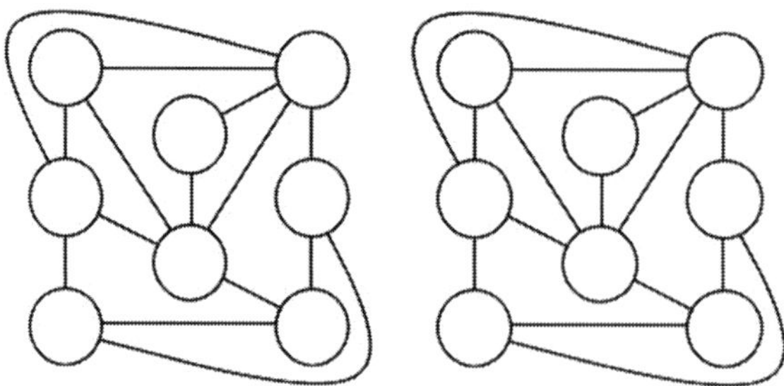

3. Draw and label a graph corresponding to the map below. Add edges to triangulate the graph. Now draw and label a map corresponding to the MPG.

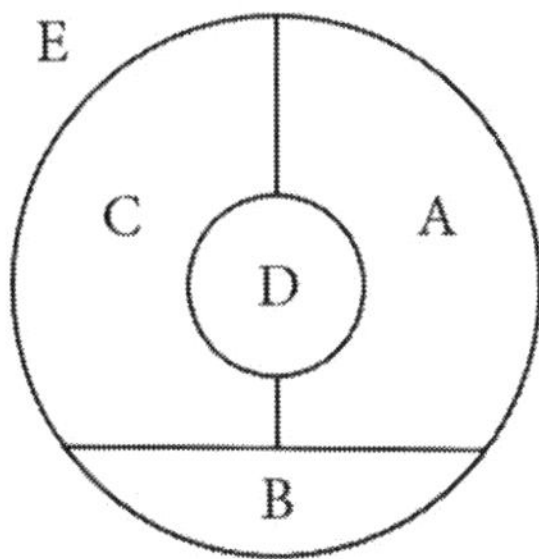

4. Are any of the graphs below isomorphic? (Recall that two graphs are isomorphic if the graphs are structurally equivalent in terms of edge-sharing.) If so, which ones?

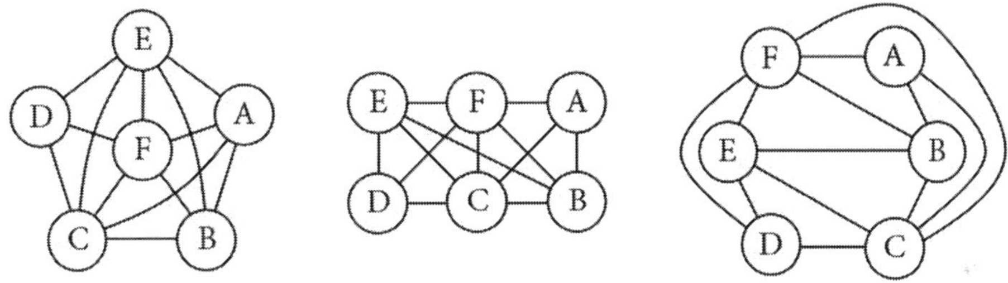

5. Are any of the graphs below isomorphic? If so, which ones?

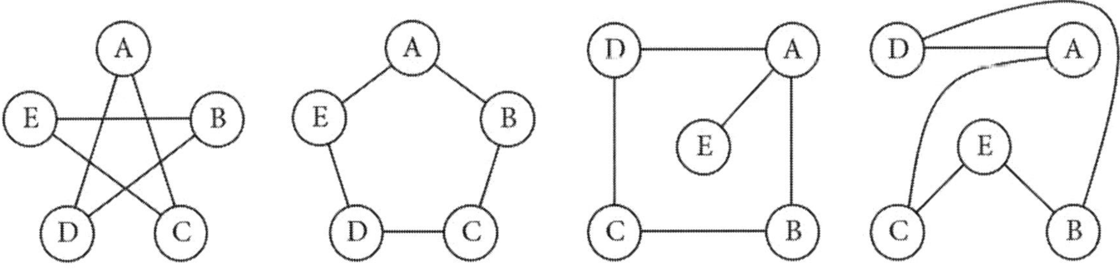

6. Looking at any map, what can you look at visually that will tell you whether or not the corresponding graph will be triangulated?

4 Euler's Formula

The number of vertices, edges, and faces of a map or a graph are related by Euler's formula, provided that we consider the unbounded surrounding area as one of the faces. We define the symbols V, E, and F as follows:

- V is the number of vertices.
- E is the number of edges.
- F is the number of faces. Remember to count the unbounded surrounding area.

According to Euler's formula [Ref. 4], for a map or a graph (or even a polyhedron):

$$V + F = E + 2$$

The number of vertices plus the number of faces is two more than the number of edges.

For the map shown above:

- V = 20. On a map, vertices are where the edges intersect. We marked the vertices with small dots (•) on the diagram above to help you count them. There are 5 for the inner pentagon, 10 for the decagon, and another 5 for the outer pentagon.
- E = 30. On a map, an edge is any line segment (or part of one) or curve that separates two regions (or which separates a region from the unbounded surrounding area). The inner pentagon has 5, there are 5 connecting the inner pentagon to the decagon, there are 10 along the decagon, there are 5 connecting the decagon to the outer pentagon, and there are 5 along the outer pentagon.
- F = 12. On a map, the faces are regions. The twelve regions are A, B, C, D, E, F, G, H, I, J, K, and L. This includes the unbounded surrounding area as a face.

Check the formula for the map on the left: V + F = 20 + 12 = 32 and E + 2 = 30 + 2 = 32. Since V + F and E + 2 both equal 32 for this map, we see that Euler's formula agrees with it.

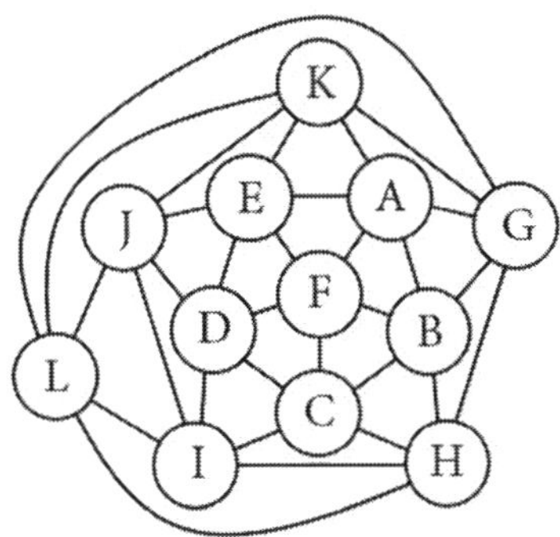

For the graph shown above:

- V = 12. On a graph, the vertices are regions where lines intersect. These are regions A, B, C, D, E, F, G, H, I, J, K, and L.

- E = 30. On a graph, each edge is a line or curve that connects a pair of regions. The edges are AB, AE, AF, AG, AK, BC, BF, BG, BH (but don't count BA since that's the same as AB, which was already counted), CD, CF, CH, CI, DE, DF, DI, DJ, EF, EJ, EK, GH, GK, GL, HI, HL, IJ, IL, JK, JL, and KL.

- F = 20. On a graph, the faces are areas formed between edges. The faces are ABF, ABG, AEF, AEK, AGK, BCF, BCH, BGH (but don't count BAF or BAG since they are the same as ABF and ABG, which were already counted), CDF, CDI, CHI, DEF, DEJ, DIJ, EJK, GKL, HIL, IJL, JKL, and the infinite area lying outside of the graph.

Note that the graph shown above corresponds to the map on the previous page, where we included region L explicitly on the graph and still counted the infinite area lying outside of the graph as its own face. On the map, L represented the unbounded surrounding area and was counted only as a face. If you are wondering if this may be inconsistent, there is a reason for drawing the map and its corresponding graph as we have done here. The values of V and F are swapped for the map compared to the graph when we draw them this way. Doing so emphasizes the fact that the graph is considered to be a dual representation of the map. (We'll elaborate on what we mean by a dual representation in the solution to Problem 1.)

For the graph, V + F = 12 + 20 = 32 and E + 2 = 30 + 2 = 32, agreeing with Euler's formula. Note that the map and graph give the same values of V, F, and E as a dodecahedron (a 12-sided polyhedron) and its dual polyhedron, which is an icosahedron (a 20-sided polyhedron). If you imagine cutting the *map* along the five outer radial edges and then folding the map up, you might be able to visualize how it can fold into the shape of a dodecahedron.

Note that if you remove region L from the previous graph, Euler's formula will still work. In that case, you would have one less vertex (region) so V = 11, five fewer edges (since L connects to G, H, I, J, and K) so E = 25, and four fewer faces (you wouldn't have GKL, HIL, IJL, or JKL) so F = 16 (note that we still include the infinite area outside the graph as a face). In this case V + F = 11 + 16 = 27 and E + 2 = 25 + 2 = 27, still satisfying Euler's formula.

When we include region L in the graph, note that 2E = 3F. That is, two times the number of edges is equal to three times the number of faces. When we include region L, 2E = 2(30) = 60 and 3F = 3(20) = 60. The formula 2E = 3F applies to any MPG. (Recall that MPG stands for "maximal planar graph" and that for a MPG every face in the graph has three "sides;" we'll call each edge a side whether it appears straight or curved.) To get the number of faces F from the number of edges for a MPG, divide E by 3 because each face has 3 edges and then multiply this by 2 because each edge is shared by 2 faces: F = 2E/3, from which it follows that F/E = 2/3. In the previous graph, for example, note that edge AE is part of both triangles AEF and AEK. The ratio of F to E is a maximum for a MPG. In this case, the ratio is 2/3, which is approximately 0.67 when rounded to two decimal places. (Note that if region L isn't included in the graph, it would be a PG, not a MPG. In that case, the ratio is slightly smaller: 16/25 = 0.64.) In contrast, if every face were a pentagon, the ratio would be 2/5 = 0.40.

Now we will do a little arithmetic to obtain a classic result related to the four-color theorem. This is generally done in a more formal, technical manner, but we'll try to keep it simple. Note that the formulas from this chapter apply if V is at least 3.

Consider the special case where every vertex of the graph has the same degree, meaning that the same number of edges intersect at every vertex. If every vertex has the same degree and if we define D to be the degree of each vertex (equal to the number of edges intersecting at each vertex), then $DV/2 = E$, which is equivalent to $DV = 2E$. This well-known result is referred to as the handshaking lemma. If V people each shake hands with D other people, $DV/2$ equals the number of handshakes that will occur. We divide by two in order to avoid double counting. For example, if there are 8 people at a gathering and each person shakes hands with 7 other people, there are $8(7)/2 = 56/2 = 28$ handshakes. If their names are Mr. A, Mr. B, Mr. C, etc., thru Mr. H, then we divide by 2 because 56 counts Mr. A shaking hands with Mr. B and Mr. B shaking hands with Mr. A separately, and similarly for all other pairs (like Mr. A and Mr. C which is the same as Mr. C with Mr. A).

What if the vertices don't all have the same degree? This will be the case with most graphs. We can still use the formula $DV = 2E$ provided that we interpret D as the *average* degree of the vertices.

If we multiply both sides of Euler's formula by 2, we get $2V + 2F = 2E + 4$. Substitute the equation $DV = 2E$ into this equation to obtain $2V + 2F = DV + 4$. Multiply both sides by 3 to get $6V + 6F = 3DV + 12$.

For a MPG, we noted that $2E = 3F$. Combine $DV = 2E$ with $2E = 3F$ to see that $DV = 3F$. Multiply both sides by 2 to get $2DV = 6F$. Substitute this into the last equation from the previous paragraph to obtain $6V + 2DV = 3DV + 12$. Subtract $3DV$ from both sides to get $6V - DV = 12$. Factor out the V to obtain $V(6 - D) = 12$. We finally obtain the formula $V = 12/(6 - D)$ for a MPG (recall that every face of a MPG has three sides). Recall that V is the number of vertices (which are regions) and that D is the average number of edges intersecting at each vertex on the graph.

When each vertex has degree D = 2, we get V = 12/(6 – 2) = 12/4 = 3. Every vertex on the MPG will connect to 2 edges if there are exactly 3 vertices.

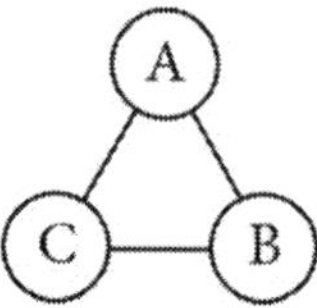

When each vertex has degree D = 3, we get V = 12/(6 – 3) = 12/3 = 4. Every vertex on the MPG will connect to 3 edges if there are exactly 4 vertices.

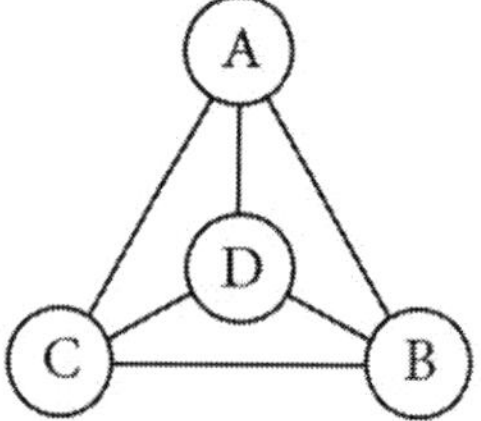

When each vertex has degree D = 4, we get V = 12/(6 – 4) = 12/2 = 6. Every vertex on the MPG can connect to 4 edges if there are exactly 6 vertices.

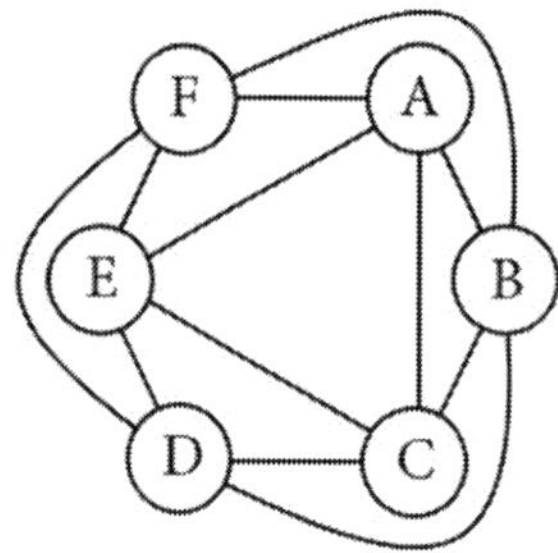

When each vertex has degree D = 5, we get V = 12/(6 – 5) = 12/1 = 12. Every vertex on the MPG can connect to 5 edges if there are exactly 12 vertices.

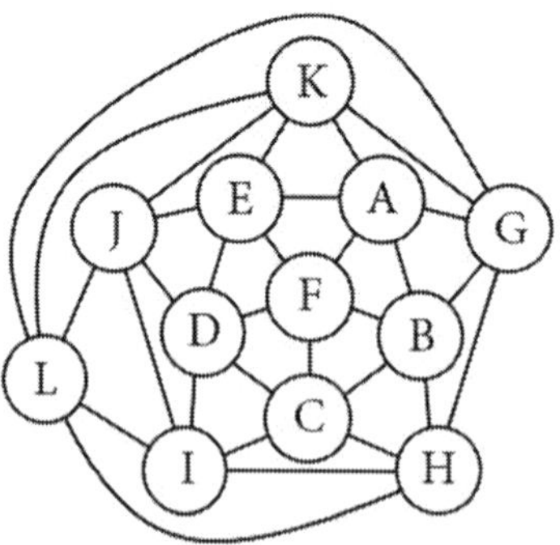

According to the formula $V = 12/(6 - D)$, there doesn't exist a solution where *every* vertex can connect to 6 edges. It isn't possible to draw a MPG (or any type of PG, as we'll see) where every vertex connects to 6 edges.

In the formula $V = 12/(6 - D)$, the only way to get a positive number for V is if D is less than 6. This means that the average degree of the vertices must be less than 6. On average, there need to be fewer than 6 edges connecting to each vertex. There can be vertices on a graph that connect to more than 6 edges, provided that the average is under 6. This shows that there must be at least *one* vertex on *any* MPG with a degree less than 6.

What if the PG isn't triangulated? In that case, the ratio of F to E will be smaller, as we noted earlier in this chapter. Let's look at a specific case with numbers. Suppose every face is a quadrilateral (a polygon with 4 sides) instead of a triangle. For quadrilateral faces, $2E = 4F$ (instead of $2E = 3F$ for triangular faces). This time, we combine $DV = 2E$ with $2E = 4F$ to get $DV = 4F$. Return to the equation $2V + 2F = DV + 4$, multiply both sides by 2 to get $4V + 4F = 2DV + 8$, and plug in $DV = 4F$ to obtain $4V + DV = 2DV + 8$. Subtract $2DV$ to get $4V - DV = 8$, which we can write as $V(4 - D) = 8$. If every face is a quadrilateral, then we find that the average degree has to be less than 4. This restriction is even more severe than it was for a MPG. Since the ratio of F to E is maximum for a MPG, the case of triangulation results in more edges per vertex. Thus, if there must be at least one vertex with a degree of less than 6 for a MPG, we can conclude that there must be at least one vertex with a degree of less than 6 for *any* PG (triangulated or not) [Ref. 2].

Returning to MPG's, it will be useful in later chapters to relate the number of edges (E) to the number of vertices (V):

- $2E = 3F$ for a MPG.
- $V + F = E + 2$ according to Euler's formula.

- Multiply both sides by 3 to get $3V + 3F = 3E + 6$.
- Plug $2E = 3F$ into the previous equation to get $3V + 2E = 3E + 6$.
- Subtract $2E$ from both sides to get $3V = E + 6$.
- Subtract 6 from both sides to see that the number of edges is $3V - 6 = E$, which is equivalent to $E = 3V - 6$. This formula applies if V is at least 3. (For very small values of V, it shouldn't be necessary to bother with the formula.)

This shows that a MPG with V vertices has $E = 3V - 6$ edges. Following are some examples.

- A MPG with $V = 3$ vertices has $E = 3(3) - 6 = 3$ edges.
- A MPG with $V = 4$ vertices has $E = 3(4) - 6 = 6$ edges.
- A MPG with $V = 5$ vertices has $E = 3(5) - 6 = 9$ edges.
- A MPG with $V = 6$ vertices has $E = 3(6) - 6 = 12$ edges.
- A MPG with $V = 7$ vertices has $E = 3(7) - 6 = 15$ edges.
- A MPG with $V = 8$ vertices has $E = 3(8) - 6 = 18$ edges.
- A MPG with $V = 9$ vertices has $E = 3(9) - 6 = 21$ edges.
- A MPG with $V = 10$ vertices has $E = 3(10) - 6 = 24$ edges.

Since every PG has an average degree value less than 6 (and thus has at least one vertex with a degree of 5 or less), it follows that:

- If every vertex of a MPG has an even degree number, there will be at least one vertex with degree 4. (See Challenge Problem 4 in Chapter 11.)
- If a MPG doesn't have any vertices with a degree less than 5, it must have at least one edge that joins together two vertices with degree 5 or at least one edge that joins together vertices with degree 5 and 6.

Chapter 4 Exercises

1. For the map below on the left, determine the number of vertices, edges, and faces. For the graph below on the right, determine the number of vertices, edges, and faces. Are this map and graph duals to one another? Explain. Verify Euler's formula for the map. Also verify Euler's formula for the graph. Determine the ratio of F to E for the graph. What can you conclude from the ratio of F to E for this graph?

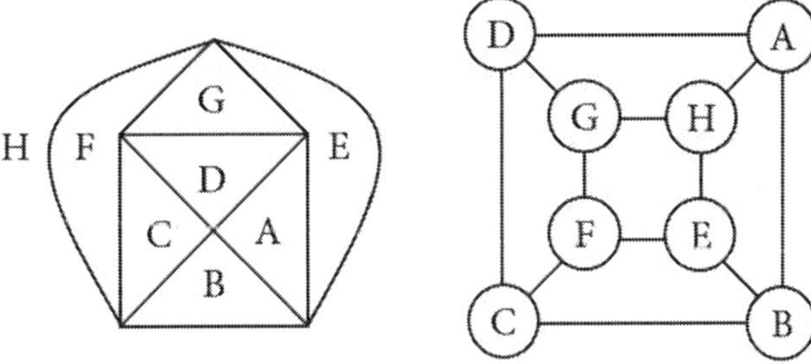

2. For the graph below, determine the number of vertices, edges, and faces. Verify Euler's formula for the graph. Determine the ratio of F to E for the graph. What can you conclude from the ratio of F to E for this graph? If the graph is not a MPG, add edges to triangulate the graph, and determine V, E, F, and F/E for the MPG.

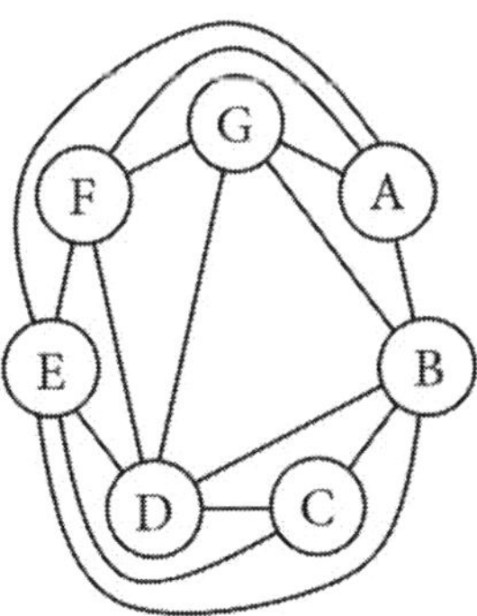

3. For the graph below determine the number of vertices, edges, and faces. Verify Euler's formula for the graph. Determine the ratio of F to E for the graph. What can you conclude from the ratio of F to E for this graph? If the graph is not a MPG, add edges to triangulate the graph, and determine V, E, F, and F/E for the MPG.

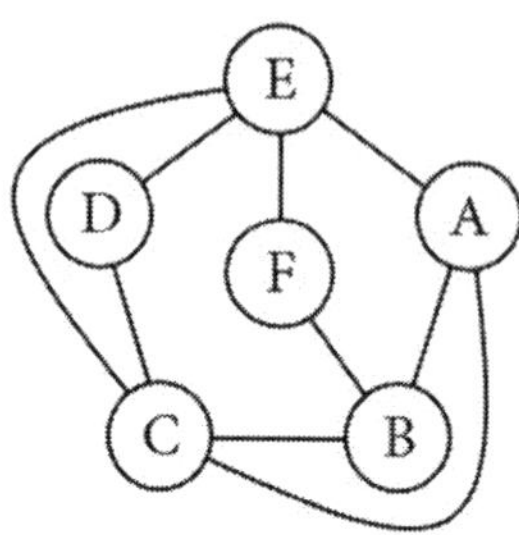

4. If every face of a PG is a quadrilateral, what is the ratio of F to E? Is this ratio larger or smaller than the ratio of F to E for a MPG?

5. If a MPG has 15 vertices, how many edges and how many faces does it have?

6. If a MPG has 10 faces, how many vertices and how many edges does it have?

7. If a MPG has 21 edges, how many vertices and how many faces does it have?

8. If a graph has 18 vertices and 48 edges, could it be a MPG? Could it be a PG?

9. If a graph has 32 vertices and 54 faces, could it be a MPG? Could it be a PG?

10. If a graph has 96 edges and 72 faces, could it be a MPG? Could it be a PG?

11. If every vertex of a MPG has degree five, how many vertices, edges, and faces does it have?

12. If one-half of the vertices of a MPG have degree five and the other half have degree six, how many vertices, edges, and faces does it have?

13. If a MPG has 2 vertices with degree 8, 2 vertices with degree 3, and each of the remaining vertices have degree 4, how many vertices, edges, and faces does it have?

5 Complete Graphs and Bigraphs

In a **<u>complete graph</u>**, every vertex connects to each of the other vertices. The notation K_V represents a complete graph with V vertices. For example, the complete graph shown below, which has $V = 8$ vertices, is represented by K_8.

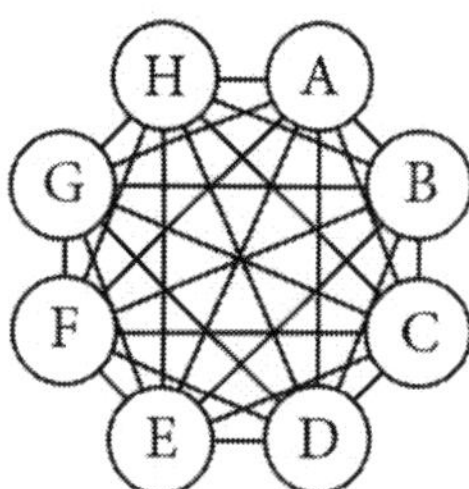

A complete graph can be used to represent the following classic handshaking problem. If V people are in a room and each person shakes hands once with each of the $V - 1$ other people, how many handshakes will there be all together? The answer is given by the **<u>handshaking lemma</u>**: $V(V - 1)/2$ equals the number of handshakes. Why? Multiply V by $(V - 1)$ and then divide by 2 to correct for double counting. To better understand the formula associated with the handshaking lemma, let's consider the K_6 graph shown below.

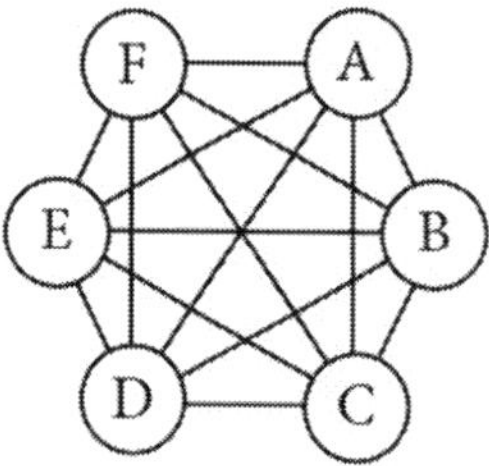

Let's list the handshakes:

- A's handshakes include AB, AC, AD, AE, and AF. That's five.
- B's handshakes include BA, BC, BD, BE, and BF. That's also five, but note that BA is the same as AB.
- C's handshakes include CA, CB, CD, CE, and CF. That's also five, but CA and CB are the same as AC and BC.

- D's handshakes include DA, DB, DC, DE, and DF. That's another five, but DA, DB, and DC are the same as AD, BD, and CD.
- E's handshakes include EA, EB, EC, ED, and EF. That's five more, but EA, EB, EC, and ED are the same as AE, BE, CE, and DE.
- F's handshakes include FA, FB, FC, FD, and FE. That's five again, but all of these are repeats. They are the same as AF, BF, CF, DF, and EF.

The previous graph, K_6, has $V = 6$ vertices. According to the handshaking lemma, there are $V(V-1)/2 = 6(5)/2 = 15$ handshakes. Look at our list. For A thru F, there are 5 handshakes each, giving us $6(5) = 30$ handshakes, but 15 of these are repeated so that there really are only $30/2 = 15$ handshakes. That's what we meant about dividing by 2 to correct for double counting. For example, BC and CB are the same handshake because they involve the same two people. A complete graph is really asking: given V vertices, how many ways are there to make pairs of them. The answer is $V(V-1)/2$.

Complete graphs with 1 to 10 vertices are illustrated on the following page. A couple of noteworthy complete graphs concerning the four-color theorem include:

- K_4 is the largest complete graph that is a MPG. K_4 is also the largest complete graph that can be colored such that the coloring satisfies the four-color theorem. As we will explore in later chapters, MPG's with K_4 subgraphs tend to have more restrictive coloring (meaning that there tend to be fewer ways to color the graphs compared to graphs that lack K_4's), vertices with degree three tend to take part in K_4 subgraphs, and even K_4 subgraphs without any vertices with degree three have separating triangles. K_4 is sometimes referred to as the **tetrahedral graph**.
- K_5 is the smallest complete graph that isn't a MPG. K_5 is also the smallest complete graph that isn't four-colorable. As we will explore in Chapter 6, K_5 plays an important role in determining whether or not a graph is a PG or if it is nonplanar. K_5 is sometimes referred to as the **pentatope graph**.

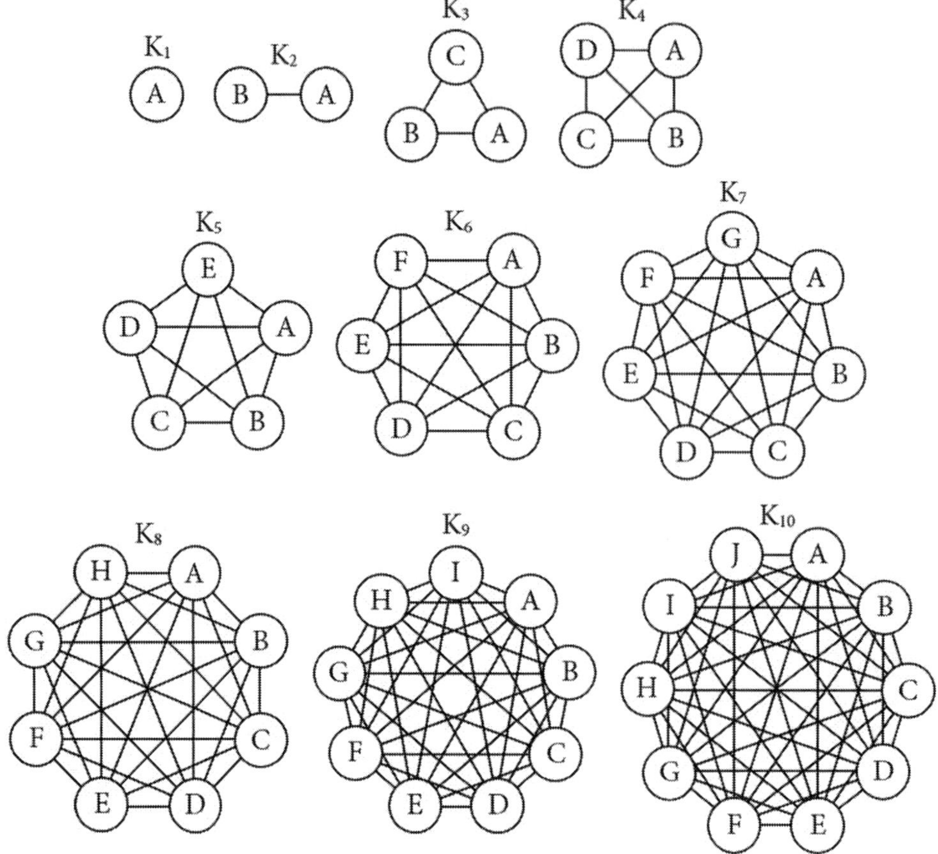

The number of edges on a complete graph with V vertices is given by the formula from the handshaking lemma: $E = V(V - 1)/2$. For example, you can verify that the K_6 graph shown above with $V = 6$ vertices has $E = 6(6 - 1)/2 = 6(5)/2 = 30/2 = 15$ edges.

A **complete bipartite graph** has two sets of vertices where every vertex of one set connects to every vertex of the other set. The notation $K_{X,Y}$ denotes a complete bipartite graph where set P has X vertices and set Q has Y vertices.

A complete bipartite graph has $E = XY$ edges because each of the X vertices in set P connects to each of the Y vertices in set Q. For example, the complete bipartite graph $K_{3,4}$ shown on the following page has $X = 3$ vertices in set P, $Y = 4$ vertices in set Q, and $E = XY = 3(4) = 12$ edges.

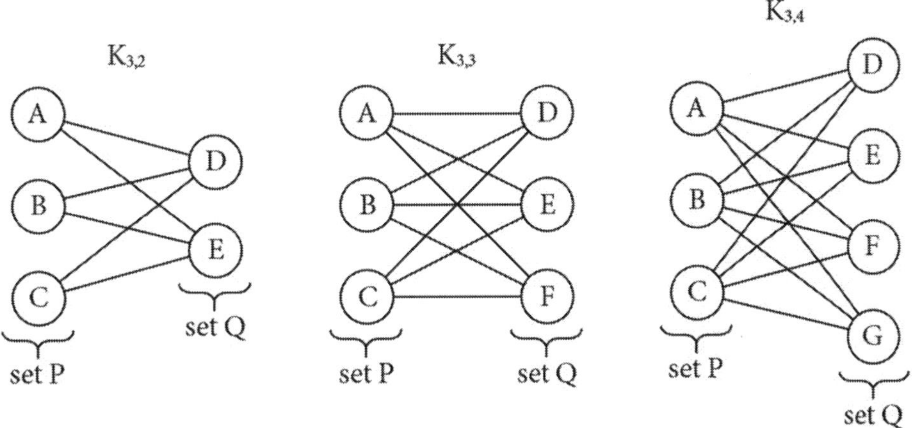

A bipartite graph is also called a **bigraph**. A bipartite graph (without the word "complete") is defined to have two sets of vertices where no two vertices of a single set connect to one another. A *complete* bipartite graph has every vertex of one set connected to every vertex of the other set and vice-versa.

The $K_{3,3}$ graph shown above is noteworthy as it concerns PG's. In Chapter 6, we will see that $K_{3,3}$ is nonplanar and that $K_{3,3}$ and K_5 play an important role in determining whether or not a graph is a PG or if it is nonplanar. The $K_{3,3}$ graph is also referred to as the **utility graph**, based on the following puzzle [Ref. 5]. Can you connect three cottages (A, B, and C) to three utilities (D, E, and F) without crossing the lines? Since the utility graph isn't planar (as we will show in Chapter 6), the answer is no. Recall that we are defining PG to include any graph that *can* be drawn in the plane without crossings (even if the graph happens to be drawn in a form that has avoidable crossings).

Observe that complete bipartite graphs are two-colorable. Since the vertices of set P don't connect to one another, every vertex in set P can be one color (such as red). Similarly, since the vertices of set Q don't connect to one another, every vertex in set Q can be another color (such as blue). For example, in the $K_{3,4}$ graph shown above, A, B, and C can each be red and D, E, F, and G can each be blue.

Chapter 5 Exercises

1. A total of 12 people attend a conference. If every person at the conference shakes hands once with every other person, how many handshakes will occur in total?

2. A research project is started with 9 female mathematicians and 7 male mathematicians. If each female shakes hands once with all of the males, each male shakes hands once with all of the females, no female shakes hands with another female, and no male shakes hands with another male, how many handshakes will occur in total?

3. Draw K_4 as it was drawn in this chapter. Now redraw K_4 to show that it is a PG. Is K_4 a MPG? How can you tell?

4. Draw K_5. Now redraw the graph with one of its edges removed. Is this new graph (with one edge removed) planar? Is it a MPG? How can you tell?

5. Draw K_6. How many edges must be removed from K_6 in order to for the new graph (with edges removed) to be a MPG? Draw the new graph (with edges removed), showing that it can be a MPG. Draw a second graph with the same number of edges removed, which is also a MPG, but where the vertices have different degrees from the first MPG. Now draw a third graph with the same number of edges removed, but which isn't a MPG.

6. How many edges must be removed from a complete graph with 9 vertices in order to make a MPG with 9 vertices? Will the resulting graph necessarily be a MPG?

7. Show that if you remove $(V - 3)(V - 4)/2$ edges from a complete graph with V vertices that the resulting graph will have the right number of edges to be a MPG with V vertices. Will the resulting graph necessarily be a MPG?

8. Draw K_5. How many edges must be removed from K_5 in order to for the new graph (with edges removed) to be $K_{2,3}$? Draw $K_{2,3}$ by removing this number of edges from K_5.

9. Draw K_6. How many edges must be removed from K_6 in order to for the new graph (with edges removed) to be $K_{3,3}$? Draw $K_{3,3}$ by removing this number of edges from K_6. Can you find more than one way to remove edges from K_6 to draw a graph that is isomorphic to $K_{3,3}$?

10. If V is an even number, show that $V(V-2)/4$ edges need to be removed from K_V in order for the new graph (with edges removed) to be $K_{N,N}$ with $N = V/2$. Will the resulting graph necessarily be $K_{N,N}$?

6 Maximal Planar Graphs

Recall from Chapter 3 that a MPG (which stands for "maximal planar graph") is a PG that is triangulated, meaning that every face is a triangle (keeping in mind that one of its edges may be curved), including the infinite area outside of the graph. A MPG is "maximal" in the sense that if another edge were added to anywhere to the MPG, then the graph would no longer be a PG (there would be an unavoidable crossing).

This chapter focuses on the important question, "How can you tell whether or not a graph is a MPG?" That is, if a graph is drawn with a crossing, is the crossing avoidable or not?

The simplest test is an exclusion test. What is an exclusion test? If a graph fails the exclusion test, then it isn't a MPG. However, if the graph passes the exclusion test, we will need more information before we can determine whether or not it is a MPG. This exclusion test is useful because it can rule many graphs out very quickly, but the exclusion test has limited use because when a graph passes the exclusion test, another test is still needed. To perform the exclusion test, count the number of vertices and edges and see if these values satisfy Euler's formula (or related formulas from Chapter 4) for a MPG. If a graph with at least 3 vertices is a MPG, the number of edges must equal $E = 3V - 6$. If E is greater than $3V - 6$, the graph isn't MPG (and it isn't even a PG).

For example, consider the graphs on the next page which have $V = 6$ vertices. A MPG with 6 vertices should have $E = 3(6) - 6 = 18 - 6 = 12$ edges.

Counting edges can get tricky when a graph has numerous vertices, but fortunately there is a simple trick to make this easy. The number of edges equals the sum of the degrees of the vertices divided by two.

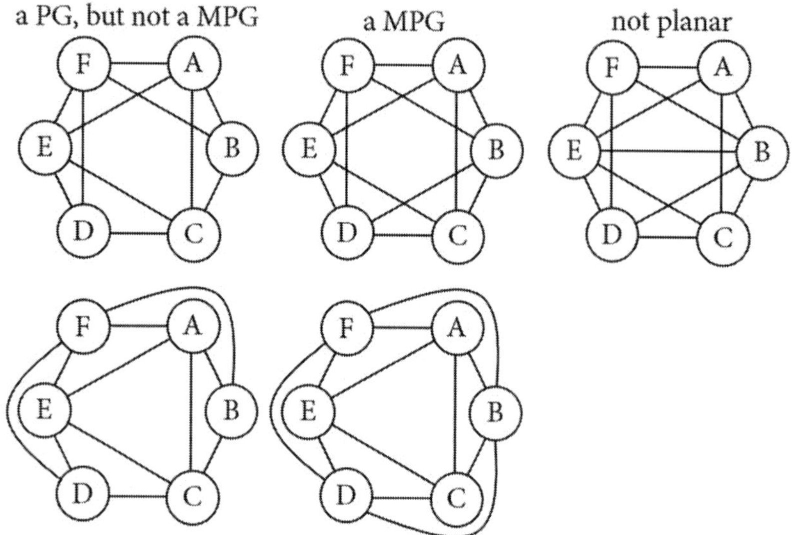

The diagrams above show how the exclusion test can determine that the right graph above isn't a MPG. As mentioned previously, a MPG with $V = 6$ vertices should have $E = 12$ edges.

- The sum of the degrees for the left graph is $4 + 3 + 4 + 3 + 4 + 4 = 22$, so it has $22 \div 2 = 11$ edges. Since 11 is less than 12, the left graph isn't a MPG. Euler's formula doesn't prove that it's a PG; this we were able to determine by redrawing the graph without crossings.

- The sum of the degrees for the middle graph is $4 + 4 + 4 + 4 + 4 + 4 = 24$, so it has $24 \div 2$ edges. Since this equals 12, the middle graph could be a MPG. Euler's formula doesn't prove that it's a MPG, but it doesn't exclude this graph from being one. We were able to determine that it was a PG by redrawing the graph without crossings. The redrawn graph plus Euler's formula then tells us that it's a MPG.

- The sum of the degrees for the right graph is $4 + 5 + 4 + 4 + 5 + 4 = 26$, so it has $26 \div 2 = 13$ edges. Since 13 is greater than 12, the right graph isn't planar (it has too many edges to be a MPG, and a MPG has the maximum number of edges for a PG). We don't need to try to redraw this graph to see whether or not it has unavoidable crossings. Since Euler's formula excludes the right graph from being a MPG (or a PG), we know that it can't be redrawn without at least one crossing.

Now let's look at an example where Euler's formula doesn't help. Both graphs below have $V = 6$ vertices. Both graphs agree with Euler's formula for a MPG: $E = 3V - 6 = 3(6) - 6 = 18 - 6 = 12$ edges. Yet the left graph is a MPG, whereas the right graph isn't planar. In this example, the exclusion test doesn't help.

- The sum of the degrees for the left graph is $4 + 5 + 3 + 4 + 5 + 3 = 24$, so it has $24 \div 2$ edges. Since this equals 12, the left graph could be a MPG. Euler's formula doesn't prove that it's a MPG, but it doesn't exclude this graph from being one. We were able to determine that it was a PG by redrawing the graph without crossings. The redrawn graph plus Euler's formula then tells us that it's a MPG.

- The sum of the degrees for the right graph is $4 + 4 + 4 + 4 + 5 + 3 = 24$, so it has $24 \div 2$ edges. Since this equals 12, the right graph could be a MPG. Euler's formula doesn't prove whether it's a MPG, but it doesn't exclude this graph from being one. If you try to redraw the graph with the same edge-sharing, it will have at least one crossing. Why is the left graph planar, but not the right graph? In the left graph, we were able to separate the inside edges into a group of three inside edges and another group of three outside edges without crossings. In the right graph, if you attempt to do this, it won't work because AD, BE, and CF triple cross (whereas AD, AE, and BF do *not*), such that at least one pair of these edges will cross inside or outside the polygon.

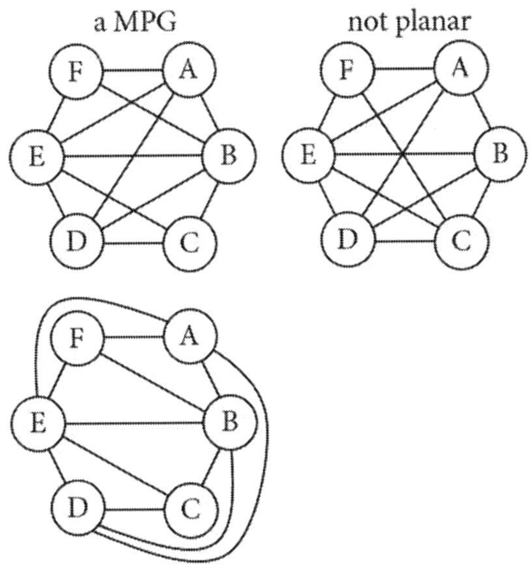

Fortunately, there are other tests besides using Euler's formula as an exclusion test.

One of these tests we have already been using: the redrawing test. If it is possible to redraw a graph without any crossings (meaning any crossings previously shown were avoidable), then the graph is planar. If a graph is planar and is also fully triangulated (which is the case if it satisfies Euler's formula for a MPG), then the graph is a MPG. On the other hand, if it isn't possible to redraw a graph without at least one crossing, the graph isn't planar.

The redrawing test is inconvenient, especially for a graph with a large number of vertices (and thus a large number of edges, too). Sometimes, there is a way to redraw a graph without crossings that isn't easy to think of.

The redrawing test is simpler when all of the vertices lie at the corners of a closed polygon. It's important that the polygon be closed; if it's missing an edge, the following rule won't apply. As we'll explore in Chapter 13 (regarding Hamiltonian cycles), any MPG can be drawn with all of its vertices on the corners of a closed polygon unless it has separating triangles (which we'll define in Chapter 12), so the polygon version of the redrawing test will actually apply to the most important examples concerning the four-color theorem. For a graph where all of the vertices lie at the corners of a closed polygon, it is a MPG if all of these apply:

- It has $E = 3V - 6$ edges, as required by Euler's formula for a MPG.
- V edges form the outline of a closed polygon.
- $V - 3$ edges can be drawn inside the polygon without crossing.
- $V - 3$ different edges can be drawn outside the polygon without crossing.

It's interesting to note that the two sets of $V - 3$ edges are interchangeable; you can put the inside edges outside and vice-versa. We will explore this more fully in Chapter 14. For now, we will focus on how this helps us determine whether or not a graph is a MPG. (Note that our polygon redrawing test is focused on possible MPG's, not more general PG's.)

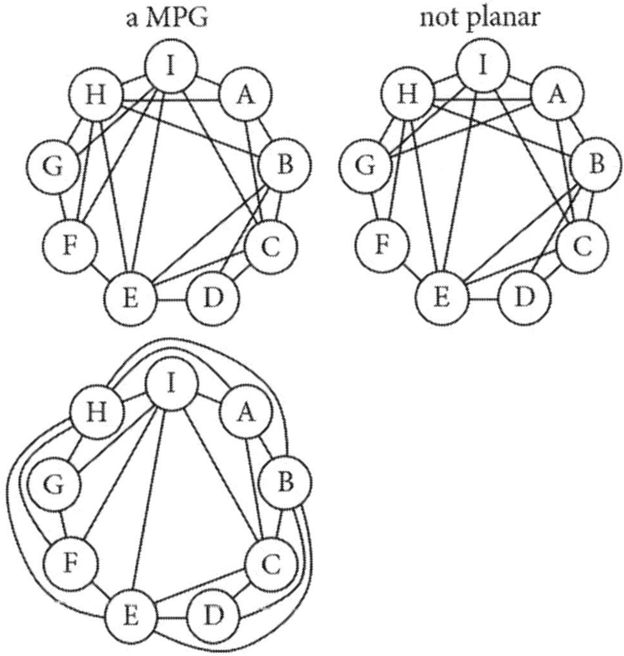

Both graphs above have V = 9 vertices. Both graphs agree with Euler's formula for a MPG: E = 3V − 6 = 3(9) − 6 = 27 − 6 = 21 edges. Yet the left graph is a MPG, whereas the right graph isn't planar. This is another example where the exclusion test doesn't help. Since each graph has all of its vertices arranged in a closed polygon, we can use the polygon version of the redrawing test. There will be V − 3 = 9 − 3 = 6 inside edges and 6 outside edges.

- In the left graph, we divided the 12 original inside edges into two groups of 6 edges that don't cross inside or outside. This makes a graph without crossings. Since this graph also has the right number of edges for a MPG, this graph is a MPG.

- In the right graph, although it has the right number of edges for a MPG, this isn't a planar graph. One way to see this is that AG, BH, and EI triple cross. By this we mean that any pair of these three lines intersect. (As a counterexample, AG, BH, and EH don't meet our definition of what a "triple cross" is because BH and EH don't cross; only AG/BH and AG/EH cross as opposed to all three pairs.) While such a triple cross in a closed polygon renders a graph nonplanar, it is important to note that a graph may still be nonplanar if it doesn't have a triple cross; the triple cross concept is useful, but it isn't sufficient.

While the polygon version helps to simplify the redrawing test, dividing the edges into two groups still becomes a challenge when there are numerous vertices (and thus many edges). Two more ways to determine whether or not a graph is a MPG apply theorems that relate to the two graphs shown below (the middle graph and right graph are isomorphic).

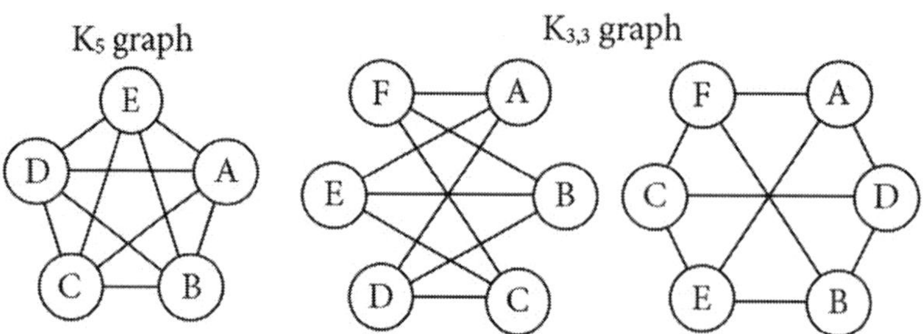

Recall from Chapter 5 that K_5, shown above on the left, is a complete graph with 5 vertices, and that $K_{3,3}$, shown above on the middle and right, is a complete bipartite graph with two sets of 3 vertices.

- K_5 has $5(4)/2 = 10$ edges. If we use the exclusion test, we find that a MPG with $V = 5$ vertices should have $E = 3(5) - 6 = 15 - 6 = 9$ edges. Since K_5 has 10 edges, it isn't planar. K_5 is an important graph because it has the fewest vertices of any nonplanar graph that isn't four-colorable; since all 5 vertices connect to one another, K_5 requires five-coloring. It's not a MPG though, so it doesn't contradict the four-color theorem; K_5 can't be mapped in the plane. Note that if you remove one edge from K_5, it will have 9 edges, be a MPG, and be four-colorable.

- $K_{3,3}$ has 9 edges. If we use the exclusion test, we find that a MPG with $V = 6$ vertices should have $E = 3(6) - 6 = 18 - 6 = 12$ edges. Since $K_{3,3}$ has 9 edges, it isn't a MPG. The exclusion test doesn't tell us whether or not $K_{3,3}$ is planar; just that $K_{3,3}$ isn't a MPG. You can use the polygon version of the redrawing test to see that $K_{3,3}$ isn't planar. Looking at the right graph above, edges AE, BF, and CD triple cross. If you put two of these edges on the inside and one on the outside, the two inside edges will cross, and if you put two of these edges on the outside and one on the inside, the two outside edges will cross. The $K_{3,3}$ graph is the smallest complete bipartite graph that isn't planar. Observe that $K_{3,3}$ can actually be colored using just two colors: since A, B, and C don't connect to one another, these three can be one color; similarly, since D, E, and F don't connect to one another, these three can be a second color.

To explain how the K_5 and $K_{3,3}$ graphs can be used to determine whether or not a graph is a PG, we need to define the following terms:

- A **subgraph** is formed using a subset of the vertices and edges of a graph.
- A **subdivision** of a graph inserts a new vertex along an edge (such that what had previously been a single edge becomes two different edges).
- **Contracting** an edge means to merge two vertices together, effectively deleting the edge that connected them. If another vertex had been connected to both vertices before they merged together, one of those two now-duplicate edges is also removed.
- A **minor** is formed by contracting the edges of another graph.

The diagrams below and on the following page help to illustrate visually what these terms mean.

the left graph below is a subgraph of the right graph

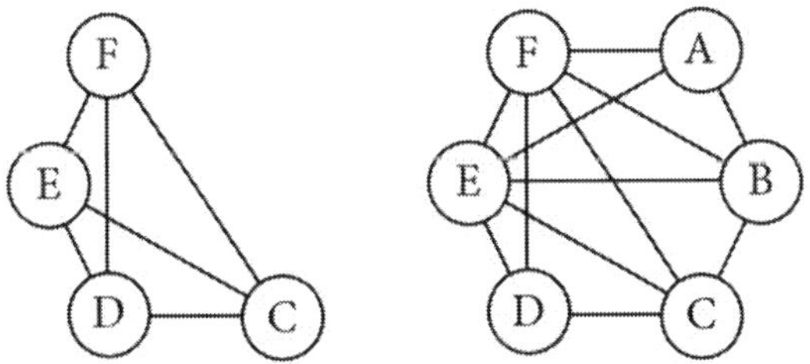

the left graph below is a subdivision of K_5

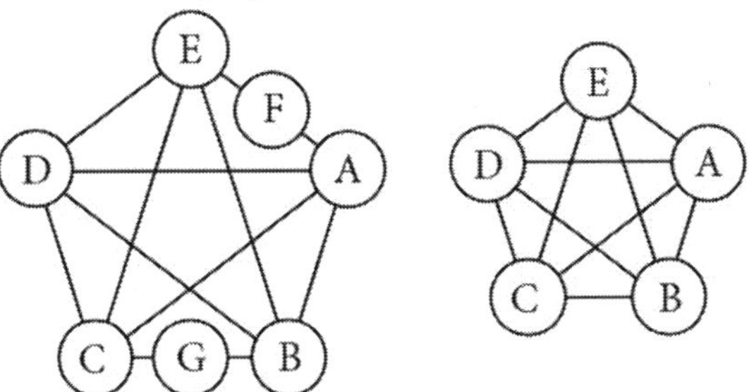

the left graph below formed by contracting edge AB in the right
graph; the left graph is thus a minor of the right graph

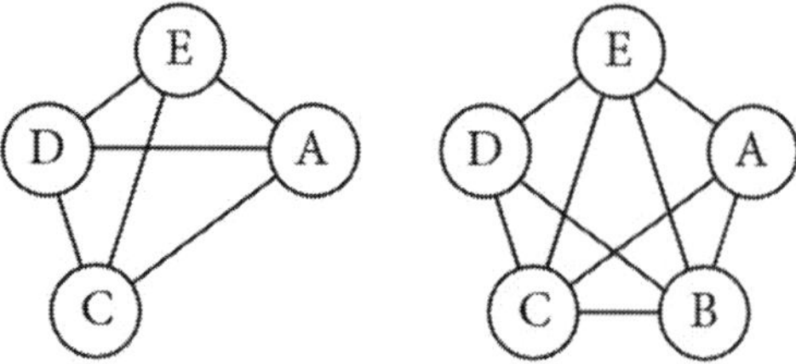

A few points regarding the previous diagrams are worth noting:

- A subgraph is contained in a larger graph. All you need to do is identify it; no subdivisions or contractions are involved. In the top right graph on the previous page, you can find the subgraph CDEF just by inspection.

- When you make a subdivision, it looks (and is!) trivial. When you apply it though, it's different. You would be looking at graphs more complicated than K_5 or $K_{3,3}$, and you would be trying to ignore certain edges such that the non-ignored edges form a subdivision of K_5 or $K_{3,3}$. We'll see an example of this later.

- When we contracted edge AB in the graph above, we merged vertices A and B, we removed edge AB (or shrank it down to zero), edge BD became edge AD (since the edge connecting B and D must remain), and similarly edge BC became edge AC.

Two similar theorems relating to K_5 and $K_{3,3}$ can be used to determine whether or not a graph is a PG. One theorem involves subdivisions; the other involves minors. (If a K_5 or $K_{3,3}$ subgraph can simply be identified without subdivision or contraction, as mentioned in the first bullet point, that's a special case of these theorems.)

- If a graph contains K_5 or $K_{3,3}$ as a subgraph, that graph isn't planar. This is a special case of Kuratowski's theorem.

- According to **Kuratowski's theorem**, a graph is planar if and only if the graph doesn't contain a subgraph that is a subdivision of K_5 or $K_{3,3}$ [Ref. 6].

- According to **Wagner's theorem**, a graph is planar if and only if the graph's minors don't include K_5 or $K_{3,3}$ [Ref. 7].

Following are a few examples.

the right graph below isn't planar because it contains a K_5 subgraph

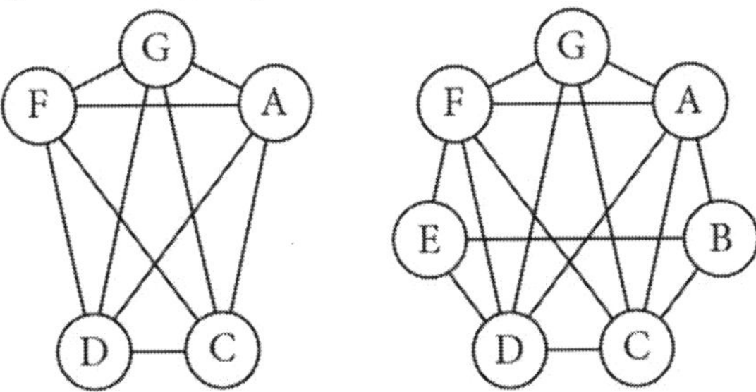

the right graph below isn't planar because it

contains a subgraph that is a subdivision of K_5

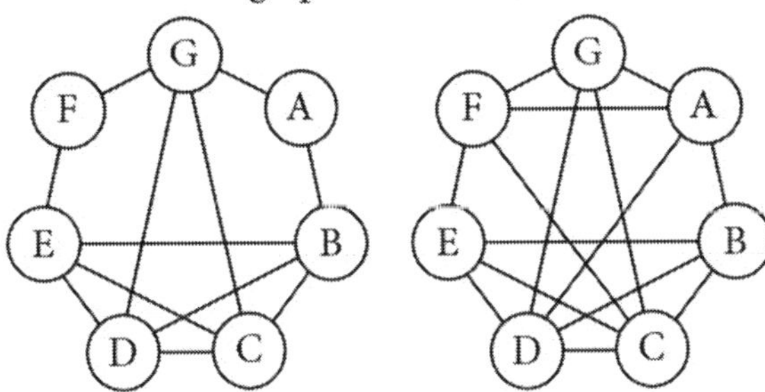

The left graph above is a subgraph of the right graph, and it is also a subdivision of K_5. To see this, imagine first drawing K_5 with vertices B, C, D, E, and G. Now subdivide edge BG into AB + AG and subdivide edge EG into EF + FG to get the left graph above. (The fact that edges BG and EG were "bent" in the process is irrelevant; bending doesn't matter.)

the right graph below isn't planar because the minor

formed by contracting E and F includes a $K_{3,3}$

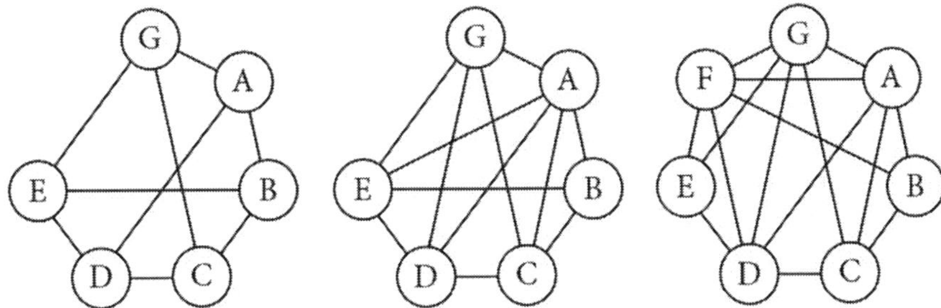

In the diagrams above, we contracted vertices E and F. Visualize vertex F being pushed down until it merges with vertex E. Doing so, edge BF will become edge BE and edge AF will become edge AE. Note that DF becomes a duplicate of DE and that FG becomes a duplicate of EG; in order to avoid double edges, one of each of the duplicates is removed. (Alternatively, you can think of DE and DF merging into one edge and similarly for FG and EG.) To see that the minor on the left is $K_{3,3}$, look at A, C, and E and then look at B, D, and G.

A graph is a PG if it satisfies Kuratowski's theorem (which is equivalent to Wagner's theorem). A graph is a MPG if it satisfies Kuratowski's theorem and also passes the exclusion test (that comes from applying Euler's formula to a MPG). Alternatively, a graph is a MPG if it satisfies Kuratowski's theorem and it is also fully triangulated (in this case, remember to check that the face corresponding to the infinite area outside is also a triangle; only 3 vertices should touch the infinite area outside).

Important reminder: We are defining PG to include any graph that *can* be drawn in the plane without crossings (even if the graph happens to be drawn in a form that has avoidable crossings). With this meaning, PG doesn't quite stand for "planar graph." If you want to be more precise, you should think of PG as "planar graph or any graph that is isomorphic to a planar graph" or "planar graph or any graph that can be redrawn in the plane without any crossings." Our definition of MPG thus has a similar interpretation.

Grammar note: Are you wondering if we should write "an MPG" instead of "a MPG"? The answer depends on how you read it. When you look at the abbreviation MPG, if in your mind you think "em pee gee" (the names of the letters), then you're right, we should write "an MPG" instead of "a MPG." However, if like the author of this book, you look at the abbreviation MPG and think "maximal planar graph" (the actual name), then we should write "a MPG" instead of "an MPG." When typing, the abbreviation MPG is handy because it saves space and makes formatting simpler. When reading and thinking, the phrase maximal planar graph is more informative than the abbreviation.

Chapter 6 Exercises

1. For each graph below and on the following page, first apply the exclusion test using Euler's formula to determine one of the following:

- The graph is definitely nonplanar.
- The graph could be a MPG.
- The graph could be a PG, but not a MPG.

Now determine if each graph is a MPG, PG but not a MPG, or nonplanar. If a graph is a MPG or PG, redraw the graph to show that the graph is a PG. If a graph is nonplanar, either apply Kuratowski's theorem or Wagner's theorem to show that the graph is nonplanar.

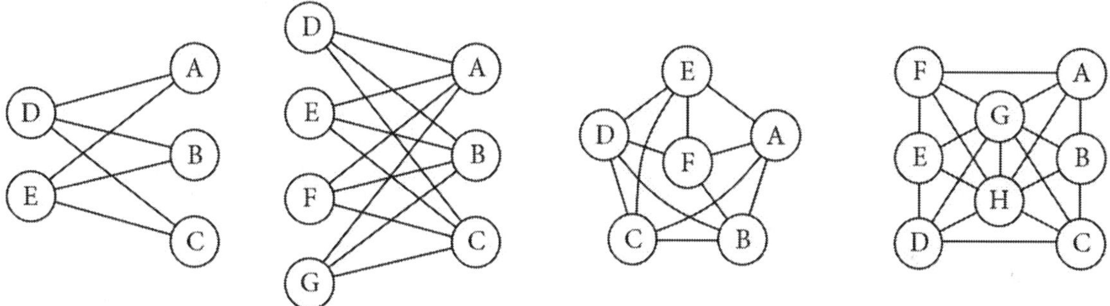

Note: It may help to review our definitions of PG and MPG on the previous page, as we are using these terms a little differently than the standard usage of "planar graph" or "maximal planar graph."

Note: This problem is continued on the next page. (There is space below to redraw graphs.)

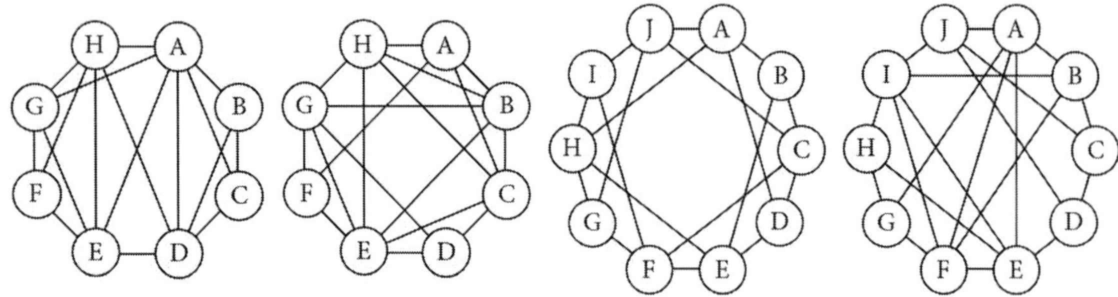

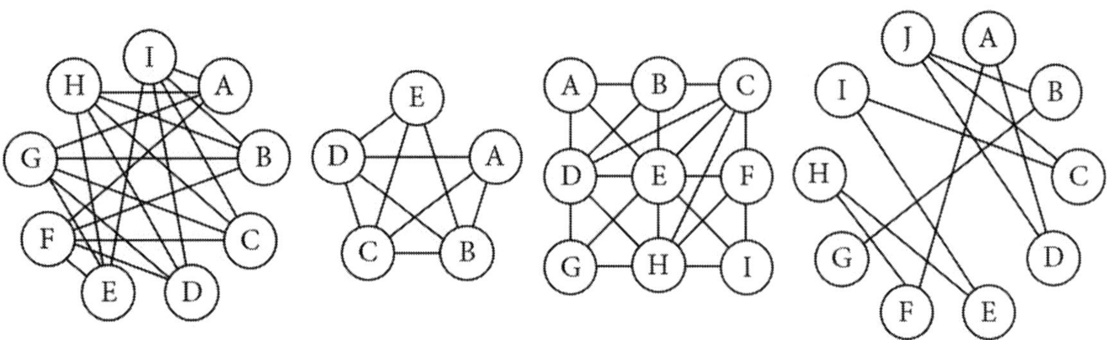

Challenge problem 1: Since $K_{3,3}$ is two-colorable, whereas K_5 isn't four-colorable, it may seem natural to wonder if either Kuratowski's theorem or Wagner's theorem may be applied in some way to prove the four-color theorem. For example, one might wonder if every graph that doesn't contain a subgraph that is a subdivision of K_5 may be four-colorable, or if every graph whose minors don't contain K_5 is four-colorable. Some of these graphs wouldn't be MPG's, but that's not a problem. If every graph that meets the specified criteria can be shown to be four-colorable and if the set of graphs includes all possible MPG's, it doesn't matter if the set of graphs also includes some non-planar graphs as well.

Do one of the following:

- Show that such an idea doesn't work by providing a counterexample (such as a graph that doesn't contain a subgraph that is a subdivision of K_5 which isn't four-colorable).
- Explain why it would be impossible or very difficult to prove the four-color theorem with this approach.
- Formulate a proof of the four-color theorem using this idea. (Before you choose this option, consider that Kuratowski's theorem has been known for nearly a century, but as of the publication of this book no attempts to prove the four-color theorem *by hand* have been accepted by the mathematics community; the only accepted proof involves computer calculations.)

Note: The answer key doesn't include answers to the challenge problems. These problems are intended to encourage you to think about the ideas. However, you may want to consider how this problem relates to the challenge problems from Chapters 10 and 16, and how this problem relates to Chapter 27.

7 Kempe Chains

In 1879, Alfred Kempe recognized that pairs of colors in PG's make strings called **Kempe chains** [Ref. 2]. For example, the MPG below shows R-Y Kempe chains. Most of the shaded vertices below participate in one very long section of the R-Y Kempe chains. (If a chain appears to end near the top, right, bottom, or left, it may continue along one of the "outside" edges.) There are also a few short sections of R-Y Kempe chains in the figure below:

- Near the bottom center, there is a short chain with just one R and one Y.
- Near the top left, there is a lone R surrounded by blues and greens.
- Toward the right, a couple of rows up is a lone Y surrounded by blues and greens.
- At the bottom, a few columns from the left is another lone Y.

Note that the R-Y at the bottom right is actually part of the main, very long section.

There are also B-G Kempe chains that complement the R-Y Kempe chains. If you focus on the non-shaded vertices, you will see the B-G Kempe chains. There are three lengthy sections of B-G Kempe chains and one short section at the top right in the MPG below.

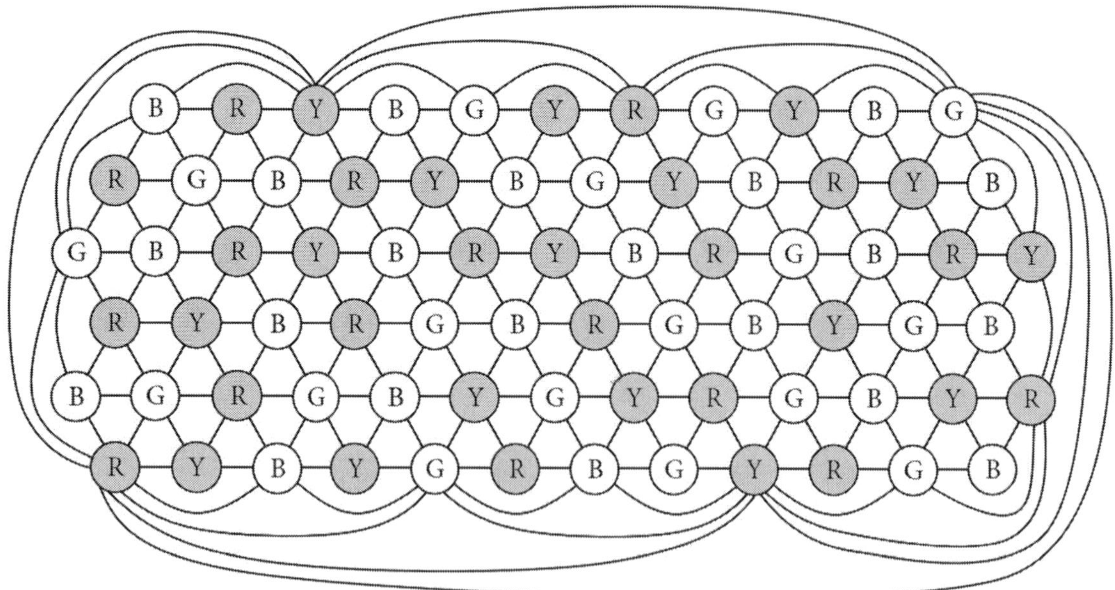

There are three different ways to visualize Kempe chains in any PG. In the previous MPG, the shading helped to visualize the R-Y and B-G Kempe chains, whereas the same MPG below helps to visualize the G-R and B-Y Kempe chains.

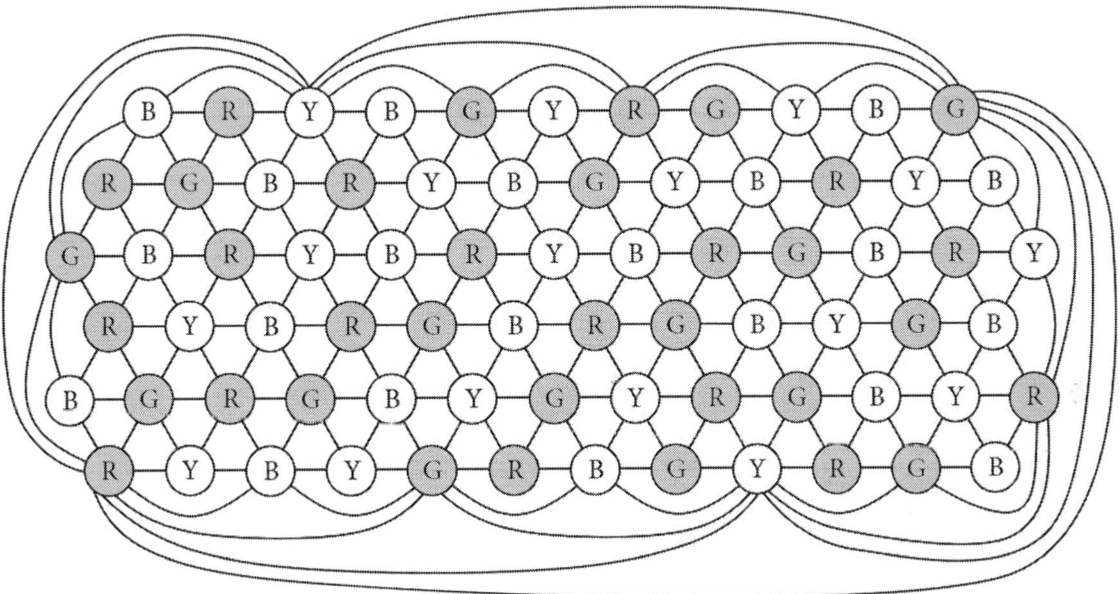

The same MPG is shown once again below, this time helping to visualize the G-Y and B-R Kempe chains. In all three cases, this MPG has multiple chain sections for each color pair.

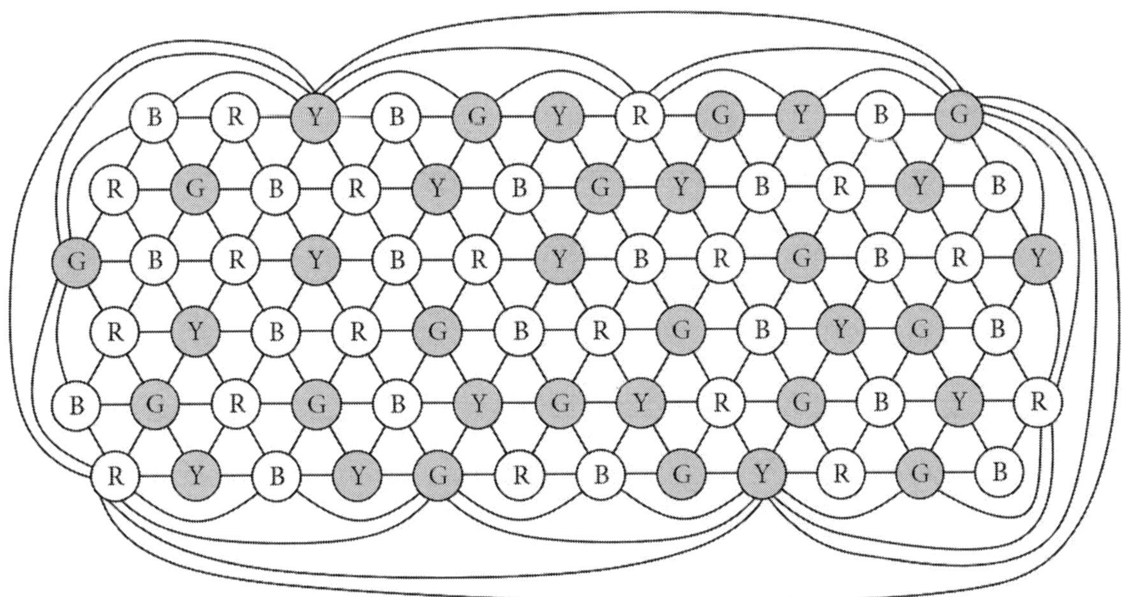

Each section of a Kempe chain is isolated from the other sections of the same color pair. For example, examine the B-G Kempe chains on the first graph of this chapter, which has three lengthy sections and one short section. Note how these four sections of B-G Kempe chains are isolated from one another by the R-Y Kempe chains. (The sections of R-Y Kempe chains are similarly isolated from one another by the B-G Kempe chains. There is a similar relationship between the two color pairs of the other two figures. This relationship is characteristic of all PG's.)

Since each section of a Kempe chain is isolated from the other sections of the same color pair, the colors of any section of a Kempe chain may be reversed and still satisfy the four-color theorem. This is an important and useful concept.

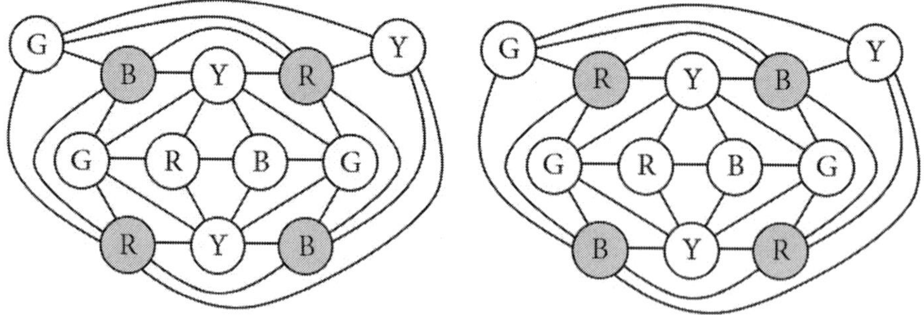

The shading of one section of the B-R chains above illustrates how the colors of any section of any Kempe chain may be reversed. Note that we reversed the colors of one section of the B-R chains, but didn't reverse the colors of the section in the center. Each section of the same chain may have its colors reversed independently of the other sections of that chain.

Why do PG's have Kempe chains? It's easy to understand why MPG's have Kempe chains. (Since a PG is formed by removing edges from a MPG and since the coloring that works for the MPG also works for the PG, it follows that PG's also have Kempe chains.)

- A MPG is triangulated. It consists of faces with three edges and three vertices.
- The three vertices of each face must be three different colors.

- Each edge is shared by two adjacent triangles, which form a quadrilateral.

- Each quadrilateral will have 3 or 4 different colors. It has 3 colors if the two vertices opposite to the shared edge happen to be the same color.

- For each quadrilateral, at least 1 vertex and at most 3 of the four vertices have colors for any color pair. For example, a quadrilateral with R, G, B, and G has 1 vertex with R-Y and 3 vertices with B-G, or you could think of it as 1 vertex with B-Y and 3 vertices with G-R, or you could think of it as 2 vertices of B-R and 2 vertices of G-Y. In the latter case, the 2 G's are not consecutive colors of the same chain.

- As you combine more triangles together (the quadrilateral only combined two) and consider the possible colorings, you will see sections of Kempe chains emerge. We will see how these Kempe chains emerge in Chapter 21.

It's also easy to see how one color pair (like R-Y) will border its counterpart (B-G):

- Draw one R vertex and one Y vertex connected by an edge.

- If a new vertex connects to each of these, it must be B or G.

- If a new vertex connects to the R but not the Y, it may be Y, B, or G.

- If a new vertex connects to the Y but not the R, it may be R, B, or G.

- Either the R-Y chain will continue to grow, or it will get surrounded by B and G.

- If you focus on the B and G, you will draw a similar conclusion for its chain.

- If one chain becomes completely surrounded by its counterpart, a new section of the chain may emerge on the other side of its counterpart.

Kempe demonstrated that all vertices with degree four (those which are connected to exactly four other vertices) are four-colorable [Ref. 2]. For example, consider the center vertex below.

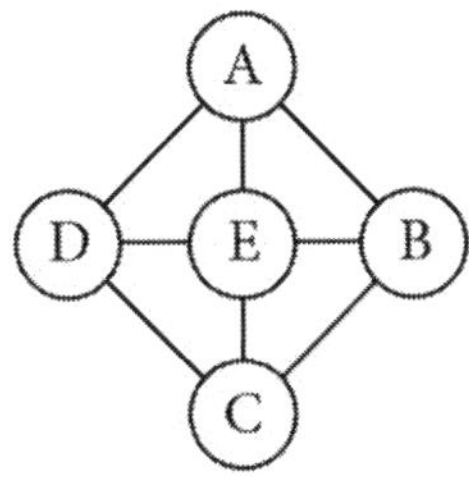

In the previous figure, vertex E is degree four since it is connected to four other vertices. Kempe showed that vertices A, B, C, and D can't be forced to be four different colors, such that vertex E can always be colored without violating the four-color theorem, regardless of how the rest of the MPG may look outside of the portion shown on the previous page.

- A and C are either part of the same section of an A-C Kempe chain, or they each lie on separated sections of A-C Kempe chains. (If A and C are red and yellow, for example, then an A-C chain is a red-yellow chain.)

- If A and C each lie on separated sections of an A-C Kempe chain, the colors of one of the sections could be reversed, which effectively recolors C to match A's color.

- If A and C are part of the same section of an A-C Kempe chain, B and D must each lie on separated sections of B-D Kempe chains because the A-C Kempe chain will block any B-D Kempe chain from reaching D from B. (If B and D are blue and green, for example, then a B-D Kempe chain is a blue-green chain.) Since B and D each lie on separated sections of B-D Kempe chains in this case, the colors of one of the sections of B-D Kempe chains could be reversed, which effectively recolors D to match B's color.

- Therefore, either C can be made to have the same color as A or D can be made to have the same color as B by reversing a separated section of a Kempe chain.

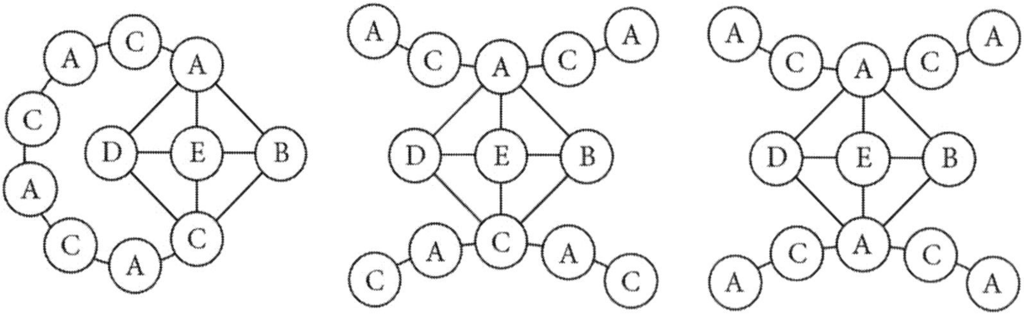

The graphs above are incomplete. These figures only show a vertex with degree four (vertex E), its nearest neighbors (A, B, C, and D), and segments of A-C Kempe chains. The entire graphs would also contain several other vertices (especially, more colored the same as B or

D) and enough edges to be MPG's. The left figure has A connected to C in a single section of an A-C Kempe chain (meaning that the vertices of this chain are colored the same as A and C). The left figure shows that this A-C Kempe chain prevents B from connecting to D with a single section of a B-D Kempe chain. The middle figure has A and C in separate sections of A-C Kempe chains. In this case, B could connect to D with a single section of a B-D Kempe chain. However, since the A and C of the vertex with degree four lie on separate sections, the color of C's chain can be reversed so that in the vertex with degree four, C is effectively recolored to match A's color, as shown in the right figure. Similarly, D's section could be reversed in the left figure so that D is effectively recolored to match B's color.

Kempe also attempted to demonstrate that vertices with degree five are four-colorable in his attempt to prove the four-color theorem [Ref. 2], but his argument for vertices with degree five was shown by Heawood in 1890 to be insufficient [Ref. 3]. Let's explore what happens if we attempt to apply our reasoning for vertices with degree four to a vertex with degree five.

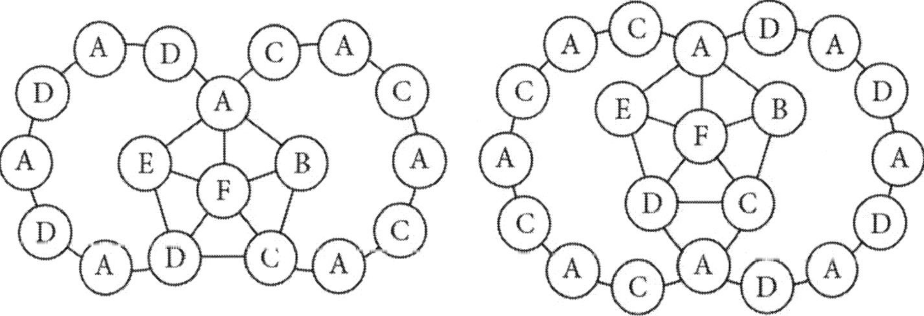

In the figures above, note how A can connect to both C and D with two different Kempe chains. Also note how these Kempe chains may cross as illustrated on the right. Note how the A-C Kempe chain separates the B and D which connect to F. It is tempting to use this separation to reverse the colors of B and D in a segment of a Kempe chain on one side of the A-C Kempe chain, such that the B connecting to F changes to D. Similarly, note how the A-D Kempe chain separates the C and E which connect to F. It is tempting to use this separation to reverse the colors of C and E in a segment of a Kempe chain on one side of the A-D Kempe chain, such that the E connecting to F changes to C. If we could make both of these changes, the vertices connected to F would be A, D, C, D, and C (since B switched to D and since E switched to C). Now F is only connected to three colors, showing that F is four-colorable. However, there is a problem with this argument. See if you can figure out what the problem is before you read on.

To help see the problem, let's draw the graph where the A-C and A-D Kempe chains cross using the numbers 1-4 for four different colors (such as green, blue, red, and yellow). In our previous argument, the color reversals would turn the two 2's into 4's or 3's.

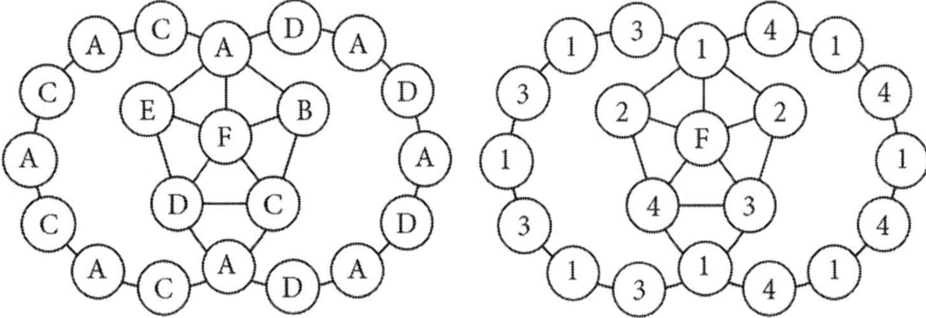

One reversal swaps 2 and 4 on B's side of the 1-3 chain. When we perform this reversal, in the crossed-chain diagram above some of the 4's in the 1-4 chain may turn into 2's. Another reversal swaps 2 and 3 on E's side of the 1-4 chain. When we perform this reversal, in the crossed-chain diagram above some of the 3's in the 1-3 chain may turn into 2's.

Color reversals are valid when we perform a single reversal at a time. The problem in the previous diagram with crossed chains is that after performing one of the reversals, the other chain may be broken, such that the second reversal is no longer isolated. That is, if we first swap 2 and 4 on B's side of the 1-3 chain and some of the 4's in the 1-4 chain turn into 2's, then when we swap 2 and 3 for E, how can we guarantee that C (which is a 3) won't change into a 2 at the same time that E changes into a 3? We can't, as the diagrams on the following page illustrate.

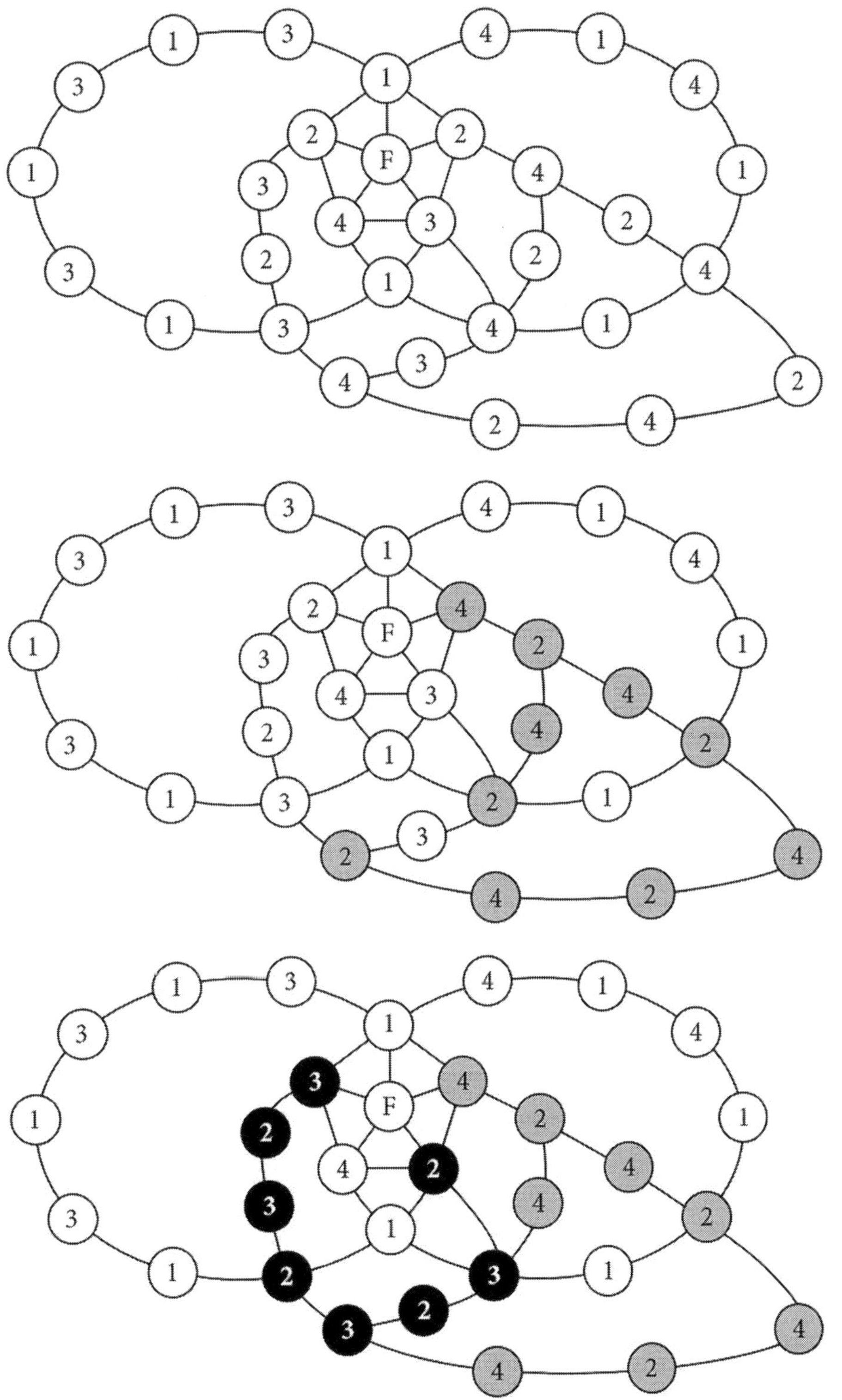

The previous diagrams show that when the two color reversals are performed one at a time in the crossed-chain graph, the first color reversal may break the other chain, allowing the second color reversal to affect the colors of one of F's neighbors. When we performed the 2-4 reversal to change B from 2 to 4, this broke the 1-4 chain. When we then performed the 2-3 reversal to change E from 3, this caused C to change from 3 to 2. As a result, F remains connected to four different colors; this wasn't reversed to three as expected.

Unfortunately, you can't perform both reversals "at the same time" for the following reason. Let's attempt to perform both reversals "at the same time." In this crossed-chain diagram, when we swap 2 and 4 on B's side of the 1-3 chain, one of the 4's in the 1-4 chain may change into a 2, and when we swap 2 and 3 on E's side of the 1-4 chain, one of the 3's in the 1-3 chain may change into a 2. This is shown in the following figure: one 2 in each chain is shaded gray. Recall that these figures are incomplete; they focus on one vertex (F), its neighbors (A thru E), and Kempe chains. Other vertices and edges are not shown.

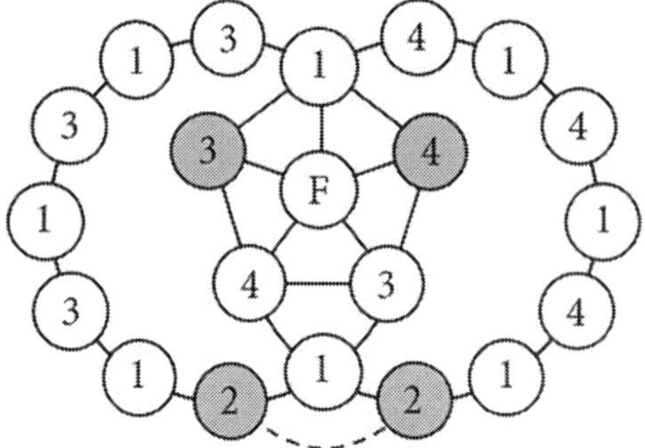

Note how one of the 3's changed into 2 on the left. This can happen when we reverse C and E (which were originally 3 and 2) on E's side of the 1-4 chain. Note also how one of the 4's changed into 2 on the right. This can happen when we reverse B and D (which were originally 2 and 4) outside of the 1-3 chain. Now we see where a problem can occur when attempting to swap the colors of two chains at the same time. If these two 2's happen to be connected by an edge like the dashed edge shown above, if we perform the double reversal at the same time, this causes two vertices of the same color to share an edge, which isn't allowed. **We'll revisit Kempe's strategy for coloring a vertex with degree five in Chapter 25.**

Chapter 7 Exercises

1. Color the graph below on the right by reversing the longer section of the two sections of the G-Y Kempe chains from the left graph. Does the coloring of the new graph still satisfy the four-color theorem? Are there any sections of Kempe chains (B-G, B-R, B-Y, G-R, G-Y, or R-Y) that can't be reversed in the new graph without causing the new coloring to no longer satisfy the four-color theorem? If so, which sections of which chains? If the coloring of a graph satisfies the four-color theorem and sections of Kempe chains are reversed one at a time, will the new coloring always satisfy the four-color theorem?

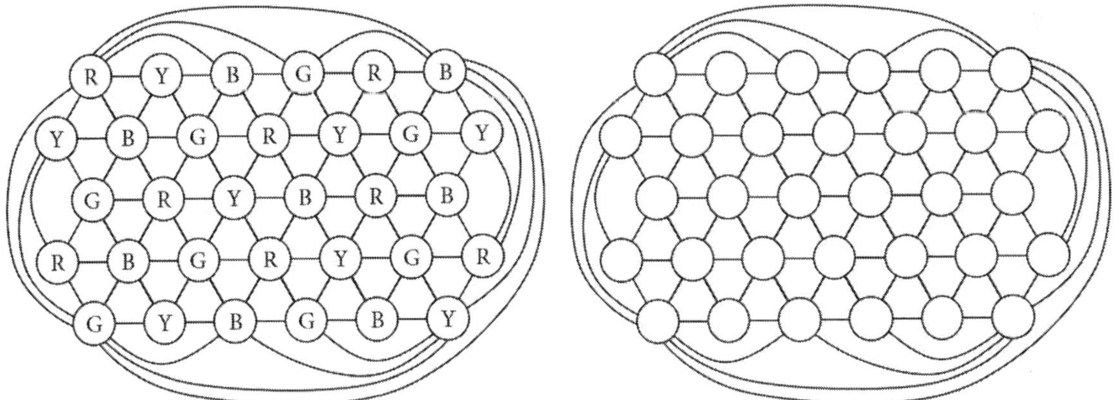

2. The graph below shows a vertex with degree four and two Kempe chains. There are other vertices and edges which aren't shown. If either Kempe chain A-B or C-D is reversed, E is still surrounded by 4 colors. How can a graph like this can satisfy the four-color theorem?

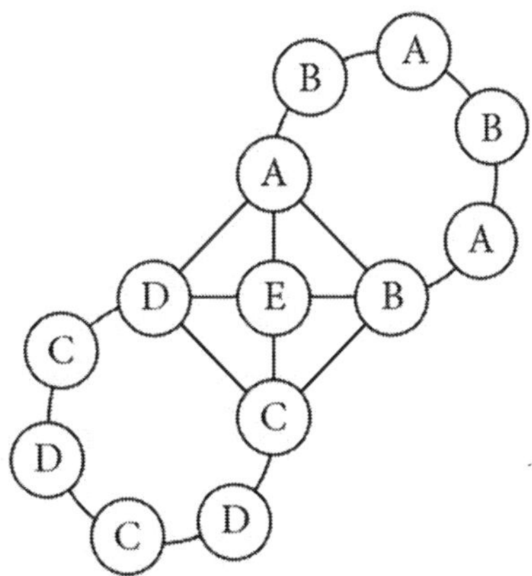

Challenge problem 1: Every MPG is triangulated, meaning that each edge is shared by two triangles. For any MPG, if you choose any of its faces and two other faces that each shares an edge with the first face, you will obtain a structure like that shown below. The partial graph below shows face BCE and two other faces (ABE and CDE) that share an edge with it. (The third face that shares an edge with BCE is not shown. There are also many other vertices and edges in the MPG that are not shown.) Apply Kempe chains to prove that, regardless of what the rest of the MPG looks like, the five vertices shown below can *always* be colored using no more than four different colors. Since we may apply this argument to any face in any MPG and two of that face's neighboring faces, does this prove the four-color theorem? Explain.

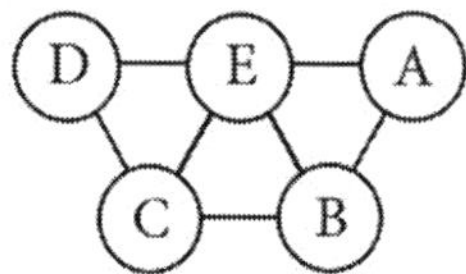

(Note how this differs from Kempe's argument for a vertex with degree five – even though every vertex in the diagram above may be degree five or higher – in that there isn't a central vertex connecting to all five of these vertices. We only need to color the five vertices shown above using *four* colors, whereas in Kempe's argument for a vertex with degree five we need to color the five surrounding vertices using *three* colors. This problem is simpler than Kempe's problem of the vertex with degree five, since we only need to recolor a single vertex.)

Note: The answer key doesn't include answers to the challenge problems. These problems are intended to encourage you to think about the ideas.

Challenge problem 2: The partial graph for a MPG below shows vertices X and Y with degree five. There are also many other vertices and edges in the MPG that are not shown. Can you apply the concept of Kempe chains to prove that X and Y can always be chosen so that both vertices are always four-colorable? Either prove this, or explain why this is impossible or very difficult. Could this proof (if it can be done) be used to prove the four-color theorem? Explain. (The main idea is this: Can you apply Kempe chains to A thru F so that X and Y are always four-colorable?)

Note that if you chain A to D and chain A to E, you can reverse F, but don't need to worry about reversing Y (at least for the first part of the problem, it is free to be chosen as desired). You should also not only consider the possibility of using Kempe chains to force two vertices connected to either X or Y to be different, but should also consider the possibility of using Kempe chains to force one vertex connected to X and another vertex connected to Y to be the *same* color. For example, connecting B to F and C to E, you can force two neighbors of X to be the same colors as two neighbors of Y. (Note that there is also the special case where one or both of E and F could be the *same* vertex as B or C, for example.)

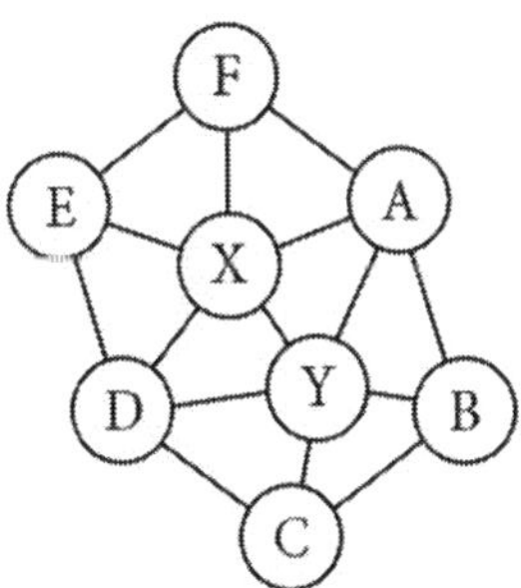

Note: The answer key doesn't include answers to the challenge problems. These problems are intended to encourage you to think about the ideas. We will consider a similar concept again in one of the challenge problems for Chapter 18.

Challenge problem 3: The MPG below is "nearly four-colored." One vertex has a fifth color, X. If the edge connecting the two shaded vertices is contracted (see Chapter 6), the graph would then be properly four-colored.

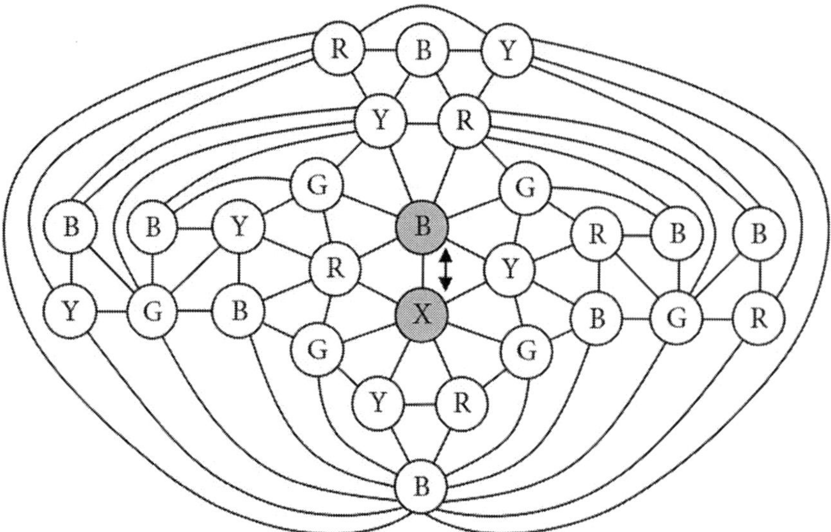

Show that X can be moved down to the left onto the vertex currently colored Y by reversing the colors of one section of a Kempe chain, allowing the graph to be four-colored. If we try to move X to the left onto the vertex currently colored G, the analogous color reversal poses a problem. Explain. Is it possible to move X onto any desired vertex in the entire MPG? Can you prove that such a "nearly four-colored" MPG can always be four-colored by moving X onto a new vertex? Can you use this to prove the four-color theorem?

Note: The answer key doesn't include answers to the challenge problems. These problems are intended to encourage you to think about the ideas. You may wish to review your ideas for this solution when you read about VS3 in Chapter 18.

8 A Few Notable Planar Graphs

The **triangle graph** has 3 vertices and 3 edges. The triangle graph has the fewest vertices of any MPG, is three-colorable, and its edges make a single complete cycle (a closed chain). It is both a cycle graph (with its edges forming a closed chain) and a complete graph (every vertex connects to all of the other vertices), which is why it may be called C_3 or K_3.

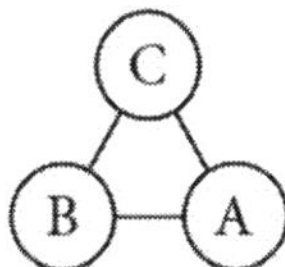

The **tetrahedral graph** has 4 vertices and 6 edges. The tetrahedral graph has the most vertices that a complete graph can have and also be a MPG, and has the most vertices that a complete graph can have and be four-colored.

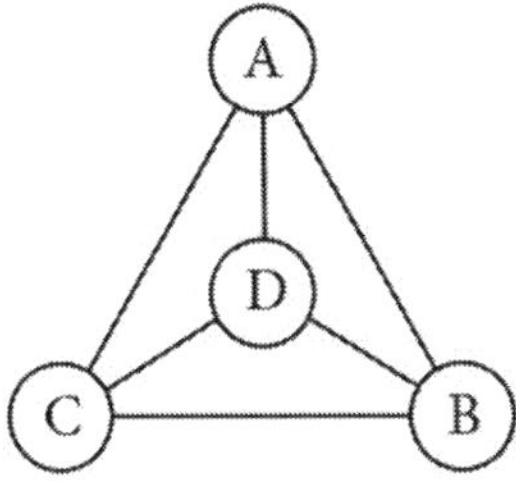

A **pentahedral** MPG has 5 vertices and 9 edges. It is the dual of the square pyramid. Recall from Chapter 4 that the dual representation swaps the roles of the vertices and faces between graphs and maps (see the solution to Problem 1 in Chapter 4). If you add edge AC to the pentahedral graph shown below, it would become the complete graph K_5.

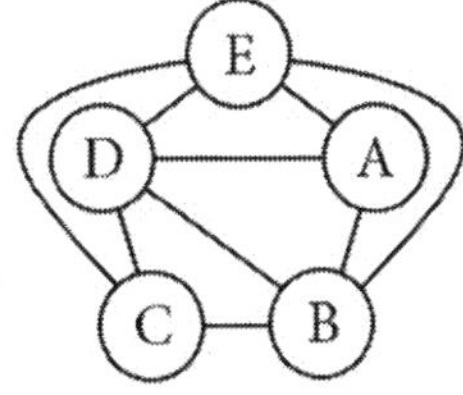

There are two structurally different **hexahedral** MPG's which have 6 vertices and 12 edges. One of these is the **octahedral** MPG. (Here, the prefix "octa," meaning 8, refers to the faces, not the vertices. An octahedron is a polyhedron formed by joining two square pyramids at their square base, such that it has 6 vertices, 12 edges, and 8 triangular faces. It is the dual polyhedron to the cube, which has 8 vertices, 12 edges, and 6 faces. See the solution to Problem 1 in Chapter 4.) The octahedral MPG is the only MPG where every vertex has degree 4. The other hexahedral MPG has two vertices with degree 5, two with degree 4, and two with degree 3. (In the case of hexahedral, the prefix "hexa" indicates 6 vertices, and includes the octahedral graph.)

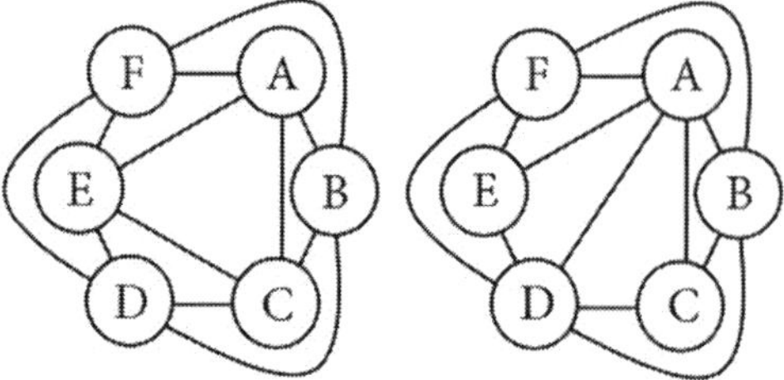

The number of structurally different MPG's grows quickly with the number of vertices:

- All MPG's with 5 vertices are structurally equivalent.
- There are 2 structurally different MPG's with 6 vertices.
- There are 5 structurally different MPG's with 7 vertices.
- There are 14 structurally different MPG's with 8 vertices.
- There are 50 structurally different MPG's with 9 vertices.
- There are 233 structurally different MPG's with 10 vertices.
- There are 1249 structurally different MPG's with 11 vertices.

The **icosahedral** MPG has 12 vertices and 30 edges. It is unique in that it is the only MPG where every vertex has degree 5. (The icosahedron is the dual to the dodecahedron, which has 20 vertices, 30 edges, and 12 faces. See the solution to Problem 1 in Chapter 4.)

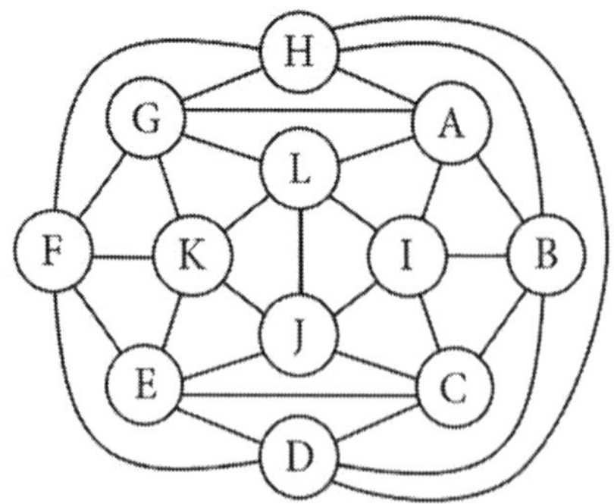

The Fritsch MPG uses just 9 vertices to provide a counterexample to Kempe's argument (Chapter 7) for four-coloring a vertex with degree 5 [Ref. 8]. (We will explore how in the exercises at the end of the chapter. We will also explore Kempe chains in the Fritsch MPG in Chapter 25.) In comparison, Heawood's graph (not shown) did this with 25 vertices [Ref. 3]. The Soifer graph (not shown) also illustrates this problem with 9 vertices, though you must add an edge to the Soifer graph to turn it into a MPG [Ref. 8].

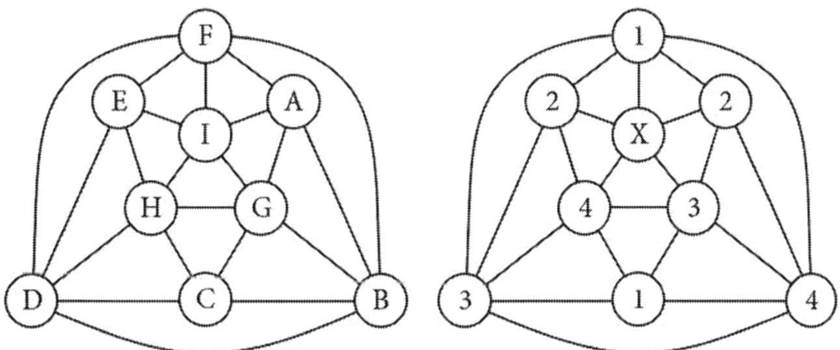

The Errera MPG with 17 vertices illustrates the problem with Kempe's argument without any vertices with a degree less than 5 [Ref. 9]. You will be able to see Kempe chains in the Errera MPG in an exercise in Chapter 25.

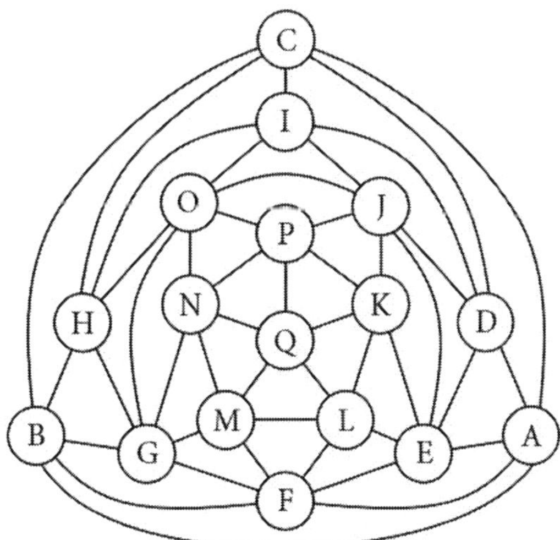

Chapter 8 Exercises

1. There are 5 structurally different heptahedral MPG's. These have 7 vertices. Draw one of each in the space below.

2. There are 14 structurally different MPG's with 8 vertices. It would be a tedious exercise to draw all 14. Instead, we challenge you to draw one MPG with 8 vertices that doesn't have a single vertex with degree 3, and to draw another MPG that has exactly two vertices with exactly degree 6 and exactly two more vertices with exactly degree 3.

3. There are 50 structurally different MPG's with 9 vertices. It would be a tedious exercise to draw all 50. Instead, we challenge you to draw one that has exactly 3 vertices with degrees exactly equal to 6.

4. In the Fritsch MPG below, vertex I is a vertex with degree five. We colored A, G, H, E, and F like Kempe colored the vertices connected to his vertex with degree five. We colored D so that there would be a 1-3 Kempe chain connecting F to G. We colored B so that there would be a 1-4 Kempe chain connecting F to H. Observe that these two chains cross at C.

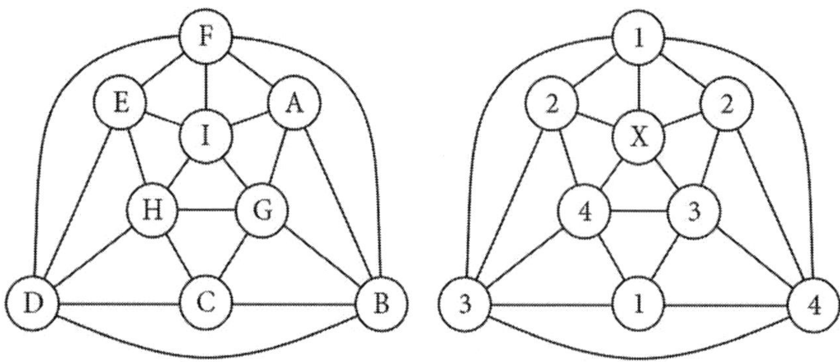

Below on the left, recolor the graph above by reversing the colors of the 2-4 chain involving vertex A. Below on the right, recolor the graph below on the left by reversing the colors of the 2-3 chain involving vertex E. Is vertex I (marked with an X) now four-colorable?

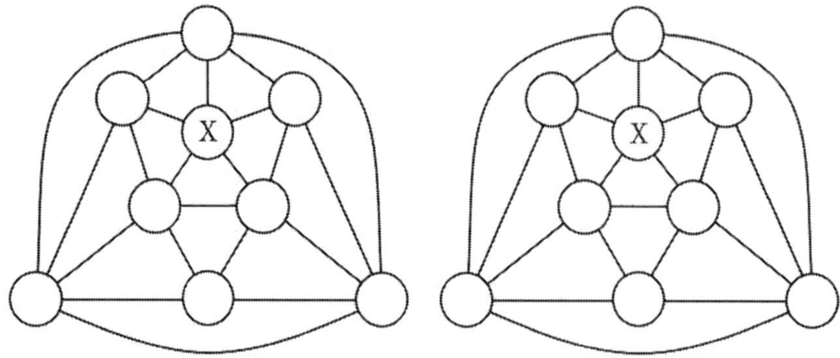

Note: This exercise is continued on the next page.

On the graph below, recolor the graph from the top of the previous page by simultaneously reversing the colors of the 2-4 chain involving vertex A and the 2-3 chain involving vertex E. Is vertex I (marked with an X) now four-colorable? Is the entire graph now properly four-colored?

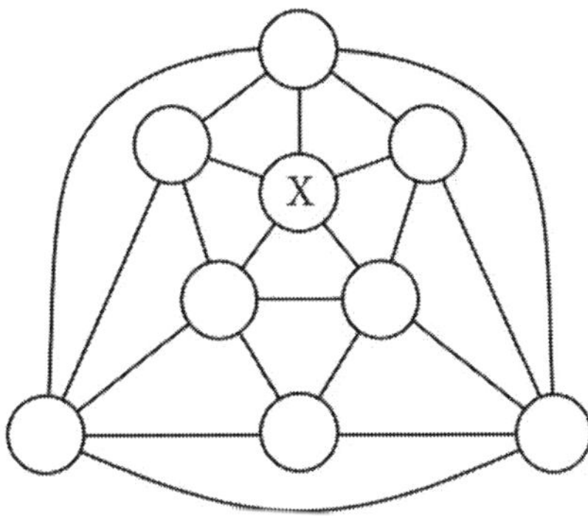

Can the Fritsch graph be four-colored? Does this graph disprove the four-color theorem? Show or explain each answer.

Note: We will consider the Fritsch graph again in Chapter 25.

9 Counting Ways

There is more than one way to color a MPG so that it satisfies the four-color theorem. For example, the following table lists 24 different ways to color the graph shown below. Note that R = red, B = blue, C = green, and Y = yellow. The graphs of this chapter include dashed lines to help show you that these are MPG's. Imagine moving the dashed lines "outside."

BGRGY	BGYGR	BRGRY	BRYRG	BYGYR	BYRYG
GBRBY	GBYBR	GRBRY	GRYRB	GYBYR	GYRYB
RBGBY	RBYBG	RGBGY	RGYGB	RYBYG	RYGYB
YBGBR	YBRBG	YGBGR	YGRGB	YRBRG	YRGRB

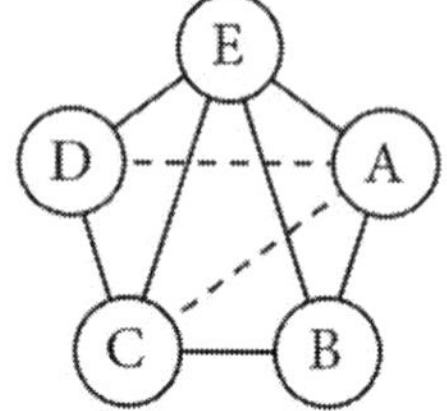

Note that these 24 ways are really just variations of a single way (ordered ABCDE). If you swap blue and green, for example, BGRGY becomes GBRBY, and then if you swap blue and yellow GBRBY becomes GYRYB. We can get all 24 ways by color swapping.

We can reduce the number 24 down to 1 if we ask a slightly different question. Instead of asking, "How many ways are there to color the graph?" we can ask, "After fixing the colors of one triangle, how many ways are there to color the remaining vertices?" You don't want to first color three random vertices because if it turns out that two of those colors needed to be the same in order to four-color the graph, you'll run into a problem. By first coloring three vertices that lie on one triangle, you "know" that those vertices must be different colors.

Looking at triangle ABC, we could choose A to be red, B to be blue, and C to be green. It then follows that D is blue and E is yellow. This results in the single answer RBGBY.

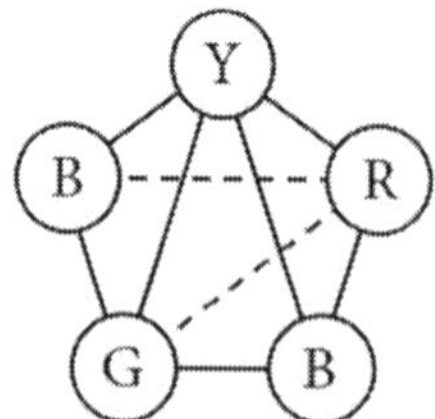

Now let's compare how many ways there are to color two different MPG's with 6 vertices such that they satisfy the four-color theorem. The left graph below can be colored 24 ways, and the right graph below can be colored 96 ways. The dashed lines can be moved "outside."

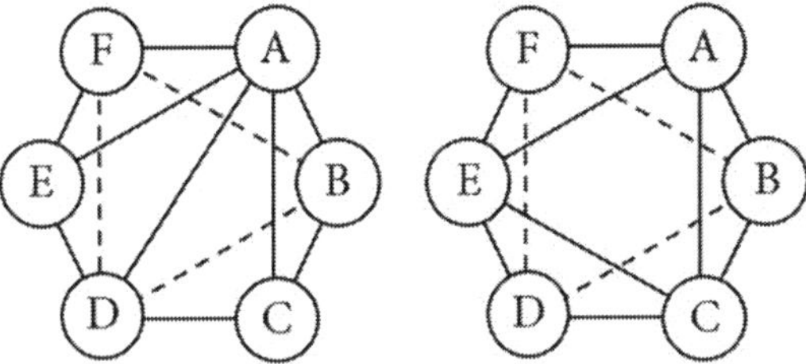

Rather than list dozens of ways to color each graph, as we did with the previous graph, let's alter the question in order to reduce the total number of ways. Let's choose to color A red, B blue, and G green, which we may do because in both cases ABC happens to be a triangle (which guarantees that these three vertices will have different colors). After coloring A red, B blue, and G green, there is only one way to color the left graph (RBGYBG), whereas there are four ways to color the right graph (RBGRBG, RBGRBY, RBGRYG, and RBGYBG).

- In the left graph, since D connects to A, B, and C, we must color D yellow once A, B, and C are set. Similarly, since F connects to A, B, and D, we must color F green once A, B, and D are set. The only color remaining for E is blue.

- The right graph is actually three-colorable (RBGRBG). Here, D can be the same as A (red) or it can be yellow, E can be the same as B (blue) or it can be yellow, and F can be the same as C (green) or it can be yellow. However, since D, E, and F connect to one another, only one of these can be yellow. This gives four possible colorings.

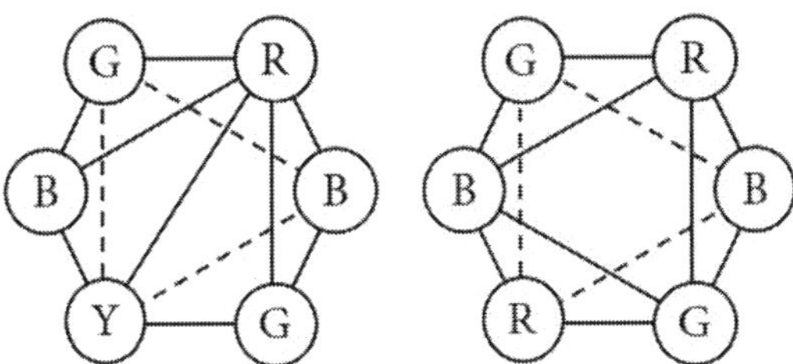

The left graph is more restrictive because A, B, C and D form a K_4 subgraph and because A, B, D, and F form another K_4 subgraph. The right graph is less restrictive because it doesn't have any K_4 subgraphs. A K_4 graph is a complete graph with 4 vertices (see the next page).

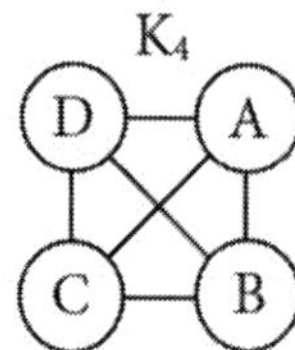

In the K_4 graph, every vertex connects to all three of the other vertices. The first graph of this chapter (with 5 vertices) has K_4 subgraphs (like ABCE) and one of the graphs with 6 vertices has K_4 subgraphs (such as ABCD). These two graphs could be colored 24 ways; or after first coloring A, B, and C, there was one way to color the remaining vertices. The other graph with 6 vertices doesn't have any K_4 subgraphs. This graph could be colored 96 ways; or after first coloring A, B, and C, there are still four ways to color the remaining vertices.

Let's examine some more MPG's and see if K_4 plays a similar role in restricting the number of ways that the graphs can be colored and satisfy the four-color theorem. The following MPG's have 7 vertices. As usual, the dashed lines help you visualize which of the edges can be moved "outside" of the polygon to convince you that these *are* MPG's. Each of these has triangle ABC, so we may choose to color A red, B blue, C green, and count how many ways there are to color the remaining vertices such that the coloring satisfies the four-color theorem.

- The left MPG can be colored one way: RBGYGBG (ordered A thru G, as usual).
- The second MPG from the left can be colored one way: RBGBYGY.
- The third MPG from the left can be colored four ways: RBGRBGY, RBGRBYG, RBGRYBG, and RBGYBYG.
- The right MPG can be colored five ways: RBGRBYG, RBGRYBG, RBGYBYG, RBGYRBG, and RBGYRYG.

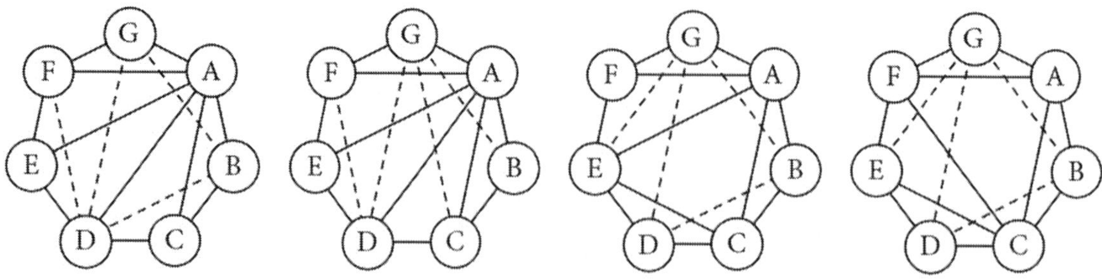

Now let's compare the number of ways that each of the previous MPG's can be colored with its K_4 subgraphs. (We're looking for subgraphs that can be found without making any subdivisions and without contracting any edges.)

- The left MPG can be colored one way. Its K_4 subgraphs are ABCD, ABDG, ADEF, and ADFG. All 7 vertices participate in K_4 subgraphs.
- The second MPG from the left can be colored one way. Its K_4 subgraphs are ABCG, ACDG, ADEF, and ADFG. All 7 vertices participate in K_4 subgraphs.
- The third MPG from the left can be colored four ways. Its only K_4 subgraph is AEFG. Only 4 of its vertices participate in K_4 subgraphs.
- The right MPG can be colored five ways. It doesn't have any K_4 subgraphs. None of its vertices participate in K_4 subgraphs.

The following graphs show a half dozen examples of the many MPG's that can be drawn with 12 vertices. These are among the more extreme examples in terms of restrictiveness. Imagine moving the dashed lines "outside" of the polygon to see that these *are* MPG's.

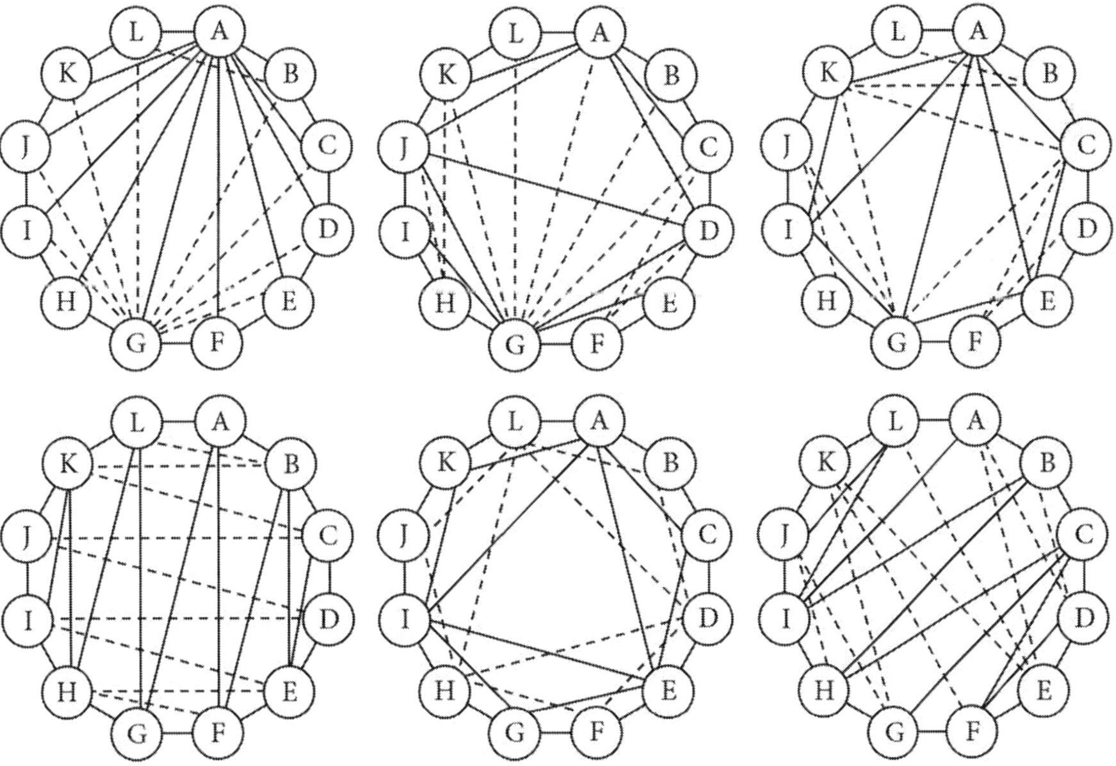

Not all of the previous graphs share the same triangles, so we will set the first three colors differently for each case.

- For the top left MPG, we will first color B blue, G green, and L yellow. The graph can then be colored one way: A red, C yellow, D blue, E yellow, F blue, K blue, J yellow, I blue, and H yellow. This gives the coloring RBYBYBGYBYBY (from A thru L, as usual). Its K_4 subgraphs are ABCG, ABGL, ACDG, ADEG, AEFG, AGHI, AGIJ, AGJK, and AGKL. All 12 vertices participate in K_4 subgraphs.

- For the top center MPG, we will first color A red, D blue, and G green. The graph can then be colored one way: C yellow, B blue, J yellow, K blue, L yellow, F red, E yellow, H red, and I blue. This gives the coloring RBYBYRGRBYBY. Its K_4 subgraphs are ABCG, ACDG, ADGJ, AGJK, AGKL, DEFG, and GHIJ. All 12 vertices participate in K_4 subgraphs.

- For the top right MPG, we will first color A red, C blue, and G green. The graph can then be colored one way: K yellow, B green, L blue, E yellow, F red, D green, I blue, J red, and H yellow. This gives the coloring RGBGYRGYBRYB. Its K_4 subgraphs are ABKL, ACEG, AGIK, CDEF, GHIJ, and GIJK. All 12 of its vertices participate in K_4 subgraphs.

- For the bottom left MPG, we will first color A red, B blue, and L yellow. There are many ways to color this graph, such as RBYBGYGRYRGY or RBYBRYBGYGRY. It doesn't have any K_4 subgraphs.

- For the bottom center MPG, we will first color A red, E blue, and I yellow. There are many ways to color this graph, such as RBYGBYGRYBGY or RBRYBRGBYRBG. It doesn't have any K_4 subgraphs.

- For the bottom right MPG (which is noteworthy in that all 12 vertices are degree 5), we will first color A red, B blue, and I green. There are many ways to color this graph, such as RBRYGBGYGBRY or RBYGYBGRGYRB. It doesn't have any K_4 subgraphs.

What trends have you noticed in this chapter?

Here is one trend that you may have noticed: A MPG that contains a K_4 subgraph (without making any subdivisions or contractions) tends to have fewer ways to be colored (using no more than four colors) compared to a MPG (with the same number of vertices) that doesn't contain any K_4 subgraphs.

This suggests that we might be able to add edges to less restrictive MPG's in order to make them more restrictive. Of course, if you add an edge to a MPG, it won't be a MPG anymore, but just like the concept of triangulation, extra edges don't hurt. If we can add an extra edge in such a way that it is easier to show that the new graph is four-colorable than it was to show for the old graph, adding an extra edge may be helpful.

In particular, it seems that it may be helpful to add an edge to a less restrictive MPG in a way that creates K_4 subgraphs. We will explore this idea in Chapter 15. As we will see in later chapters, K_4's can be useful in a few different ways:

- Their presence helps make coloring a MPG less restrictive.
- K_4's often include a vertex with degree 3. In Chapter 11, we'll see that some graphs with K_4's can be trivially colored by removing vertices with degree 3.
- Every K_4 contains a separating triangle (Chapter 12). In Chapters 12-13, we'll explore how such a separating triangle can be used to divide a MPG into smaller MPG's.

If we add an extra edge to a MPG to create a K_4 subgraph, we may wish to take advantage of these properties of K_4's. We will explore adding edges in Chapter 15, removing edges in Chapter 17, and the full significance of adding or removing edges in Chapter 27.

Chapter 9 Exercises

1. Each MPG below has three vertices already colored. For each MPG:

- Color the remaining vertices so that it satisfies the four-color theorem.

- Determine how many different ways there are to color the remaining vertices.

- Determine how many K_4 subgraphs there are (without making any subdivisions or contractions). List the K_4 subgraphs. How many vertices participate in K_4's?

Note: We repeated the graphs a couple of times in case it may help with scratch work.

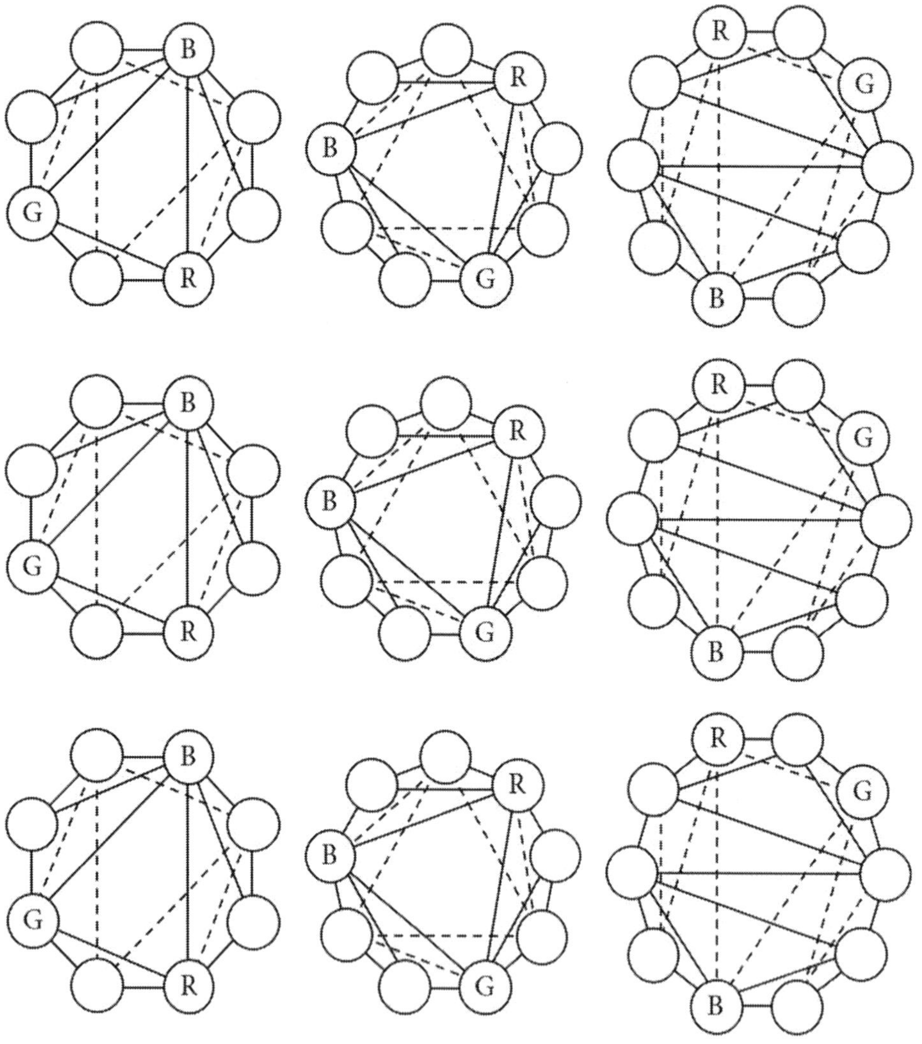

Challenge problem 1: We discussed two different methods for counting the number of ways that a MPG can be four-colored. One method is simply to count how many ways a MPG can be four-colored. A second method is to first set the colors of the vertices of one face of the MPG and then determine how many ways the MPG can be four-colored from this point onward. For this second method, does it matter which face is colored first? Either provide an example which demonstrates that this does matter or prove that it doesn't matter. How does the answer for the second method relate to the answer for the first method (if at all)?

Challenge problem 2: MPG's that contain K_4 subgraphs (without making any subdivisions or contracting edges) are either trivially four-colorable (as we'll explore in Chapter 11) or can be reduced to MPG's with fewer vertices (as we'll discuss in Chapters 13 and 16). When a MPG contains a K_4 subgraph, it either has a vertex with degree three that may be removed from the graph or has a separating triangle that lets us split the MPG into two smaller MPG's (with the K_4 no longer present), so that if we could prove that all MPG's that don't contain K_4 subgraphs are four-colorable, we could easily prove the four-color theorem. (We'll explore these ideas in later chapters.) You may have observed that MPG's that don't contain K_4 subgraphs tend to be four-colorable more than one way. If you could prove that every MPG that doesn't contain a K_4 subgraph is four-colorable more than one way (or if you could prove that it is four-colorable **<u>at least one</u>** way), then you could prove the four-color theorem. Either use this idea to prove the four-color theorem or explain why it is impossible or very challenging to prove that every MPG that doesn't contain a K_4 subgraph is four-colorable at least one way.

Note: The answer key doesn't include answers to the challenge problems. These problems are intended to encourage you to think about the ideas. However, you may wish to consider how the second challenge problem relates to Chapter 27.

Challenge problem 3: The left MPG below has one full B-Y Kempe chain and one full G-R Kempe chain. If instead the left graph is divided into B-G and R-Y Kempe chains or divided into B-R and G-Y Kempe chains, will each chain be full or will any of the chains be separated like the G-R Kempe chains shown in the right graph below? If all possible Kempe chains in the left graph are full, can you prove that if two complementary Kempe chains (like B-Y and G-R) are full in any MPG that all possible Kempe chains will also be full?

Does the number of ways that a MPG is four-colorable (without fixing any of the colors first) relate to the number of full or separated Kempe chains? If so, how? (For example, is every MPG with two complementary Kempe chains that are both full four-colorable 24 ways, and do these 24 ways correspond to 24 independent color swaps, like interchanging R and B? If there are any separated Kempe chains, does each separated segment increase the number of possible color swaps, thereby increasing the number of ways that the MPG can be colored?)

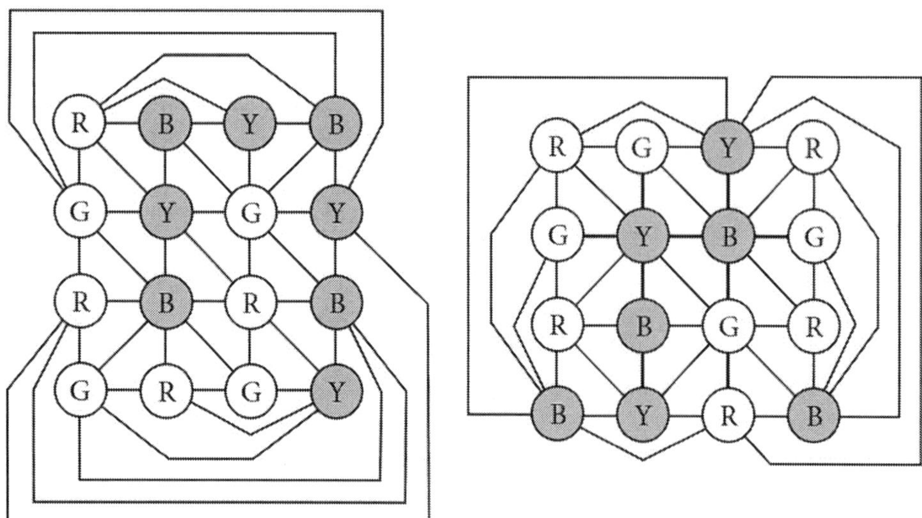

Note: The answer key doesn't include answers to the challenge problems. These problems are intended to encourage you to think about the ideas. However, once you read Chapter 22, you may wish to review your solution to this problem.

10 Logic Puzzle

Coloring a map or graph in such a way that it satisfies the four-color theorem is like solving an ill-conditioned logic problem. Why "ill-conditioned"? A typical logic problem gives you exactly the right amount of information such that if you solve the problem correctly, there is exactly one answer to the problem. When coloring an empty map or graph, the information given in the problem is insufficient, which results in multiple solutions to the problem.

A graph that is constrained such that it is four-colorable only 24 ways can be solved like a well-conditioned logic problem once you set the colors of the vertices of one triangular face. Once these three colors are set, for the most restrictive graphs there is only one way to color the remaining vertices. Logic may be applied to work out the colors of the remaining regions.

If the graph it isn't highly constrained, meaning that that there are more than 24 ways to color it (or equivalently, after setting the colors of the vertices of one triangular face, there are multiple ways to color the remaining vertices), then a logic table will be ill-conditioned. There won't be a single solution for coloring the remaining vertices.

Let's first look at a highly restrictive graph (the corresponding map is shown on the left). If we set B to be color 1, C to be color 2, and D to be color 3, there is only one solution.

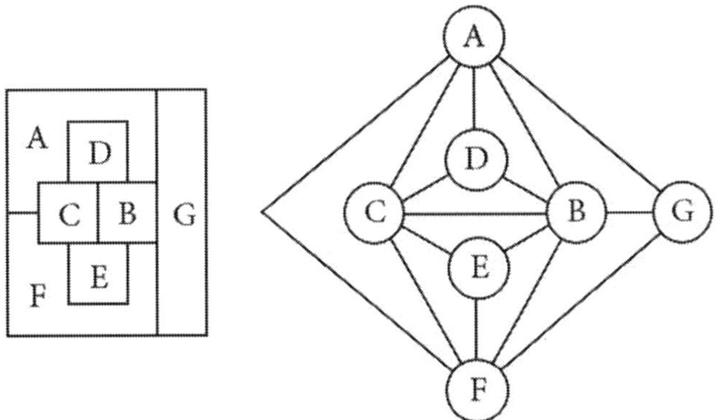

For the map and corresponding graph shown on the previous page:

- A connects to B, C, D, F, and G (but not E).

- B connects to A, C, D, E, F, and G.

- C connects to A, B, D, E, and F (but not G).

- D connects to A, B, and C (but not E, F, or G).

- E connects to B, C, and F (but not A, D, or G).

- F connects to A, B, C, E, and G (but not D).

- G connects to A, B, and F (but not C, D, or E).

We can set up an equivalent logic problem based on the sharing between regions. Since B, C, and D are the vertices of a triangular face, these three regions must be three different colors. We may assign colors 1, 2, and 3 to these regions in our logic table.

	A	B	C	D	E	F	G	1	2	3	4
A		×	×	×		×	×				
B	×		×	×	×	×	×	✓			
C	×	×		×	×	×			✓		
D	×	×	×							✓	
E		×	×			×					
F	×	×	×		×		×				
G	×	×				×					

Now we apply logic:

- A must be color 4 because it can't be the same as B, C, or D (which are 1, 2, 3).

- F must be color 3 because it can only be the same color as D.

	A	B	C	D	E	F	G	1	2	3	4
A		×	×	×		×	×				✓
B	×		×	×	×	×	×	✓			
C	×	×		×	×	×			✓		
D	×	×	×							✓	
E		×	×			×					
F	×	×	×		×		×			✓	
G	×	×				×					

- E must be color 4 because it can't be the same as B, C, or F (which are 1, 2, 3).

- G must be color 2 because it can't be the same as A, B, or F (which are 4, 1, 3).

	A	B	C	D	E	F	G	1	2	3	4
A		×	×	×		×	×				✓
B	×		×	×	×	×	×	✓			
C	×	×			×	×	×		✓		
D	×	×	×						✓		
E		×	×			×					✓
F	×	×	×		×		×			✓	
G	×	×				×			✓		

According to our solution to the logic problem, the map and graph can be colored like this:

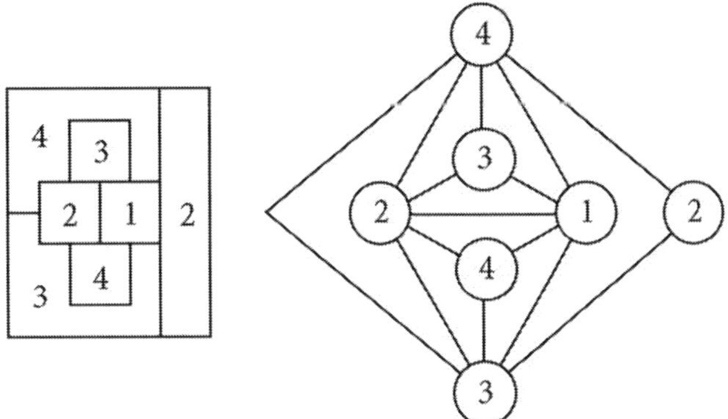

Setting the problem up with a logic table doesn't appear to have made the map or graph any easier to color. The logic table merely provides one way to show the logical decision-making involved in the four-coloring.

Now let's look at a less restrictive graph. If we set A to be color 1, C to be color 2, and E to be color 3 in the graph below, there will be multiple solutions.

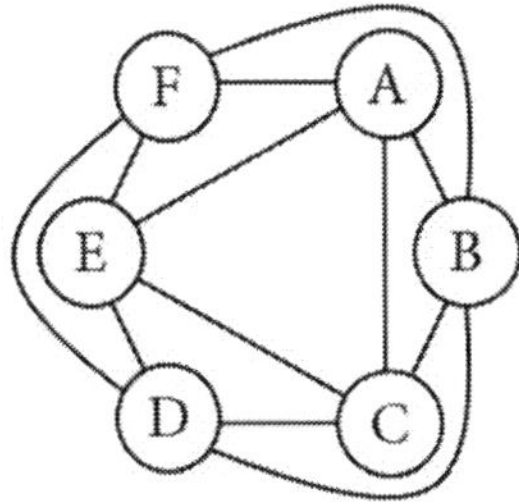

For the graph shown on the previous page:

- A connects to B, C, E, and F (but not D).
- B connects to A, C, D, and F (but not E).
- C connects to A, B, D, and E (but not F).
- D connects to B, C, E, and F (but not A).
- E connects to A, C, D, and F (but not B).
- F connects to A, B, D, and E (but not C).

We can set up an equivalent logic problem based on the sharing between regions. Since A, C, and E are the vertices of a triangular face, these three regions must be three different colors. We may assign colors 1, 2, and 3 to these regions in our logic table.

	A	B	C	D	E	F	1	2	3	4
A		×	×		×	×	✓			
B	×		×	×		×				
C	×	×		×	×			✓		
D		×	×		×	×				
E	×		×	×		×			✓	
F	×	×		×	×					

In this example, even after setting the first three colors, the problem remains ill-conditioned since the problem still has multiple solutions. For example, B can't be the same as A or C, but be could be the same as E. Therefore, B can be color 3 or 4. Similarly, D can be color 1 or 4 and F can be color 2 or 4. This allows four possible solutions. One is B = 3, D = 1, and F = 2. For the other three solutions, change one of these colors to 4. Note that two regions can't both be color 4 since regions B, D, and F are all connected to one another.

What would a logic table look like if a solution didn't exist? One example is to work out a logic table for **three-coloring** (*not* four-coloring) of the K_4 graph. In the following logic table, region D can't be colors 1, 2, or 3 because it can't be the same color as regions A, B, or C, which results in no solution (unless you introduce color 4).

	A	B	C	D	1	2	3
A		×	×	×	✓		
B	×		×	×		✓	
C	×	×		×			✓
D	×	×	×				

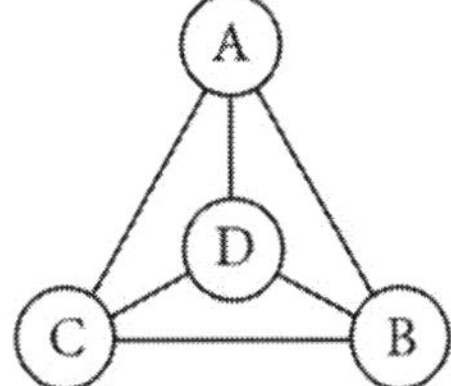

The only way to solve the logic table above is to allow region D to be a fourth color. This proves that K_4 is four-colorable, but not three-colorable.

Chapter 10 Exercises

1. Proceed to complete the logic table for the partially colored graph below. If you are able to find a unique solution to the logic table, complete the logic table and color the graph according to the completed logic table. If you aren't able to find a unique solution to the logic table, describe why not, explain what caused the problem, and interpret what this means.

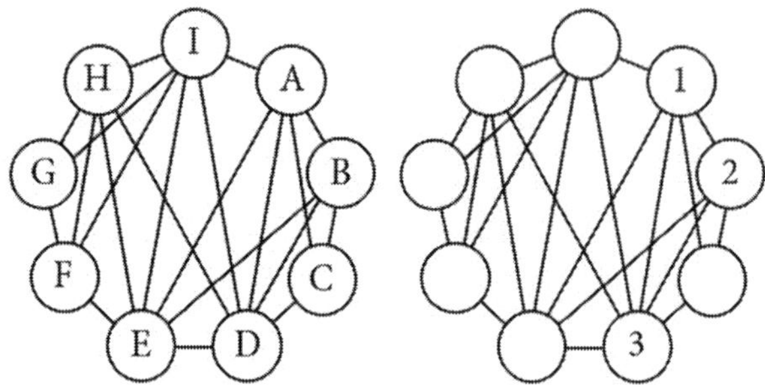

	A	B	C	D	E	F	G	H	I	1	2	3	4
A										✓			
B											✓		
C													
D												✓	
E													
F													
G													
H													
I													

2. Proceed to complete the logic table for the partially colored graph below. If you are able to find a unique solution to the logic table, complete the logic table and color the graph according to the completed logic table. If you aren't able to find a unique solution to the logic table, describe why not, explain what caused the problem, and interpret what this means.

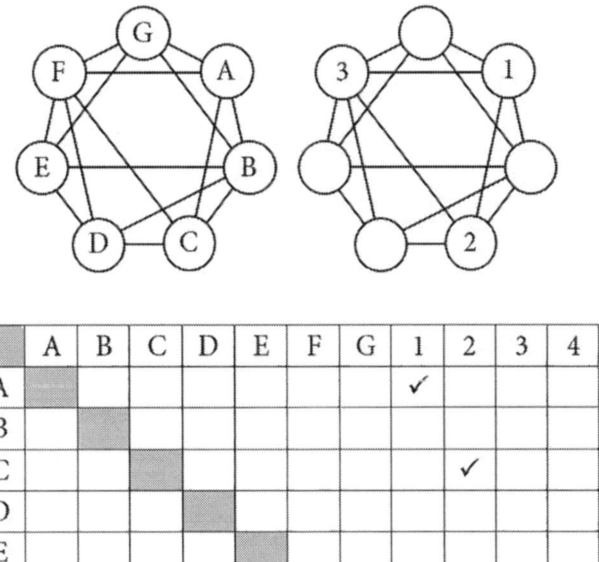

	A	B	C	D	E	F	G	1	2	3	4
A								✓			
B											
C									✓		
D											
E											
F										✓	
G											

3. Proceed to complete the logic table for the partially colored graph below. If you are able to find a unique solution to the logic table, complete the logic table and color the graph according to the completed logic table. If you aren't able to find a unique solution to the logic table, describe why not, explain what caused the problem, and interpret what this means.

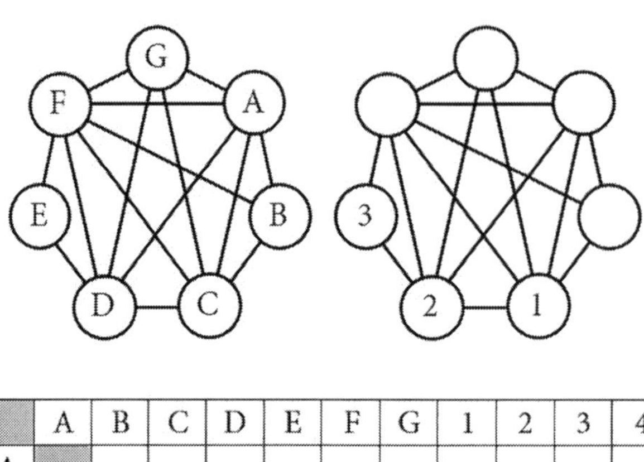

	A	B	C	D	E	F	G	1	2	3	4
A											
B											
C							✓				
D								✓			
E									✓		
F											
G											

4. Proceed to complete the logic table for the partially colored graph below. If you are able to find a unique solution to the logic table, complete the logic table and color the graph according to the completed logic table. If you aren't able to find a unique solution to the logic table, describe why not, explain what caused the problem, and interpret what this means.

Note: This logic table only has three colors. The goal for this logic table is to determine if this graph is or isn't three-colorable (not whether it is four-colorable).

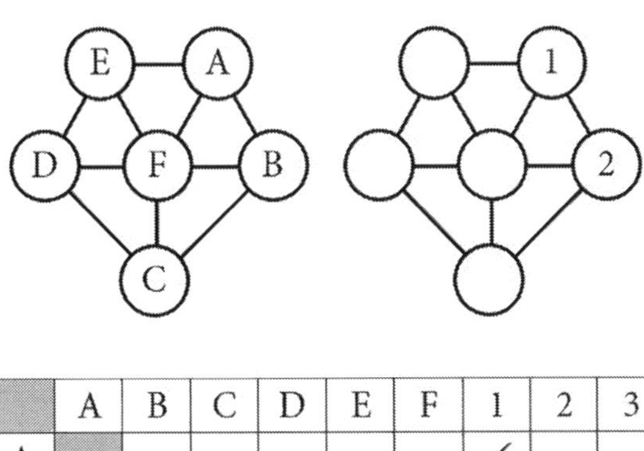

	A	B	C	D	E	F	1	2	3
A							✓		
B								✓	
C									
D									
E									
F									

Challenge problem 1: A K_4 subgraph is four-colorable, whereas a K_5 subgraph isn't four-colorable. Is it possible for a graph to contain a subgraph that is a subdivision of K_5 (which would make the graph nonplanar according to Kuratowski's theorem) to be four-colorable? Either provide an example, prove that it isn't possible, or argue why it would be very difficult to determine. Similarly, is it possible for a graph to contain a subgraph that is a subdivision of K_4 to be three-colorable? Either provide an example, prove that it isn't possible, or argue why it would be very difficult to determine.

Could the answers to these questions help to prove the four-color theorem? Explain. Are the answers to the above questions relevant to the challenge problem from Chapter 6? Explain.

Note: The answer key doesn't include answers to the challenge problems. These problems are intended to encourage you to think about the ideas. However, you may wish to consider how this problem relates to Chapter 27.

11 Trivial Four-Coloring

A vertex with a degree equal to three is connected to exactly three other vertices. A vertex with degree three is guaranteed to be four-colorable. Why? Regardless of which colors the other three vertices have, there will always be at least one color remaining that is different from the colors of those three vertices. For example, consider the diagram below, which is "zoomed in" on a vertex with degree three. The rest of the graph is irrelevant to the current discussion. Just focus on the vertex with degree three and the three vertices to which it is connected. If A is red, B is blue, and C is green, for example, then D can be yellow. If instead A is yellow, B is green, and C is yellow, then D can be red or blue. No matter which colors you choose for A, B, and C, there will always be at least one color left over for D.

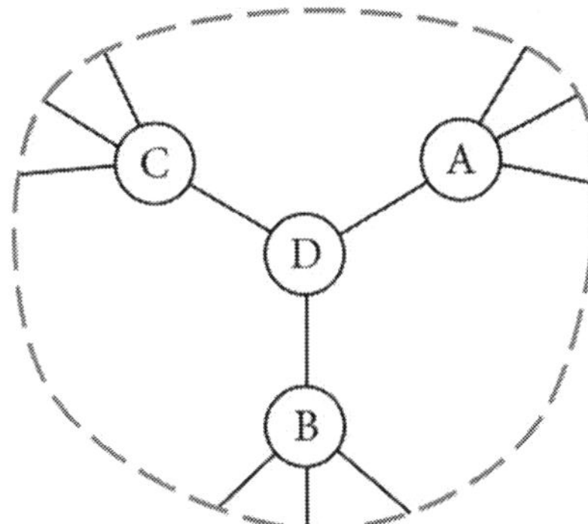

Any vertex with degree three (or less) may be removed from a graph, provided that we also remove the edges connecting it to the other vertices. If a graph is four-colorable after the vertices with degree three have been removed, it will still be four-colorable when the vertices with degree three are replaced. The examples of this chapter will illustrate this concept, including cases where vertices that originally had higher degrees may also be removed.

This means that we don't need to worry about any graphs that have at least one vertex with degree three. If we can prove the four-color theorem for all planar graphs that only have vertices with degree four and higher, it will follow that all planar graphs are four-colorable.

For example, the MPG on the left has three vertices with degree three: D, F, and I. When we remove these vertices and their connecting edges, we obtain the middle MPG. Although vertices C, G, and J had previously been degree four, after the edges connecting D, F, and I were removed, C , G, and J became degree three. We may now remove vertices C, G, and J. When we do this, even the remaining vertices (A, B, E, and H) are now degree three. Since the entire graph has unraveled through the process of removing vertices of degree three and their connecting edges, this graph is one of many graphs that are trivially four-colorable.

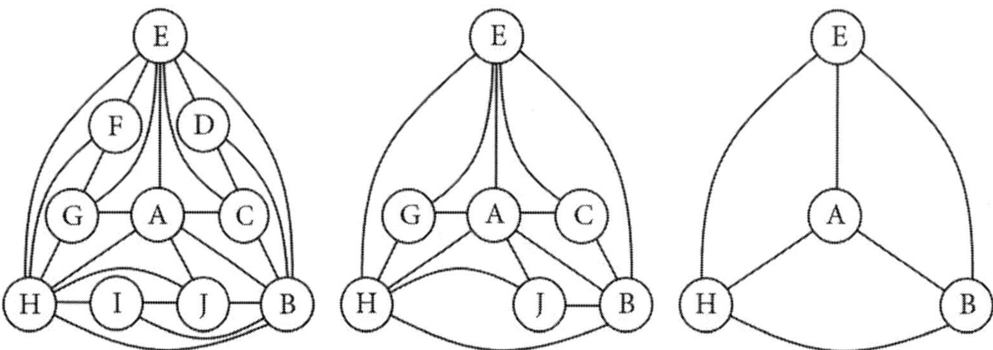

The previous MPG can be trivially colored using no more than four colors by coloring the graph in reverse. The MPG on the right only has four vertices, so it is easily four-colored. Each vertex added in the middle MPG uses the color that is unused by its three neighbors, and similarly for the MPG on the left.

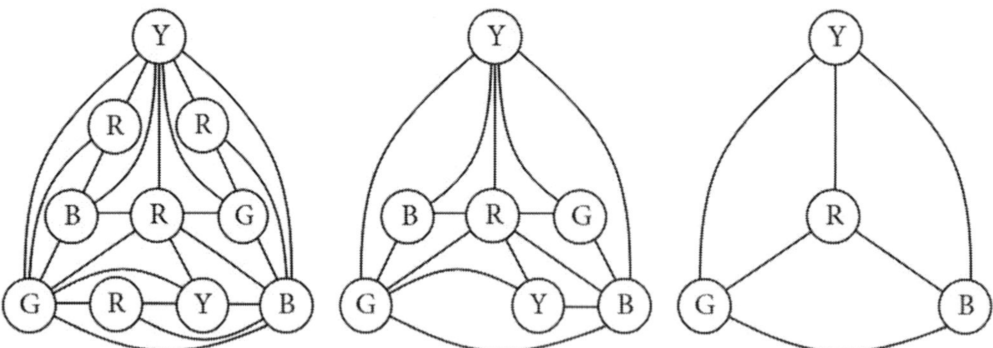

There are a great many MPG's with at least one vertex with degree three which completely unravel (and thus are trivially four-colorable) when vertices of degree three are removed from the graph. However, this is not always the case, as the next example shows.

The MPG on the left has one vertex with degree three: F. In this case, when F and its connecting edges are removed, the new MPG doesn't have any vertices with a degree less than four. In this example, only one vertex can be removed. Once the MPG on the right is shown to be four-colorable, it will follow that the MPG on the left is also four-colorable.

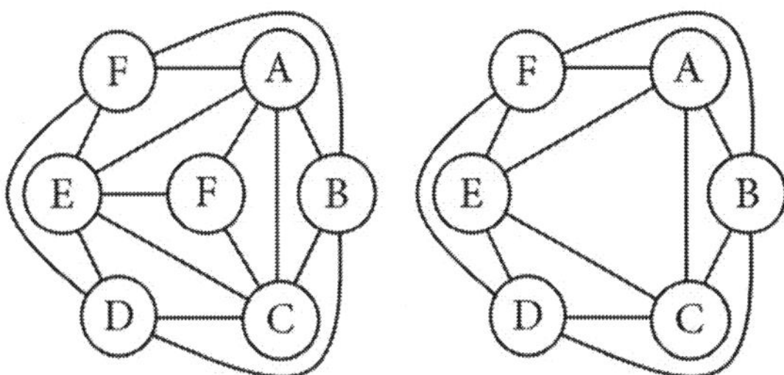

Vertices with degree three are common in MPG's that have K_4 subgraphs. For example, in the MPG on the left above, ACEF is a K_4 and F has degree three. On the previous page, BCDE, BIIJ, and EFGII are K_4's, and in each case one of the vertices has degree three. Note that the two graphs below are isomorphic: they are both K_4's. The only difference is that the left graph was drawn in planar form. If you imagine pushing edge AC outside of the graph on the right, you can see how D can lie at the center of that graph. In an isolated K_4, each vertex has degree three. When K_4 is a subgraph in a MPG, some or all of its vertices have a degree of four or higher because its vertices connect to other vertices in the MPG. If one of the K_4's vertices doesn't connect to other vertices in a MPG, that vertex will have degree three. (In the next chapter, we'll see an example of how all four vertices of a K_4 subgraph may have a degree of four or higher. In that case, although no vertex may be removed, the K_4 subgraph still serves to simplify the graph: as we'll see, such a K_4 subgraph has a separating triangle, which allows the graph to be divided into two smaller graphs. Are you wondering how D could connect to other vertices in the K_4 below? Put vertices in triangle ABD, for example.)

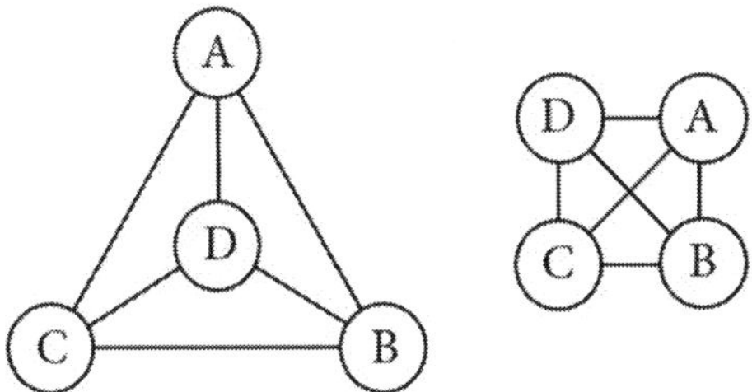

When removing vertices with degree three causes a MPG to completely unravel (and thus be trivially four-colorable), this generally happens because K_4 subgraphs make the coloring maximally restrictive. We saw that in the first example of this chapter, where three different K_4's involve 9 of the 10 vertices. Another example like this appears below.

Recall the MPG below from Chapter 9. This graph contains 9 K_4 subgraphs: ABCG, ABGL, ACDG, ADEG, AEFG, AGHI, AGIJ, AGJK, and AGKL. All 12 vertices participate in K_4 subgraphs. Presently, only two vertices have degree three: F and H. If you remove F and H and their connecting edges, E and I will be degree three. When E and I are removed, D and J will be degree three. This continues until the entire graph unravels. (If you're not seeing this, try redrawing the graph by removing one pair of vertices at a time, similar to the first example from this chapter.)

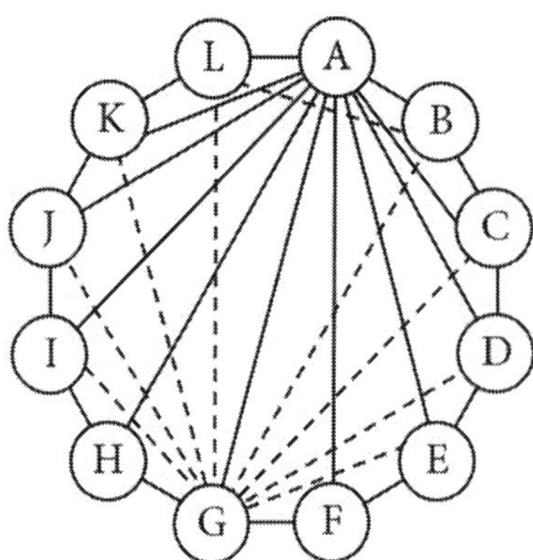

You could similarly make the case for removing vertices with degree four. How? Kempe showed how any vertex with degree four is four-colorable (Chapter 7). First remove all vertices with degree four (or lower). Then four-color the remaining vertices. Now add vertices with degree four back into the graph one at a time. Apply Kempe's method for coloring each newly added vertex. Since each color reversal would be done one at a time, we wouldn't need to worry about the problem of a double reversal associated with Kempe's argument for vertices with degree five. Many MPG's would completely unravel after removing vertices with degree four and lower, showing that they are trivially four-colorable.

There is an important difference between removing a vertex with degree four and removing a vertex with degree three along with their connecting edges. After removing a vertex with degree three and its connecting edges, the new graph is still a MPG. Since E = 3V − 6 for a MPG (Chapter 4), a MPG with one less vertex would have three fewer edges. That is, when V is reduced by 1, the quantity 3V − 6 (which equals E) is reduced by 3. (This is easy to see on the last page of Chapter 4 prior to the end-of-chapter exercises.) When we remove a vertex with degree three and its connecting edges, we're removing one vertex and three edges. Since this decreases V by 1 and decreases E by 3, the new graph is still a MPG. Removing a vertex with degree four removes one vertex and four edges (decreasing E by 4), such that the new graph isn't a MPG. That isn't a problem though. As we explained in Chapter 3, if a PG isn't a MPG, we can add edges to it as needed to triangulate the graph (turning it into a MPG). If the MPG with the added edge(s) is four-colorable, the PG without the added edge is also four-colorable. What does this mean? If you remove a vertex with degree four and its connecting edges, you just need to add one edge to triangulate the graph (so that the new graph is a MPG). In this book, our motivation for removing or adding edges revolves around the four-color theorem and various attempts to prove it.

We would still be left with graphs that have vertices with degree five and higher, even after removing vertices with degree four and lower. For example, given 12 vertices or more, you can always draw a MPG that doesn't have *any* vertices with degree four or lower, like the icosahedral graph shown below.

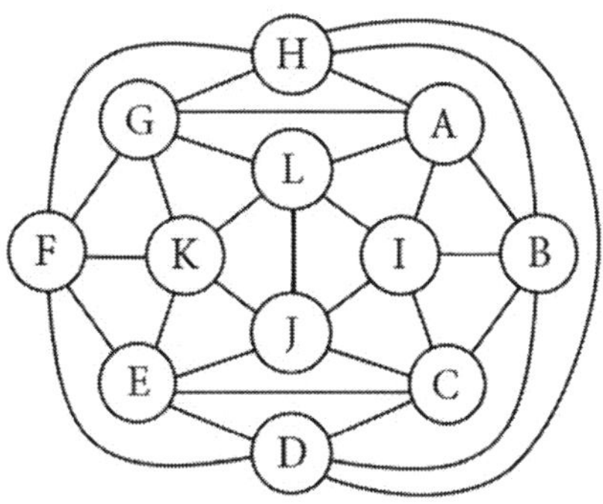

Chapter 11 Exercises

1. The same MPG is drawn twice below. On the left graph, enter the numbers 1 thru 11 to show a possible order in which the vertices could be colored so that when any single vertex is colored, at the time that it is colored that vertex doesn't share an edge with more than three other vertices that have already been colored. On the right graph, color the vertices in the order that you entered on the left to show that this MPG is indeed trivially four-colorable.

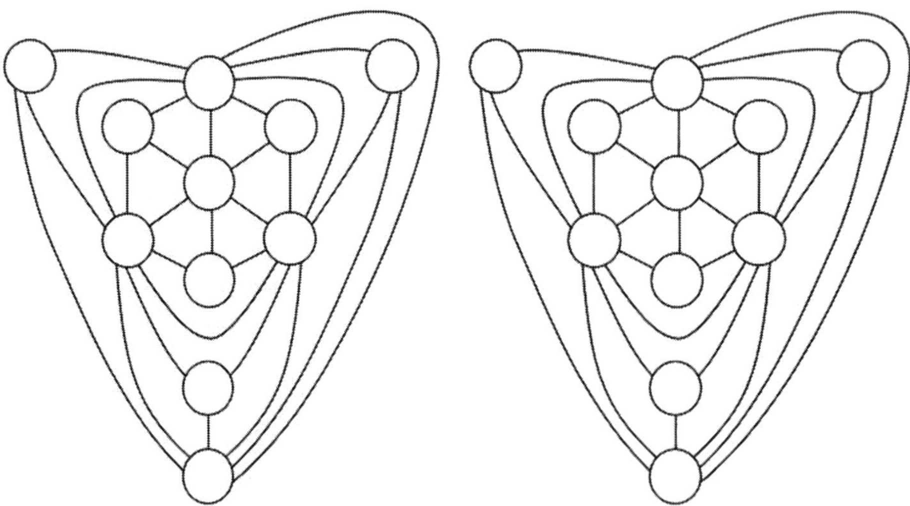

2. The same map is drawn twice below. (Note that this is a *map*, not a graph.) On the left map, enter the numbers 1 thru 10 to show a possible order in which the regions could be colored so that when any single region is colored, at the time that it is colored that region doesn't share an edge with more than three other regions that have already been colored. On the right map, color the regions in the order that you entered on the left to show that this map is indeed trivially four-colorable. Remember to count the unbounded surrounding area as a region of the map.

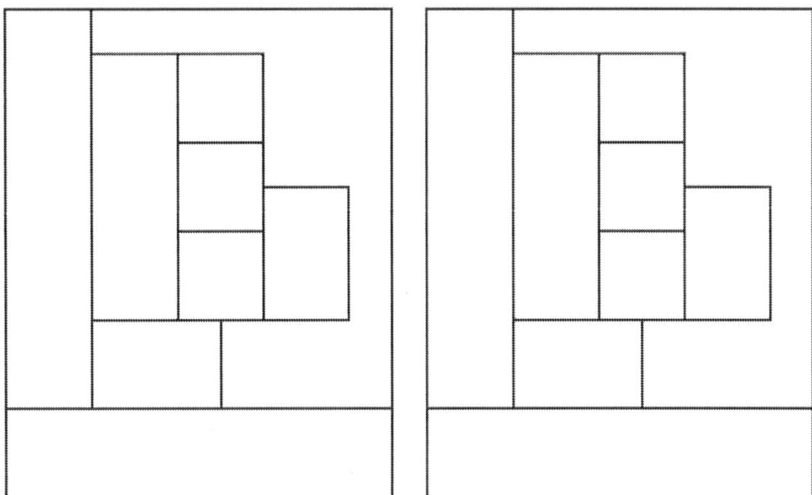

3. In the MPG shown below, 13 of the 28 vertices have degree 3. Redraw the graph without any of these vertices with degree 3. Will this graph completely unravel if we continue to remove vertices with degree 3? What if we also remove vertices with degree 4?

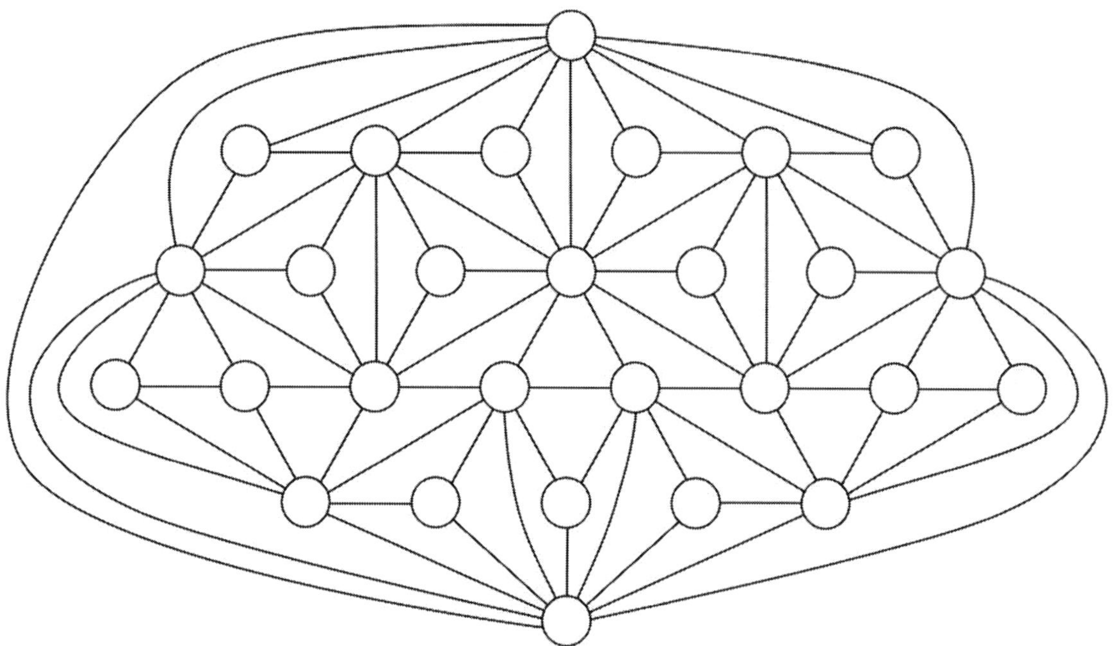

4. The same MPG appears twice below and once again on the following page. The left MPG has the vertices numbered in a specified order, which **saves two vertices with degree four until the end** (labeled 10 and 11). In the middle graph below, color the vertices in the order that you entered on the left, but please **read all of these directions before you begin**. We recommend using a pencil and eraser to label the vertices B, R, G, and Y (or 1, 2, 3, and 4). The coloring of vertices 1 thru 9 is trivial, but we have a particular goal in mind: We want at least one of the remaining vertices, 10 or 11 (which must be degree four), to connect to vertices of all four colors. If it doesn't happen this way on your first attempt, try coloring the first 9 vertices differently until it does (so that you will be able to appreciate the last part of this exercise). Once at least one of vertices 10 and 11 is connected to all four colors, **on the following page** apply Kempe's method for coloring a vertex with degree four in order to complete the four-coloring of the MPG. If both vertices 10 and 11 connect to four different colors, apply Kempe's method to one vertex at a time (so that you don't run into the problem that Kempe had in his argument with vertices with degree five which involved reversing the colors of two chains at the same time).

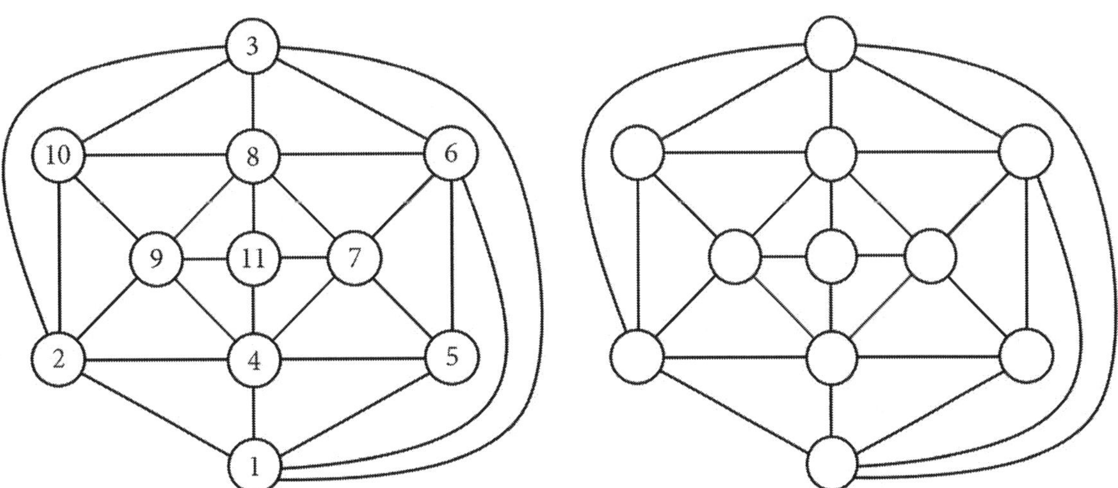

Note: This problem has another copy of this graph on the following page.

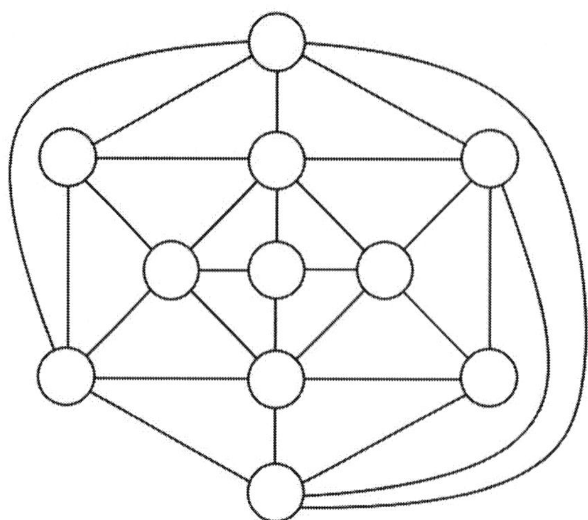

5. The icosahedral MPG below has 12 vertices with degree 5. Attempt to build as much of this graph **without assigning any colors** as possible in a trivially four-colorable way until you reach a point where it isn't possible to add another vertex in such a way that the added vertex will also be trivially four-colorable. What is the largest number of vertices that can be built in a trivially four-colorable way before you run into a problem?

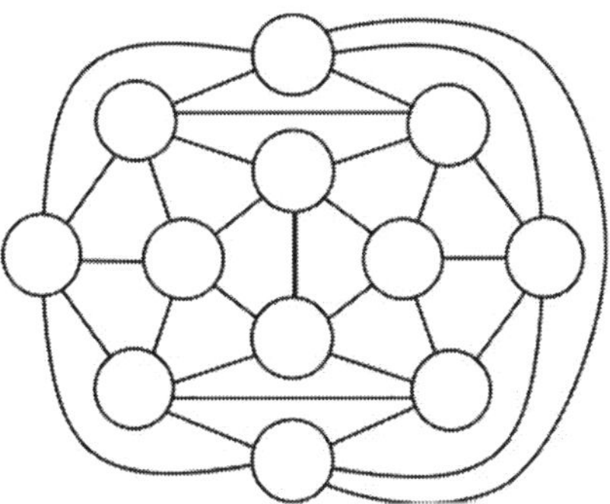

Challenge problem 1: Is it possible for a MPG (with at least 5 vertices) to have a vertex with degree three that isn't part of a K_4 subgraph? Either provide an example of this or prove that it isn't possible.

Challenge problem 2: If a MPG (with at least 5 vertices) has two or more vertices with degree three, is it possible for any pair of these vertices with degree three to connect to one another with a shared edge? Either provide an example of this or prove that it isn't possible.

Note: The answer key doesn't include answers to the challenge problems. These problems are intended to encourage you to think about the ideas. However, for the second challenge problem, you might explore Chapter 28.

Challenge problem 3: Prove that every MPG would completely unravel if you could successively remove vertices with degree five. Show that Kempe's argument (Chapter 7) for degree five vertices could be used to prove the five-color theorem (note that this is a *five*, not a four). (Heawood, in fact, proved the five-color theorem based on Kempe's attempted proof of the four-color theorem.) If you could find a way to prove that every vertex with degree five is four-colorable, you could justify removing vertices with degree five from any MPG and thus prove the four-color theorem (which is what Kempe's argument would have done if there hadn't been a subtle problem with reversing the colors of two chains at the same time).

Reconsider Challenge Problem 2 from Chapter 7, which had two vertices with degree five connected. Can the solution of that challenge problem (modified, if needed) be used to allow us to remove all vertices with degree five and thus prove the four-color theorem? Explain.

Challenge problem 4: Prove that if every vertex of a MPG has an even degree number, the MPG must have at least one vertex with degree four. Does this mean that every such MPG is trivially four-colorable? Explain.

Note: The answer key doesn't include answers to the challenge problems. These problems are intended to encourage you to think about the ideas.

12 Separating Triangles

Imagine making a MPG with a K_4 subgraph where every vertex of the K_4 has a degree of four or higher. Start by thinking about the K_4 shown below. In order for vertex D to have a degree of four or higher, we must add new vertices inside at least one of the faces.

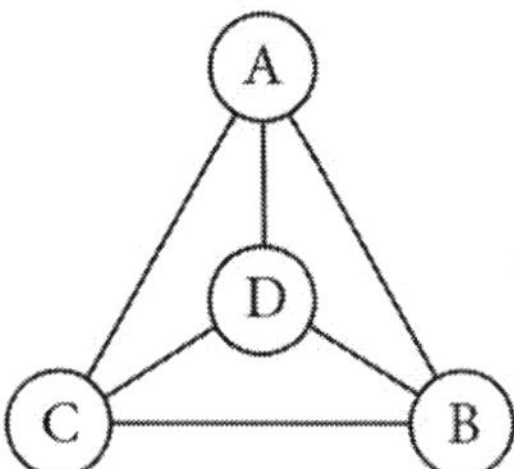

Note that K_4 has four faces: ABD, ACD, BCD, and the outside face ABC corresponding to the infinite area outside. If we add new vertices in at least two of these faces (well, for ABC it would be "out" rather than "in"), we can make a MPG where all of the vertices of the K_4 subgraph have a degree of at least four. An example is shown below.

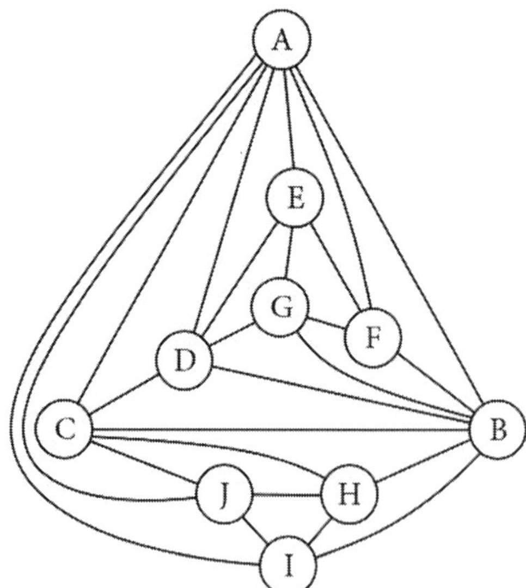

Find A, B, C, and D in the triangle above. This was our original K_4, which is now a subgraph of a MPG with 10 vertices. We added vertices E, F, and G in face ABD, and added vertices H, I, and J "in" the outside face ABC (when a face corresponds to the infinite area outside, the area of the face is "out" of the face rather than "in" it). We then added enough edges to triangulate the graph, turning it into a MPG. For V = 10 vertices, we need a total of E = 3V – 6 = 3(10) – 6 = 24 edges (arranged so that every face has three edges). The way we chose to add the needed edges, note that all 10 vertices of the MPG have a degree of at least four, including all four vertices of the original K_4 subgraph.

Every K_4 subgraph has a separating triangle. What does this mean? A **separating triangle**, which we will abbreviate ST, is a triangle consisting of three vertices and three edges, where the triangle isn't one of the faces. A ST has at least two faces inside of it and at least two faces outside of it. (Note that one of the faces outside of a ST can be the infinite area outside of the MPG.)

In the previous example, ABD is a ST because ABD isn't a face and because there are at least two faces inside and at least two faces outside of ABD. We call it a ST because vertices A, B, and D divide the MPG into two smaller MPG's, as seen in the diagram below. Observe that triangle ABD appears in both MPG's below, and now ABD is a face (whereas it wasn't in the original graph). Any MPG that has a ST can similarly be split into two separate MPG's. (There is actually a *second* ST in the MPG below and on the previous page: triangle ABC.)

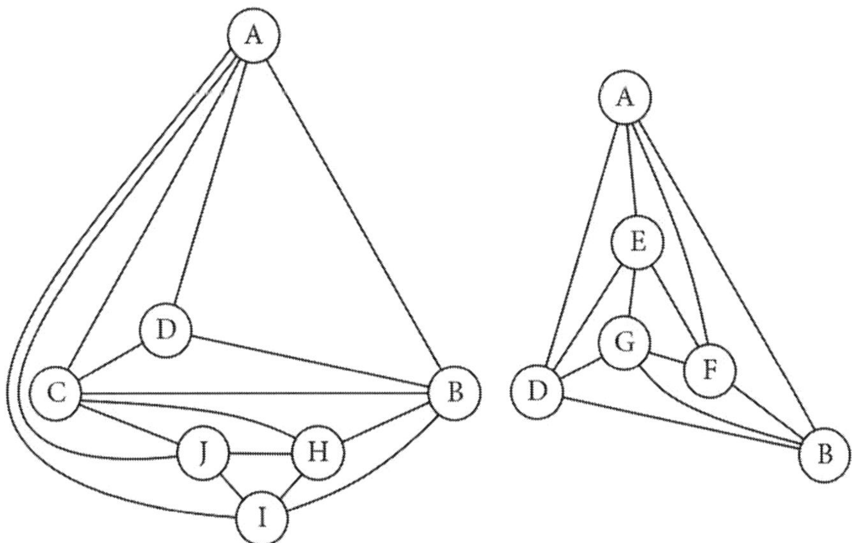

If each of the smaller MPG's is individually four-colorable, it follows that the original MPG is also four-colorable. Why? The only thing that the two smaller MPG's have in common is the ST, which has three vertices and three edges. When we color the two smaller MPG's, in each case we may begin by coloring the vertices of the ST. For example, we can choose A to be red, B to be blue, and D to be yellow for each MPG. Since A, B, and D lie on one face of each of the smaller MPG's, we know that these three vertices must have different colors. After coloring each of the smaller MPG's, when we put them together to form the original MPG, the same coloring will work for the original MPG because A, B, and D have the same colors in each of the smaller MPG's.

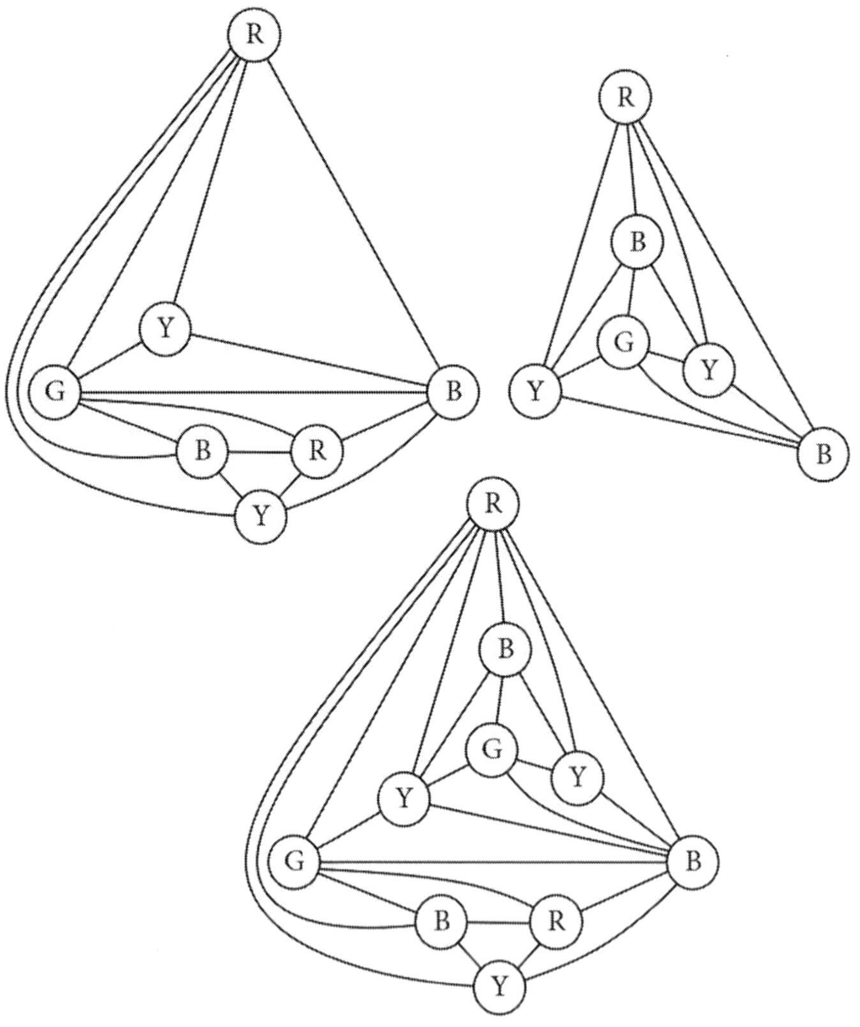

This means that we don't need to worry about any graphs that have ST's. If we can prove the four-color theorem for all MPG's that don't have ST's, it will follow that the four-color theorem also applies to MPG's that do have ST's (since a ST can be used to divide a MPG into two smaller MPG's).

Let's look at another example of a ST. Can you find a ST in the MPG below?

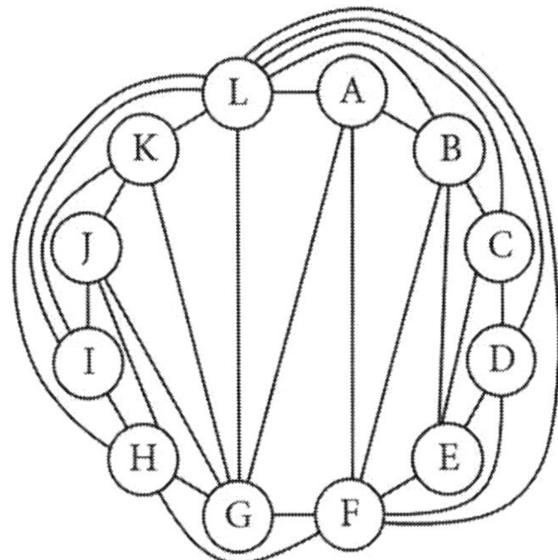

You can find the ST in the K_4 subgraph. The K_4 subgraph is AFGL. (Why is AFGL a K_4 subgraph? Each of these three vertices connects to all three of the others.) The ST is FGL. Why? FGL isn't a face. There are at least two faces inside FGL and at least two faces outside of FGL (in this case, there are many more than two faces both inside and outside of the ST; the requirement is for there to be at least two).

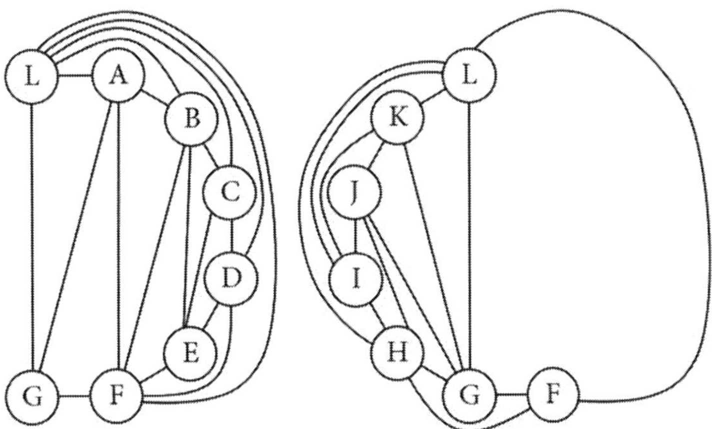

At this point, we've shown that it isn't necessary to prove the four-color theorem directly for any MPG's that contain K_4 subgraphs for the following reasons:

- If a K_4 subgraph has a vertex with degree three, the vertex with degree three may be removed from the MPG (as discussed in Chapter 11). This subgraph will no longer be a K_4. Since we don't need to prove the four-color theorem directly for any MPG that has a vertex with degree three, we don't need to consider any MPG's with K_4's with a vertex with degree three. If we prove the four-color theorem for all MPG's without vertices with degree three, it will follow that all MPG's with vertices with degree three are also four-colorable.

- Even if a K_4 subgraph doesn't have any vertices with degree three, the K_4 subgraph will have a ST. The ST may be used to divide the MPG into two smaller MPG's. If we prove the four-color theorem for all MPG's without ST's, it will follow that all MPG's with ST's are also four-colorable.

Combining these ideas together, if we prove the four-color theorem for all MPG's without K_4 subgraphs, it will follow that all MPG's with K_4 subgraphs are also four-colorable.

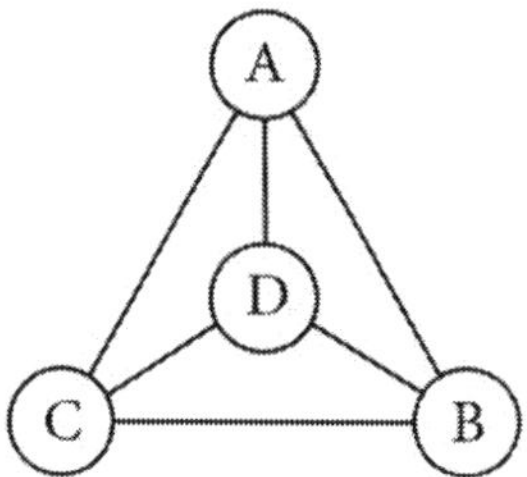

Technically, all K_4 subgraphs have ST's, even if a K_4 subgraph has a vertex with degree three. However, if there is a vertex with degree three, we may simply remove that vertex from the MPG (Chapter 11). Imagine the K_4 above as a subgraph in a MPG. Vertex D has degree three. If you make the ST ABC to divide the MPG into two graphs, all you are effectively doing anyway is removing vertex D from the MPG.

Chapter 12 Exercises

1. Identify all of the ST's in each MPG below.

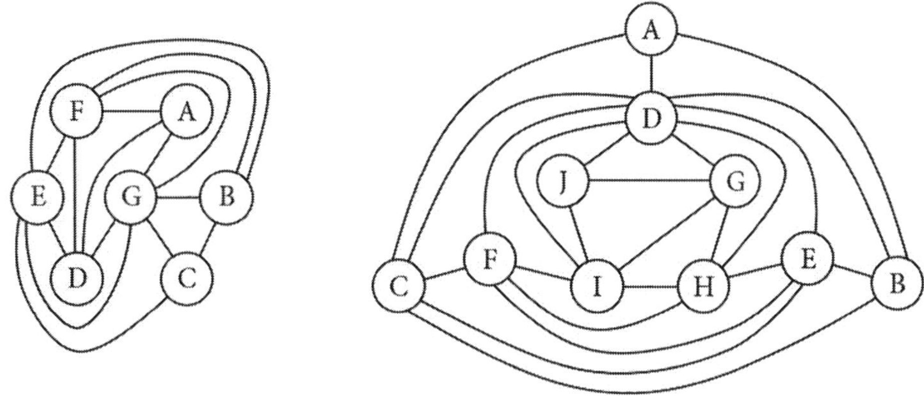

2. Identify a ST in the MPG below. Draw two smaller graphs to show how this ST may be used to divide the MPG into two separate graphs.

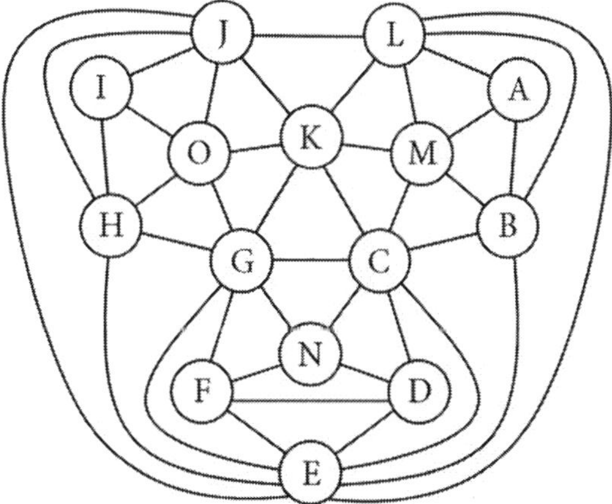

3. The MPG below has been divided into two graphs using its ST. Color the two smaller graphs below and show that this same coloring works for the original graph.

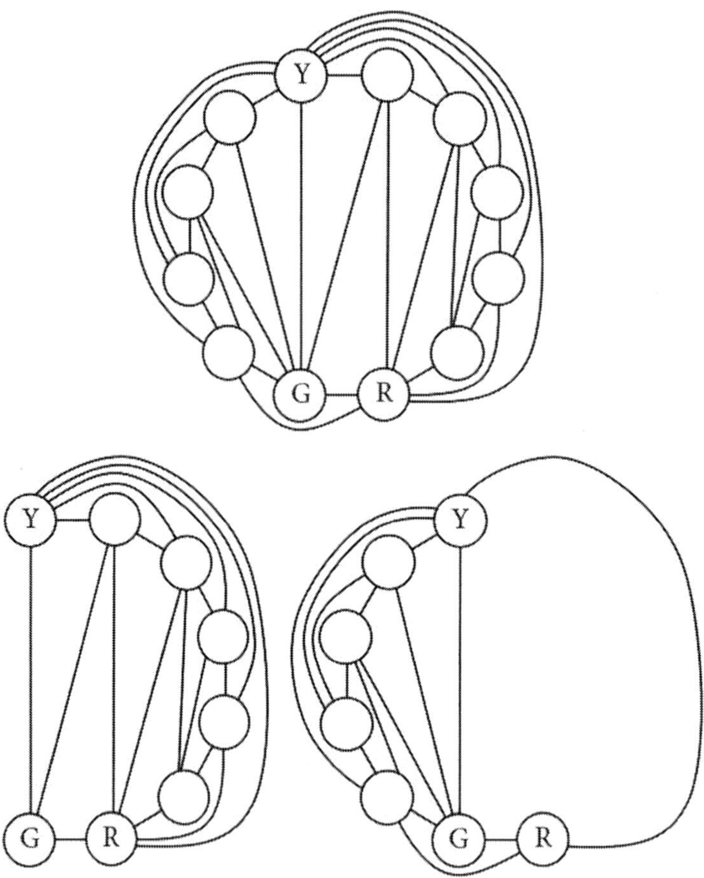

4. A MPG has a ST involving vertices E, K, and P. This ST is used to separate the MPG into two smaller graphs. One graph is four-colored with E green, K blue, and P yellow. The other MPG is four-colored with E red, K green, and P blue. Explain precisely how one or both of these graphs can be recolored such that the original MPG is properly four-colored.

5. What is the greatest number of ST's possible that a MPG with 10 vertices can have? Draw one MPG with this number of ST's with 1 red vertex, 1 blue vertex, 4 green vertices, and 4 yellow vertices. Draw another MPG with this number of ST's with 3 green vertices, 3 yellow vertices, 2 red vertices, and 2 blue vertices.

Challenge problem 1: Are all ST's (of MPG's) parts of K_4 subgraphs? Either provide an example with a ST (in a MPG) that isn't part of a K_4 subgraph or explain why all ST's are contained in K_4 subgraphs.

Challenge problem 2: Can we make separating quadrilaterals, pentagons, or any other polygons besides a ST and use them to separate a MPG into two smaller graphs? If so, will the same principle apply that if each of the smaller graphs is four-colorable the original MPG will also be four-colorable? Either provide an example and explain what allows you to do this or explain why this is only possible for a triangle.

Note: The answer key doesn't include answers to the challenge problems. These problems are intended to encourage you to think about the ideas.

13 Hamiltonian Cycles

A **circuit** refers to a path in a graph that begins and ends at the same vertex (meaning that the path is a closed loop), while a **cycle** refers to a circuit that doesn't repeat any vertices. A **Hamiltonian cycle**, which we will abbreviate HC, is a cycle that involves every vertex in a graph [Ref. 10]. Many (but not all) MPG's have a HC. If a MPG has a HC, there exists some way to begin at one vertex, travel along an edge to another vertex, travel along another edge to another vertex, continuing to do so until every vertex is visited exactly once, and then travel along one more edge to return to the original vertex. An example of a HC is shown below.

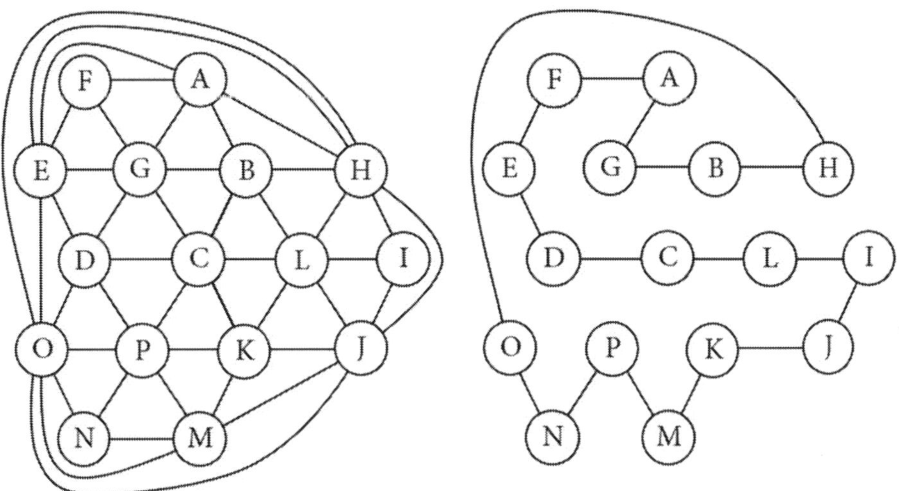

The MPG on the left includes the HC shown on the right. If you start at vertex H, for example, you can then trace the cycle HBGAFEDCLIJKMPNOH. It's important to finish on the same vertex as you start. (If there isn't an edge returning to the starting vertex from the ending vertex, then it's a Hamiltonian "path," not a "cycle.") When a HC exists for a MPG, there are often multiple HC's. For example, we could start out HBAGFE instead of HBGAFE and still obtain a HC. There are numerous HC's available for the above MPG.

An example of a MPG that doesn't have a HC is shown on the following page. This graph is known as the Goldner-Harary graph [Ref. 11].

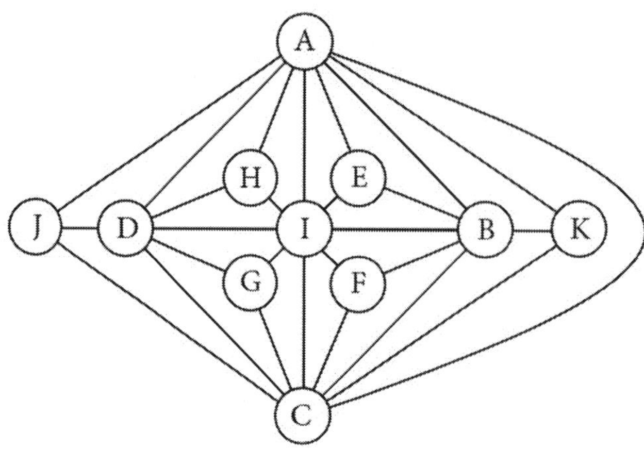

The best we can do with the above MPG is GDHAEIFBKCJ, but although we were able to use all 11 vertices going from G thru J, unfortunately there isn't an edge allowing a return from the final point (J) to the initial point (G). What if we don't start at G? Okay, let's try it again: CKBFIEAHDG. This time, we didn't even reach J (we could have after D, but then we wouldn't have reached G). Another option is JDHAEIGCFBK. We visited all 11 vertices once again, but were unable to return to the starting point (J) from the final point (K). Try as you might, no HC exists for the MPG above. (We did find a Hamiltonian *path*, but that's not the same as a Hamiltonian *cycle*.)

How can you tell whether a MPG will have a HC? It turns out that if a MPG doesn't have a ST, then it has a HC [Ref. 12]. Recall from Chapter 12 that ST stands for "separating triangle." (However, if a MPG has ST's, this doesn't mean it won't have a HC; it's possible for a MPG to have ST's and still have a HC.)

The previous graph has 4 ST's: ABI, ADI, BCI, and CDI. Each of these ST's has 3 faces inside of it and 15 faces outside of it. As far as the four-color theorem is concerned, the ST's of this particular graph are trivial, since the vertex inside of each ST is degree three. The vertices inside of the ST's are H, E, F, and G. Recall from Chapter 11 that we may remove any vertices with degree three from the graph along with their connecting edges. When we do this, we obtain the following graph, which now has a HC.

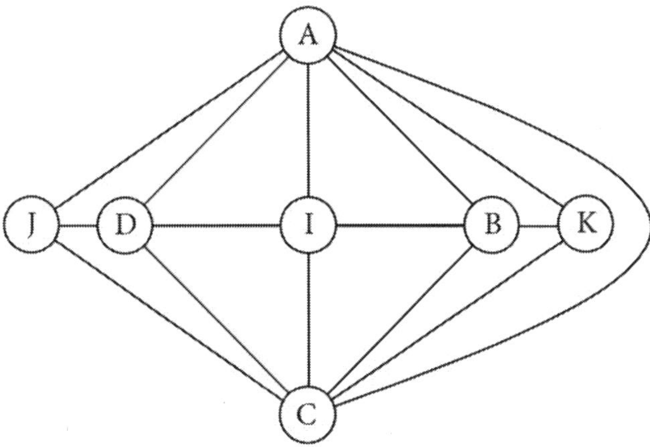

What if the ST's are more complicated? As shown in Chapter 12, we may use the ST to divide the MPG into two smaller graphs such that if the smaller graphs are four-colorable, the original MPG is also four-colorable. This means that we don't need to work with any MPG's that have ST's as far as proving the four-color theorem is concerned. All of the remaining MPG's will therefore have HC's. The MPG below shows an example with ST's that contain multiple vertices and edges.

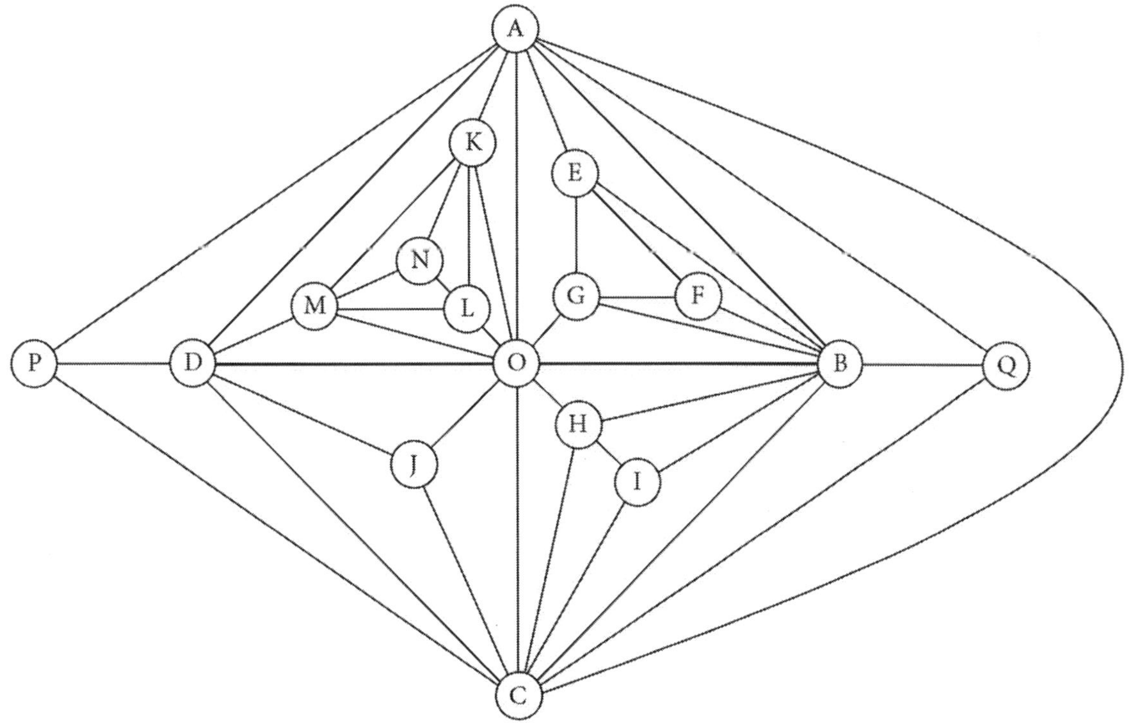

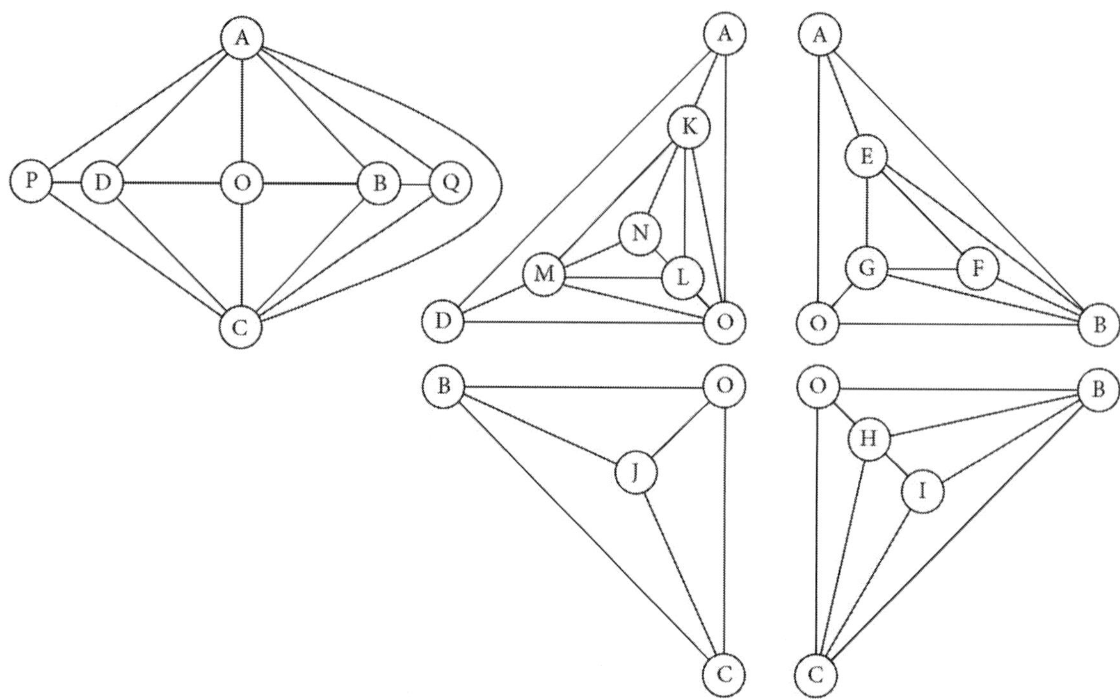

The MPG on the previous page has ST's: ABO, ADO, BCO, and CDO. Above, the MPG has been divided into one parent MPG (on the left) and four daughter MPG's (on the right). The three exterior vertices of each daughter MPG "fit" in the corresponding ST of the parent MPG. This is the sense in which these triangles are "separating."

Since the entire original graph was a MPG, both the parent graph and daughter graphs are also MPG's. In the parent MPG, the ST's are interior faces. In the daughter MPG's, each ST is an exterior face (corresponding to the infinite area outside).

Note that if the parent MPG and the daughter MPG's are each four-colorable, the original MPG in its entirety is also four-colorable. This follows from the fact that each daughter MPG has three vertices in common with the parent MPG, and that all three vertices are connected to one another in both the parent and daughter MPG's. For example, consider ADO. Since A connects to both D and O in the parent MPG and also in the daughter MPG, vertices A,

D, and O must be three different colors in each MPG. If the parent and daughters are each four-colorable, we can choose to color each MPG such that A, D, and O match colors.

This means that we don't need to worry about MPG's that don't have HC's:

- We may focus on attempting to prove that every MPG with a HC is four-colorable.
- If a MPG doesn't have a HC, it must have ST's.
- The ST's allow us to separate the MPG into parent and daughter MPG's.
- This allows us to separate a MPG that doesn't have a HC into MPG's that do have HC's. (What if a daughter MPG doesn't have a HC? Then the daughter MPG has ST's, and we can use the ST's to make granddaughter MPG's, and so on.)
- It follows that if every MPG with a HC is four-colorable, then even MPG's without HC's are four-colorable.

If we wish to do so, we may thus focus solely on MPG's that have HC's.

As we will explore in the next chapter, if a MPG has a HC, we can draw the MPG with all of its vertices arranged in the shape of a closed convex polygon. This allows us to draw every MPG in a similar form.

Another interesting cycle is the **<u>Eulerian cycle</u>**, which is somewhat different from the HC. An Eulerian cycle is a path that begins at one vertex, travels along every edge exactly once, and finishes at the original vertex. The goal of finding an Eulerian cycle is to use every edge once (whereas the goal of finding a HC is to visit every vertex once; note that the HC doesn't use all of the edges of a MPG with more than three vertices). If the degree of every vertex is even (and the graph is connected, of course), the graph has an Eulerian cycle [Ref. 13].

Chapter 13 Exercises

1. For each MPG below:

- Determine whether or not the graph has a HC. If the graph has a HC, draw one.

- Identify any ST's.

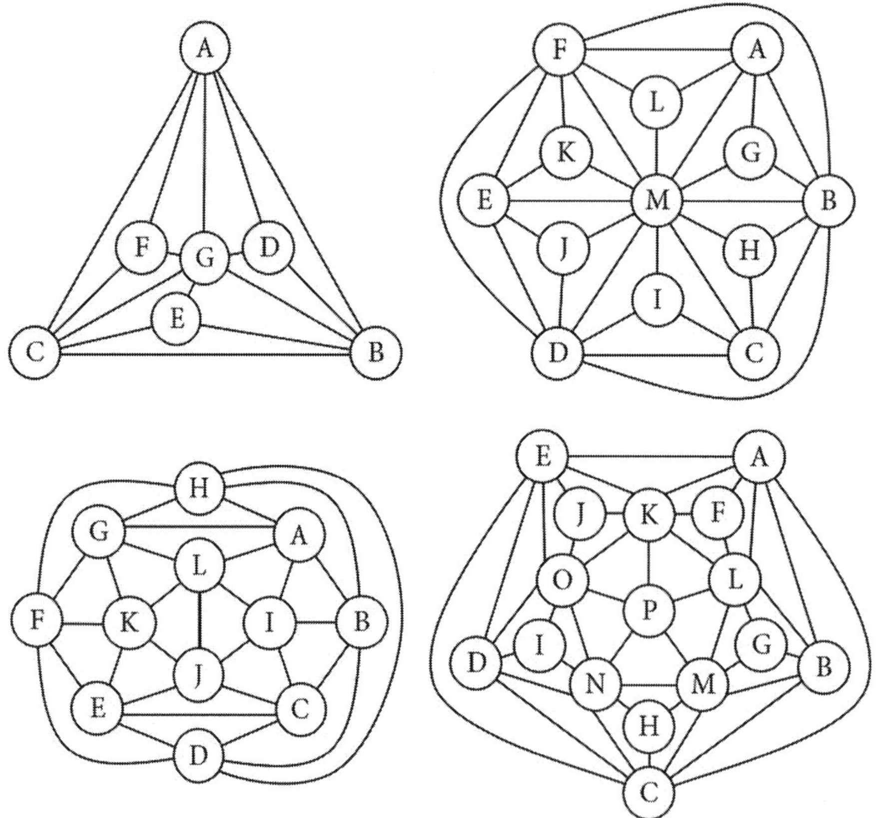

14 Polygon Graphs

As discussed in Chapter 13, as far as the four-color theorem is concerned, we only need to consider MPG's that have HC's. Any MPG that has a HC may be drawn with all of its vertices arranged in the shape of a closed convex polygon, as shown below.

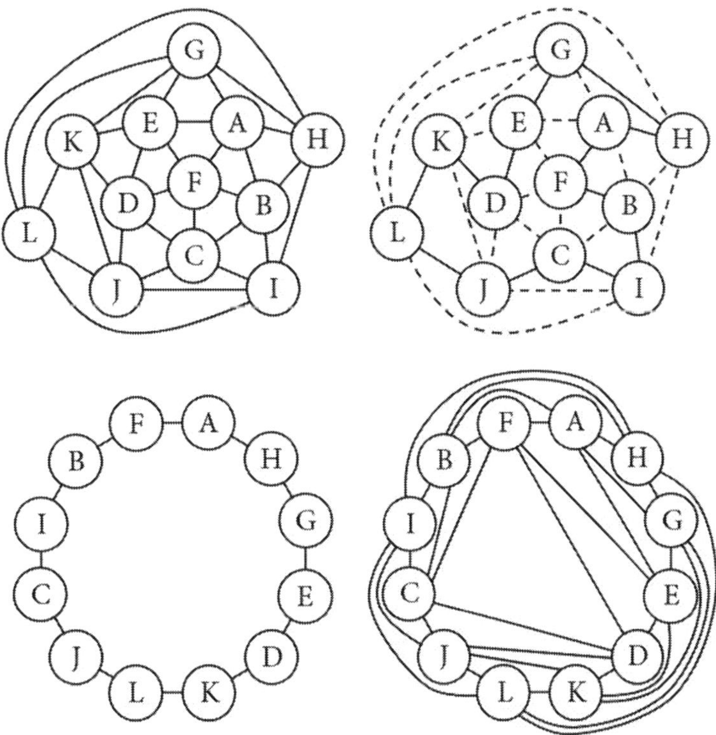

The diagrams above show how a MPG with a HC can be drawn as a closed convex polygon.

- First identify one HC. See the solid lines at the top right: AHGEDKLJCIBFA.
- Arrange the vertices and edges of the HC in order in the shape of a closed convex polygon. See the diagram at the bottom left.
- Add in the remaining edges. One-half of the remaining edges will fit inside of the polygon and one-half will fit outside of the polygon. Choose wisely and there won't be any crossings (since a MPG is a planar graph). See the diagram at the bottom right.

This allows us to draw all MPG's with HC's in a consistent form.

The MPG on the previous page has $V = 12$ vertices and $E = 3V - 6 = 3(12) - 6 = 30$ edges.

- $V = 12$ vertices are arranged in the shape of a convex polygon.
- $V = 12$ edges connect the $V = 12$ vertices to form the closed convex polygon.
- $V - 3 = 12 - 3 = 9$ edges lie inside of the polygon.
- $V - 3 = 12 - 3 = 9$ edges lie outside of the polygon.
- There are $V + 2(V - 3) = 12 + 2(12 - 3) = 12 + 2(9) = 12 + 18 = 30$ edges in total, which agrees with the formula from Chapter 4 for a MPG: $E = 3V - 6 = 3(12) - 6 = 30$.

Every MPG with a HC that is drawn in the shape of a closed convex polygon will have $V - 3$ edges inside of the polygon and $V - 3$ edges outside of the polygon. Why? This follows from Euler's formula.

- Recall from Chapter 4 that a MPG with V vertices has $E = 3V - 6$ edges.
- The HC connects the V vertices with V edges. Subtract V edges from $3V - 6$ to see that there are $2V - 6$ edges remaining.
- Divide this by 2 to get $V - 3$. There are $V - 3$ edges inside of the polygon and $V - 3$ edges outside of the polygon.
- As with any MPG, it is possible to draw all of the edges without crossing.

The HC divides the MPG in half in the following sense. There are $V - 2$ faces inside of the polygon and $V - 2$ faces outside of the polygon. This makes $F = 2(V - 2) = 2V - 4$ faces in total. This agrees with Euler's formula. From Chapter 4, for a MPG, $E = 3V - 6$ and $2E = 3F$. Multiply $E = 3V - 6$ by two to get $2E = 6V - 12$. Since $2E = 3F$, it follows that $3F = 6V - 12$. Divide both sides by 3 to see that $F = 2V - 4$.

The two sets of $V - 3$ edges can always be inverted for a MPG with a HC. The $V - 3$ edges on the "inside" of the polygon and the $V - 3$ edges on the "outside" of the polygon may be swapped, as illustrated by the following diagram. In this sense, the polygon serves as a "ring of invertibility."

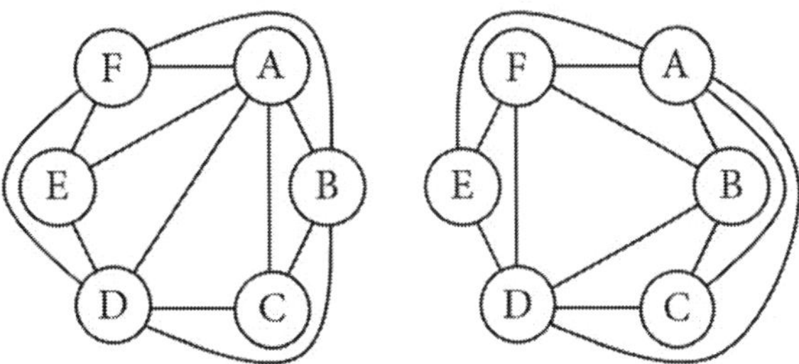

The two MPG's shown above are isomorphic: both contain edges AB, AC, AD, AE, AF, BC, BD, BF, CD, DE, DF, and EF. (Recall from Chapter 1 that isomorphic graphs are structurally equivalent in terms of edge-sharing.) As far as which vertices share edges is concerned, the two graphs shown previously may effectively be treated as the *same* MPG. The only distinction is that "inside" and "outside" of the HC have been inverted.

It is well-known that any PG can be drawn on the surface of a sphere instead of in the plane [Ref. 14]. Imagine drawing a MPG with a HC on the surface of a sphere. We may choose to place the HC along the equator of the sphere. Then the two sets of V − 3 edges that we interpreted as "inside" and "outside" edges in the plane would simply correspond to the two hemispheres of a sphere. Inverting "inside" and "outside" in the plane is simply like turning the sphere upside down.

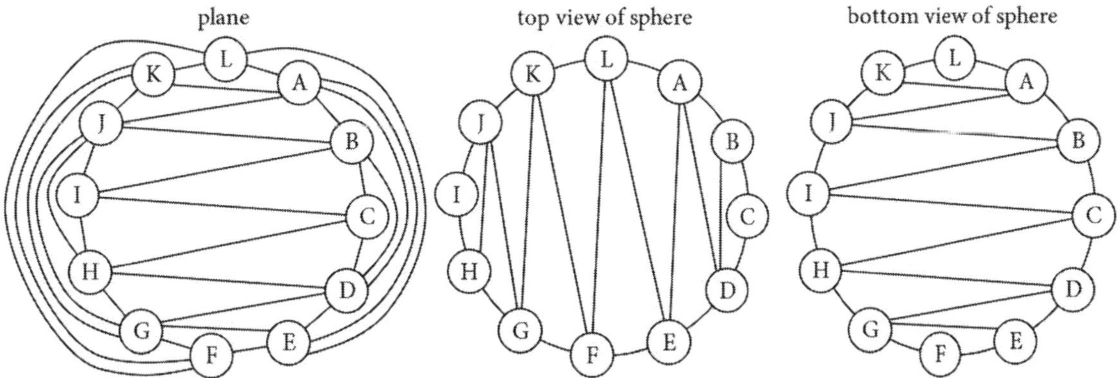

Technically, the HC around the equator would look like its mirror image in the bottom view compared to the top view. To make the comparison of the two views simpler, we mirrored the bottom view image in the diagram above so that the HC around the equator would be consistent with the top view.

Chapter 14 Exercises

1. Redraw the MPG below in the form of a closed convex polygon.

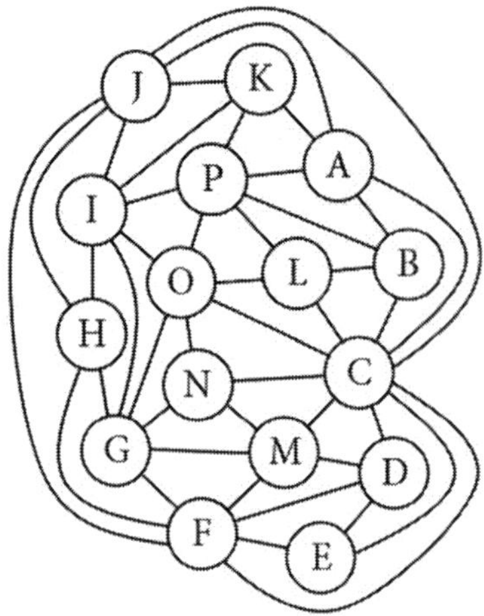

2. Redraw the MPG below with its inside and outside edges inverted.

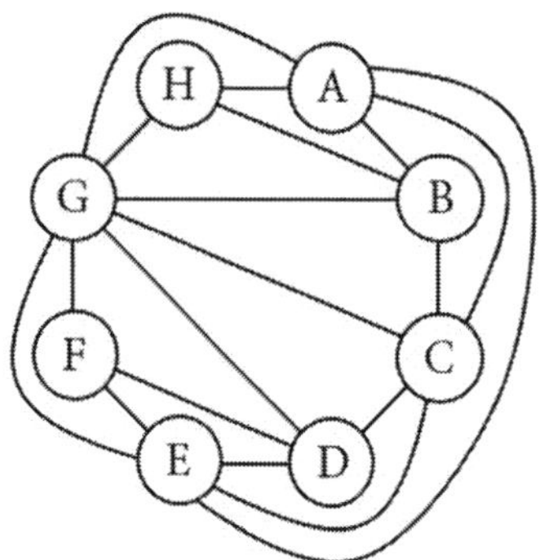

3. If a MPG with 24 vertices which has a HC is drawn in the form of a closed convex polygon, how many inside edges and outside edges will it have? How many edges does it have in total? How many inside faces and outside faces will it have? How many faces does it have in total?

Challenge problem 1: If a MPG with a HC is drawn in the form of a closed convex polygon and we remove the V − 3 outside edges, prove that the remaining graph is three-colorable. Similarly, if we remove the V − 3 inside edges (instead of removing the outside edges), prove that the remaining graph is three-colorable. Can you use these proofs to prove that when both sets of V − 3 edges (inside and outside) are included that the complete MPG is four-colorable? Either use this to prove the four-color theorem or explain what makes this task very difficult or impossible to carry out.

Challenge problem 2: Show that the faces (not the vertices) of every MPG with a HC are trivially four-colorable since the set of faces on either side of the HC is two-colorable. Show that we can actually go a step further: The faces of many MPG's with HC's are actually three-colorable. Draw a variety of examples to illustrate this. Are the faces of every MPG with a HC three-colorable? Either prove this, find a counterexample, or explain why it would be very difficult to determine. What are the prospects of using either of these results (four-coloring or three-coloring of the faces of MPG's with HC's) to prove the four-color theorem (which relates to vertices, not faces)? Now consider that the four-color theorem applies to all maps, where faces (not vertices) are colored. Can these results (regarding the coloring of the faces of MPG's with HC's) be used to prove the four-color theorem for maps? Explain.

Note: The answer key doesn't include answers to the challenge problems. These problems are intended to encourage you to think about the ideas.

15 Adding Edges

In Chapter 9, we observed that MPG's that don't contain any K_4 subgraphs tend to have more ways of being colored (using no more than four colors). This suggests that we might be able to go a step *beyond* triangulation. Specifically, we might be able to add edges to less restrictive MPG's to make them more restrictive with regard to four-coloring.

For example, consider the MPG below on the left. If we color A red, B blue, and C green, there are 4 ways to color D, E, and F (as shown in Chapter 9). If we add a new edge to connect BE, as shown below on the right, there will be just one way to color D, E, and F. Adding one edge made this graph much more restrictive. It's no longer a MPG, but that's not really a problem. We can apply the same concept that motivated triangulation in the first place (Chapter 3): if the graph with the extra edge(s) is four-colorable, the MPG made by removing the extra edge(s) will also be four-colorable.

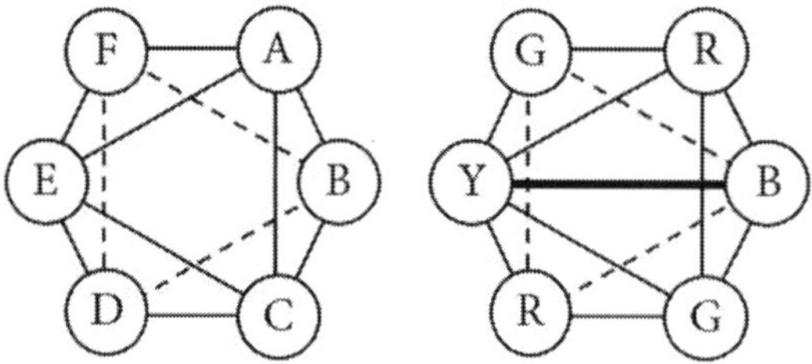

The trick is finding a way to add extra edges that allows the new graph to be four-colorable. For example, when we add new edges, we must be careful not to create any K_5 subgraphs, since K_5 isn't four-colorable; however, creating a $K_{3,3}$ subgraph isn't necessarily a problem since $K_{3,3}$ by itself is two-colorable. Recall that we discussed K_5 and $K_{3,3}$ graphs in Chapters 5-6. If we add edges to a MPG, the new graph will no longer be a MPG, so it must have a K_5 or a $K_{3,3}$ as a minor according to Wagner's theorem, as discussed in Chapter 5. If we can add edges to a less restrictive MPG so that the new graph is four-colorable, in some cases it might be easier to show that the new graph is four-colorable than it would be to show that the original MPG was four-colorable.

Note that adding edge BE to the previous graph created K_4 subgraphs, like ABEF and BCDE. All 6 vertices participate in these two K_4 subgraphs. Originally, there were no K_4 subgraphs.

In the diagrams of this chapter, the dashed lines indicate which edges could be moved outside of the polygon to help you visualize that the original graph is indeed a MPG, while the thick lines indicate which edges could be added to the MPG's so that they could be colored a single way (once three vertices of one of the triangular faces have been set).

Important reminder: We are defining PG to include any graph that *can* be drawn in the plane without crossings (even if the graph happens to be drawn in a form that has avoidable crossings). With this meaning, PG doesn't quite stand for "planar graph." If you want to be more precise, you should think of PG as "planar graph or any graph that is isomorphic to a planar graph" or "planar graph or any graph that can be redrawn in the plane without any crossings." Our definition of MPG thus has a similar interpretation.

In each of the following graphs, we set A red, B blue, and G green.

- After adding edges CE and CG to the left graph, it can be colored one way. There were no K_4 subgraphs to begin with. After adding edges CE and CG, the K_4 subgraphs include ABCG, ACFG, BCDE, BCEG, CDEF, and CEFG, involving all 7 vertices.
- After adding edge AD to the right graph, it can be colored one way. There was one K_4 subgraph originally: DEFG. New K_4 subgraphs include ABCD, ACDF, and ADFG.

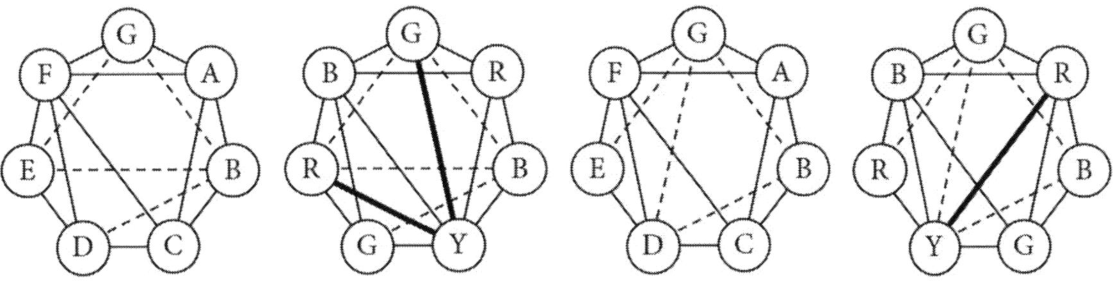

In the following graph, we added edges BE, CE, and FL so that the right graph can only be colored one way after setting A red, B blue, and D yellow. There were no K_4 subgraphs to begin with. Now the K_4 subgraphs include ABDE, BCDE, CDEF, and EFKL, involving 8 of the 12 vertices.

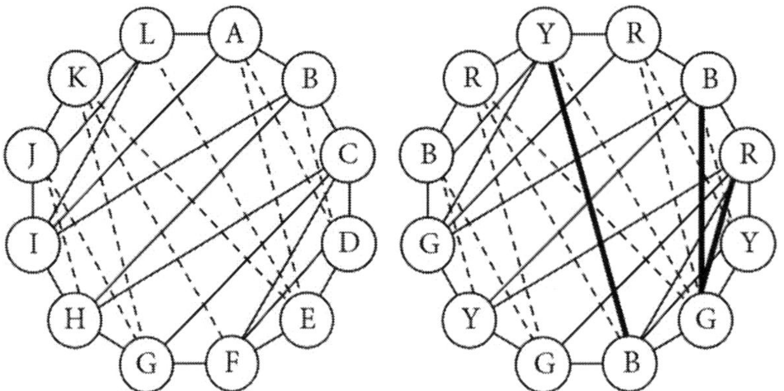

The examples that we've looked at so far have shown how to add a minimal number of edges so that the graph can be four-colored exactly one way (after setting the three colors for the vertices of one triangular face). As we'll see, it's often possible to add many more than the minimal number of edges such that the graph remains four-colorable.

The graph shown above on the left is the icosahedral MPG, which has $V = 12$ vertices and $E = 3V - 6 = 3(12) - 6 = 36 - 6 = 30$ edges. Although the previous example showed one way to add a mere 3 edges to make the graph four-colorable exactly one way (after setting three colors), it's actually possible to add a staggering 24 additional edges to the icosahedral graph so that the graph remains four-colorable. You can't put the 24 edges anywhere though. You have to arrange them so that the 12 vertices of the new graph are each degree 9 instead of degree 5. The new graph has 54 edges instead of 30, as shown on the following page. (As mentioned earlier, after adding edges to a MPG, the new graph is no longer a MPG.)

Where you add the extra edges can make a huge difference. For example, it's possible to add a mere 4 edges to the icosahedral MPG in such a way that the new graph isn't four-colorable.

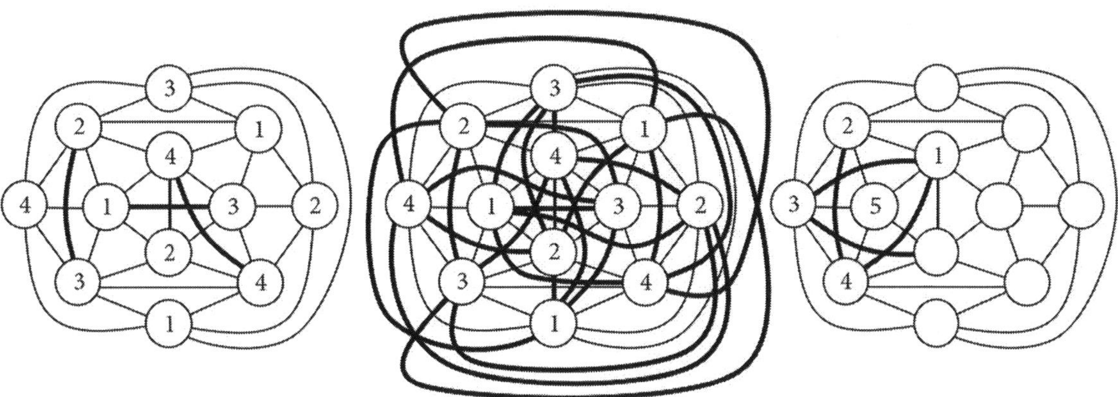

Three extreme cases of adding edges to the icosahedral graph are shown above.

- The left graph adds a minimal number (just 3) edges such that the new graph is four-colorable a single way (once three colors are set). These 3 edges make 3 strategically placed K_4 subgraphs.

- The middle graph adds a maximum number (24) edges such that the new graph is four-colorable. We call this ultimate four-coloring and will discuss graphs like this in Chapter 16 (which also presents a visually simpler way to draw such graphs). In this graph, each vertex connects to all 9 of the other vertices that aren't the same color.

- The right graph shows that it only takes 4 edges to transform the icosahedral graph into a graph that isn't four-colorable. We added all 4 edges around a single vertex (the one colored 5) in such a way as to create a K_5 subgraph, which isn't four-colorable.

How do you know where you can or can't add edges to a MPG, so that the new graph will be four-colorable? Here are a few important rules:

- If a MPG doesn't have any K_4 subgraphs, we can potentially add 3 edges to *any* vertex with degree 5 in such a way that the vertex with degree 5 will be four-colorable.

- If edges are added in such a way as to create a K_5 subgraph, the new graph won't be four-colorable. (Subdivisions and minors with K_5's could be okay. See Chapter 27.)

- Be careful not to add an edge that results in a double edge.

- Try to add new edges that create K_4 subgraphs to make the coloring more restrictive.

The partial graphs below are zoomed in on a vertex with degree five (colored 4). They show how it is possible to add three edges to any vertex with degree five, provided no pair of its neighboring vertices is already involved in a K_4 subgraph.

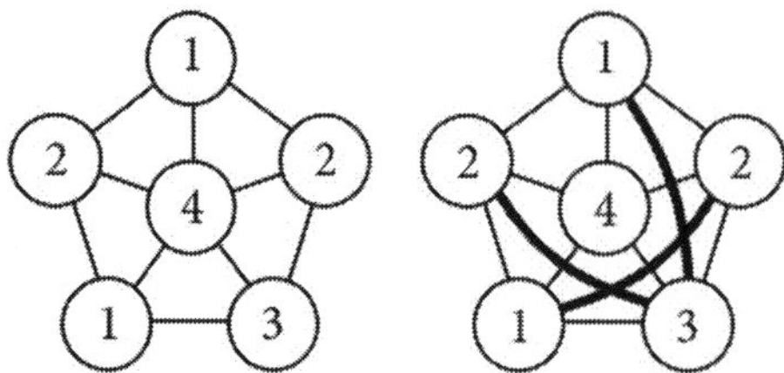

Why does it matter if a pair of neighboring vertices is part of a K_4 subgraph? If a pair of neighboring vertices is already part of a K_4 subgraph, there is already an edge connecting that pair of vertices in the MPG; in that case, you can't add another edge to connect them without creating a double edge. As shown in the partial graph below, if A and C are already part of a K_4 subgraph, this means that there is already an edge connecting them, in which case connecting them with another edge would create a double edge.

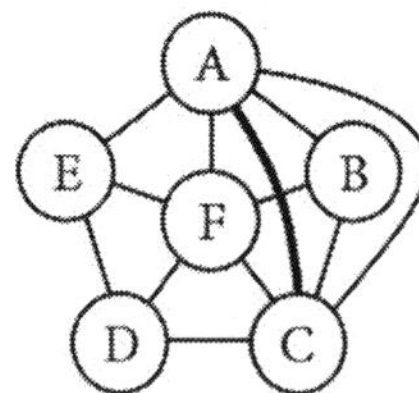

We don't really need to worry about K_4 subgraphs already in the MPG. As we discussed in Chapter 12, if we can prove that all MPG's without K_4's are four-colorable, it will follow that all MPG's are four-colorable. Of course, adding new edges to any MPG will create K_4's, but these K_4's are in graphs that aren't MPG's, and the purpose of these extra K_4's would be to help make the coloring of the graph more restrictive (so that it is easier to color).

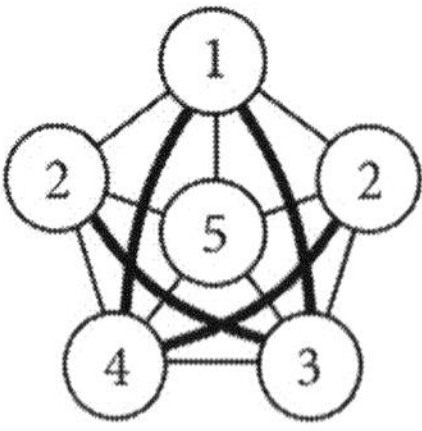

We can't add four edges to a vertex with degree five without creating a K_5 subgraph, which isn't four-colorable.

A possible problem with adding edges to a vertex with degree five is knowing which vertices can be connected by a new edge. If the four-coloring somehow requires two particular vertices to be the *same* color, connecting those vertices with a new edge will force them to be different colors, causing the new graph to be five-colorable. Rather than worry about what might happen if you add an edge, just add the edge anyway and see what happens. If you add an edge and the new graph is four-colorable, then the original MPG was four-colorable. If instead the new graph isn't four-colorable, then you know to remove the extra edge.

Chapter 17 will look at the reverse process: removing edges. The significance of adding or removing edges will be explored more fully in Chapter 27.

Chapter 15 Exercises

1. Each MPG below has three vertices already colored. For each graph:

- What is the minimum number of edges that need to be added such that the new graph is four-colorable a single way?

- Draw these edges on the graph and color the new graph.

- If the added edges are now removed, does the same coloring work for the original MPG?

- How many K_4 subgraphs are there before and after adding the new edges?

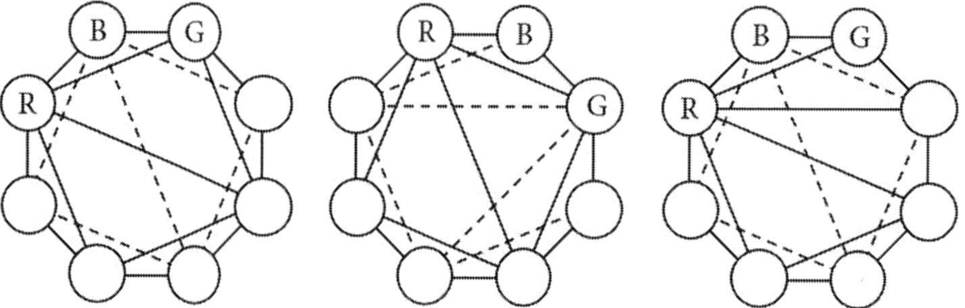

2. Each MPG below is already colored. For each graph, what is the maximum number of edges that can be added such that the current coloring works for the new graph? Draw these edges on each graph.

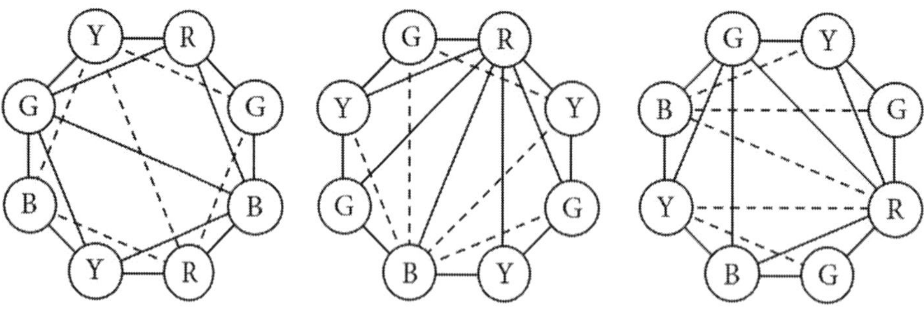

Challenge problem 1: Prove that it is possible to add 3 edges to any vertex with degree 5 in any MPG that doesn't contain K_4 subgraphs (without making any subdivisions or contracting any edges) provided that the edges connect the right pairs of vertices, even if the MPG is five-colored instead of four-colored. The left diagram below shows 3 edges that can be added.

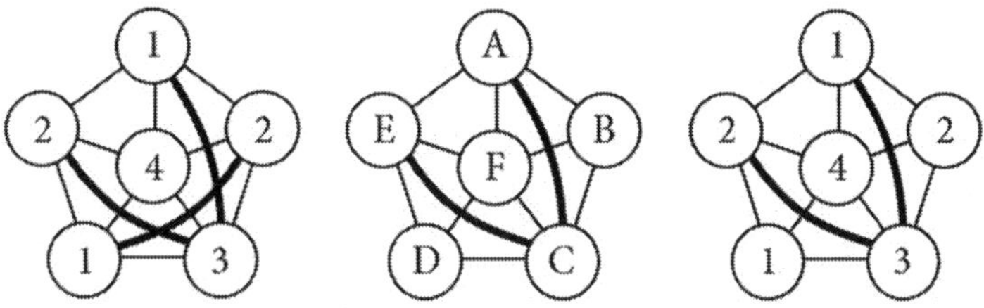

Suppose now that we choose only to add 2 of these 3 possible edges, as shown in the middle and right diagrams. Does triangle ACE meet the definition of a ST? Can we use triangle ACE to divide the new graph (which is no longer a MPG) into two smaller graphs? Either show that we can in an example MPG or demonstrate why we can't. If we can divide the graph in this manner, can we prove the four-color theorem by adding 2 edges to every vertex with degree 5 to divide MPG's into smaller graphs? Explain.

Note: The answer key doesn't include answers to the challenge problems. These problems are intended to encourage you to think about the ideas.

Challenge problem 2: If a MPG doesn't contain any K_4 subgraphs, will adding a new edge always create a K_4 subgraph in the new graph (which is no longer a MPG)? Either prove this, provide a counterexample, or explain why it would be very difficult to determine. In Chapter 12, we saw that when a K_4 is present in a MPG, the K_4 always includes a ST. Does the K_4 in the new graph made by adding an edge to a MPG similarly include a ST, or is the triangle that would normally be regarded as a ST somehow different in this case? Can we use a K_4 made by adding an extra edge to a MPG to divide the new graph into two smaller graphs? Either show that we can in the example graph below (where IL is the added edge) or demonstrate why we can't. If we can do this, can we prove the four-color theorem using a "divide and conquer" strategy as follows? Add an edge to a MPG, use the K_4 to divide the new graph into two smaller graphs, add an edge to each of these smaller graphs (if needed), use the K_4's to divide those graphs into smaller graphs, and so on.

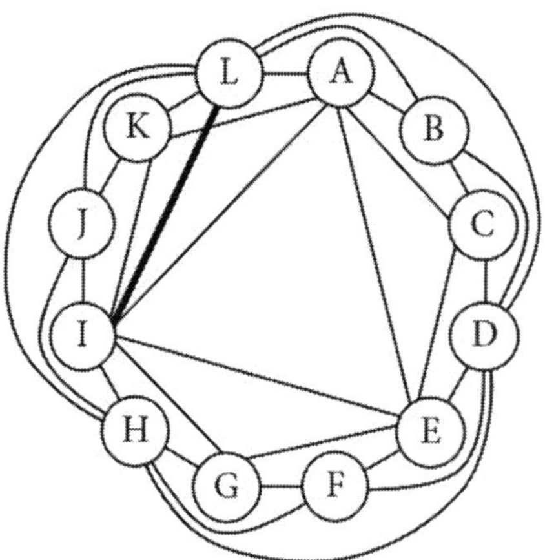

Note: The answer key doesn't include answers to the challenge problems. These problems are intended to encourage you to think about the ideas.

16 Ultimate Four-Coloring

In this chapter, we will construct and consider **ultimate four-colorable graphs**, which we will abbreviate U4CG, defined as follows:

- K vertices are blue (B). Each blue vertex connects to every vertex that isn't blue.
- L vertices are green (G). Each green vertex connects to every vertex that isn't green.
- M vertices are red (R). Each red vertex connects to every vertex that isn't red.
- N vertices are yellow (Y). Each yellow vertex connects to every vertex that isn't yellow.
- The total number of vertices is V = K + L + M + N.

An U4CG is a **complete multipartite graph** with four sets of vertices where every vertex of one set connects to every vertex of the other sets, but where no vertices of the same set are connected. (We saw an example of a complete multipartite graph with *two* sets of vertices in Chapter 5, called a complete bipartite graph. The complete multipartite graphs of this chapter are similar, except that they have four sets of vertices rather than two.)

A general U4CG is shown below, where B_1, B_2, B_3, thru B_K represent the K regions colored blue; G_1, G_2, G_3, thru G_L represent the L regions colored green; R_1, R_2, R_3, thru R_M represent the M regions colored red; and Y_1, Y_2, Y_3, thru Y_N represent the N regions colored yellow.

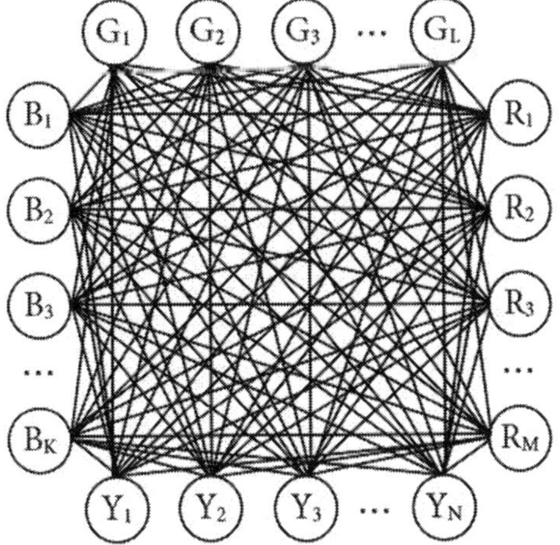

The $\cdots$ symbol in the above graph means "and so on."

Let's count how many edges an U4CG has. Recall that K, L, M, and N represent the numbers of blue, green, red, and yellow vertices, respectively.

- K blue vertices connect with L + M + N vertices that aren't blue: K(L + M + N).

- L green vertices connect with K + M + N vertices that aren't green: L(K + M + N).

- M red vertices connect with K + L + N vertices that aren't red: M(K + L + N).

- N yellow vertices connect with K + L + M vertices that aren't yellow: N(K + L + M).

Add these up and divide by two (to correct for double-counting, like we did in Chapters 4-5).

$$E = \frac{K(L + M + N) + L(K + M + N) + M(K + L + N) + N(K + L + M)}{2}$$

Apply the distributive property of algebra. For example, K(L + M + N) = KL + KM + KN.

$$E = \frac{KL + KM + KN + KL + LM + LN + KM + LM + MN + KN + LN + MN}{2}$$

We applied the commutative property of multiplication, that order doesn't matter when multiplying numbers. For example, LK = KL. Now we will combine like terms. For example, KL + KL = 2KL. Note that each term appears twice on the right-hand side.

$$E = \frac{2KL + 2KM + 2KN + 2LM + 2LN + 2MN}{2}$$

$$E = KL + KM + KN + LM + LN + MN$$

An alternative formula for the number of edges of an U4CG appears in Exercise 11 of Chapter 17. For example, an U4CG with K = 3 blue vertices, L = 4 green vertices, M = 5 red vertices, and N = 6 yellow vertices has E = 3(4) + 3(5) + 3(6) + 4(5) + 4(6) + 5(6) = 12 + 15 + 18 + 20 + 24 + 30 = 119 edges. The total number of vertices in this example is V = K + L + M + N = 3 + 4 + 5 + 6 = 18. Note that this U4CG isn't planar because it has too many edges to be a MPG, since a MPG with V = 18 vertices would have E = 3V − 6 = 3(18) − 6 = 54 − 6 = 48 edges instead of 119 edges (using Euler's formula for a MPG from Chapter 4). Most U4CG's are nonplanar because they have too many edges to be MPG's (but there are special cases where U4CG's have few enough edges to be PG's, like the case where K = L = M = N = 1).

The significance of the U4CG is that any four-colorable graph (planar or not) can be drawn by removing edges from an U4CG. The same principle that motivated triangulation (Chapter

3) applies here: Since every U4CG is inherently four-colorable, any graph that is obtained by removing edges from an U4CG is also four-colorable. If we could prove that every possible MPG can be obtained by removing edges from an U4CG, we could use that proof as a lemma to easily prove the four-color theorem.

Let's consider a specific example of an U4CG with 2 blue vertices, 2 green vertices, 3 red vertices, and 1 yellow vertex. This U4CG is shown below.

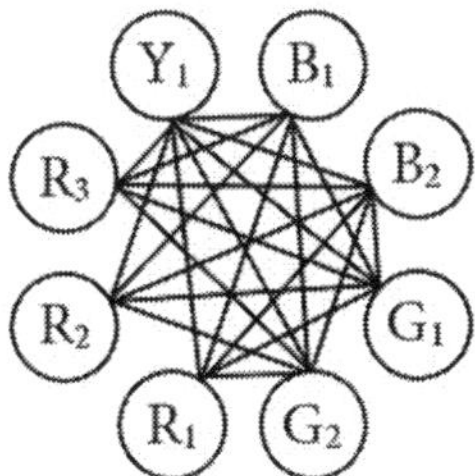

- Each blue connects to all 6 non-blue regions.
- Each green connects to all 6 non-green regions.
- Each red connects to all 5 non-red regions.
- The only yellow connects to all 7 non-yellow regions.
- $K = 2$, $L = 2$, $M = 3$, and $N = 1$.
- There are $V = K + L + M + N = 2 + 2 + 3 + 1 = 8$ vertices.
- There are $E = KL + KM + KN + LM + LN + MN = 2(2) + 2(3) + 2(1) + 2(3) + 2(1) + 3(1) = 4 + 6 + 2 + 6 + 2 + 3 = 23$ edges.
- A MPG with $V = 8$ vertices would have $E = 3V - 6 = 3(8) - 6 = 24 - 6 = 18$ edges. Thus, the U4CG shown above is nonplanar. (See Chapter 6.)

Like all U4CG's, the graph shown above is four-colorable by construction since it doesn't connect any edges to two vertices that have the same color. Any graph that can be drawn by removing edges from this (or any other) U4CG is also four-colorable.

Note that the previous graph isn't the only type of U4CG that can be drawn for a graph with 8 vertices. An U4CG with $V = 8$ vertices could have:

- 2 blues, 2 greens, 2 reds, and 2 yellows.

- 3 of one color, 3 of another color, and 1 each of the remaining two colors, such as 3 blues, 3 greens, 1 red, and 1 yellow.

- 3 of one color, 1 of another color, and 2 each of the remaining two colors, such as 3 blues, 1 green, 2 reds, and 2 yellows.

- 4 of one color, 2 of another color, and 1 each of the remaining two colors, such as 4 blues, 2 greens, 1 red, and 1 yellow.

- 5 of one color and 1 each of the remaining three colors, such as 5 blues, 1 green, 1 red, and 1 yellow.

Note that K, L, M, and N must all be positive in order for a graph to be U4CG. If one value is zero, it would be a three-colored graph instead of a four-colored graph. Observe that any graph that is three-colorable is also four-colorable (provided that the graph has at least four vertices). For example, suppose that a graph is colored using red, green, and blue only (and that no two vertices of the same color share an edge). Choose any color that has at least two vertices of that color and change exactly one of its vertices to yellow. The graph will now be four-colored (and the yellow certainly won't share an edge with any other yellows).

Let's look at two extreme types of U4CG's. The first extreme is the case of equitable coloring where V is evenly divisible by 4. In this case, there are equal numbers of vertices for each color: $K = L = M = N = V/4$. All 6 terms in the formula for the number of edges will be the same:

$$E = 6\left(\frac{V}{4}\right)\left(\frac{V}{4}\right) = \frac{6V^2}{16} = \frac{3V^2}{8}$$

For example, for V = 12 vertices, equitable coloring would have 3 blues, 3 greens, 3 reds, and 3 yellows. In this case, the number of edges is $E = 3(12^2)/8 = 3(144)/8 = 54$. Each vertex connects to all 9 of the other vertices which aren't the same color as itself. We actually saw this in an example in Chapter 15, where we added 24 edges to the icosahedral graph.

Another important extreme is the case of lopsided coloring. We don't need to concern our-selves with U4CG's where half of the vertices are a single color if our interest in U4CG's relates to the four-color theorem. Any MPG (with $V > 2$) that can be formed by removing edges from an U4CG won't have half of its vertices a single color when it is colored. (You may explore the reason for this in one of the exercises at the end of the chapter, if you wish.)

With this in mind, for an even number of vertices, the most lopsided coloring of an U4CG would have $V/2 - 1$ vertices one color, $V/2 - 1$ vertices a second color, and the remaining two vertices would be the remaining two colors. To determine the number of edges in this case, set $K = L = V/2 - 1$ and $M = N = 1$.

$$E = KL + KM + KN + LM + LN + MN$$

$$E = \left(\frac{V}{2} - 1\right)\left(\frac{V}{2} - 1\right) + \left(\frac{V}{2} - 1\right)(1) + \left(\frac{V}{2} - 1\right)(1) + \left(\frac{V}{2} - 1\right)(1) + \left(\frac{V}{2} - 1\right)(1) + (1)(1)$$

Anything times one equals itself.

$$E = \left(\frac{V}{2} - 1\right)\left(\frac{V}{2} - 1\right) + \frac{V}{2} - 1 + \frac{V}{2} - 1 + \frac{V}{2} - 1 + \frac{V}{2} - 1 + 1$$

Apply the foil method: $(W + X)(Y + Z) = WY + WZ + XY + XZ$.

$$E = \frac{V^2}{4} - \frac{V}{2} - \frac{V}{2} + 1 + \frac{V}{2} - 1 + \frac{V}{2} - 1 + \frac{V}{2} - 1 + \frac{V}{2} - 1 + 1$$

Two pairs of $V/2$'s cancel. The remaining pair equals V (since a half plus a half makes one whole). There are two $+1$'s and four -1's, which add up to -2.

$$E = \frac{V^2}{4} + V - 2$$

For example, for $V = 12$, there could be 5 blue vertices, 5 green vertices, 1 red vertex, and 1 yellow vertex. (Note that $12/2 - 1 = 6 - 1 = 5$.) In this example, $E = 12^2/4 + 12 - 2 = 144/4 + 10 = 36 + 10 = 46$ edges. (Compare this to the 54 edges for an U4CG with equitable coloring.)

It is interesting to compare these two types of extreme U4CG's to two types of extreme MPG's.

- One type of extreme MPG has a fairly equitable distribution of the degrees of its vertices. One example is the icosahedral MPG where all 12 vertices have degree 5. We saw in Chapter 15 that the icosahedral MPG can be four-colored with an equitable distribution of colors (3 blues, 3 greens, 3 reds, and 3 yellows).

- Another type of extreme MPG has a lopsided distribution of the degrees of its vertices. One example has two vertices with a degree of $V - 1$, two vertices with degree 3, and all of the remaining vertices have degree 4. Such a MPG has the maximum number of ST's (Chapter 12). This MPG has the same coloring as the U4CG with lopsided coloring that we considered: $V/2 - 1$ vertices of one color, $V/2 - 1$ vertices of another color, and 1 vertex each for the two remaining colors.

These two extreme types of MPG's can be drawn by removing edges from the corresponding types of extreme U4CG's. If we could show that all MPG's can be drawn by removing edges from U4CG's, we could use this to prove the four-color theorem.

In Chapter 17, we will discuss the idea of removing edges from one type of graph to make another type of graph. In Chapter 27, we'll explore the significance of U4CG's more fully (but in that chapter we will refer to them as "complete tetrapartite graphs").

Notation: Since K_7 represents a complete graph with 7 vertices and $K_{5,8}$ represents a complete bipartite graph with one set of 5 vertices and one set of 8 vertices (Chapter 5), we could use the notation $K_{K,L,M,N}$ to represent an U4CG (which is a complete multipartite graph with four sets of vertices) with K blue vertices, L green vertices, M red vertices, and N yellow vertices.

Grammar note: If "an U4CG" and "a MPG" seem backwards to you (that is, you think they should be "a U4CG" and "an MPG"), see the last paragraph in Chapter 6.

Chapter 16 Exercises

1. For each U4CG described below, draw the graph and also determine how many vertices and edges the graph has.

- 4 blue vertices, 3 green vertices, 2 red vertices, and 1 yellow vertex
- 3 blue vertices, 3 green vertices, 2 red vertices, and 2 yellow vertices

2. Which structurally different types of U4CG's can be formed with 12 vertices? (If a graph can be made from another by swapping colors, it isn't structurally different. For example, a graph with 3 blues, 2 greens, 1 red, and 1 yellow has the same structure as a graph with 3 reds, 2 yellows, 1 blue, and 1 green.) Determine the number of edges for each case.

3. Are any U4CG's planar? If so, which ones? Are any U4CG's also MPG's? If so, which ones. Are there any U4CG's with few enough edges to be PG's, but which aren't planar? If so, which ones?

4. Explain why any MPG that can be formed by removing edges from an U4CG won't have half of its vertices a single color when it is colored such that no two vertices that share an edge have the same color (of course, $V > 2$ for a MPG).

5. Show that any U4CG that meets the condition from Problem 4 has a HC, and thus may be redrawn as a closed convex polygon similar to the graphs of Chapter 14 (except that they will generally not be MPG's).

6. Identify a HC in the U4CG shown below and use it to redraw the graph as a closed convex polygon similar to the graphs of Chapter 14 (except that it isn't a MPG).

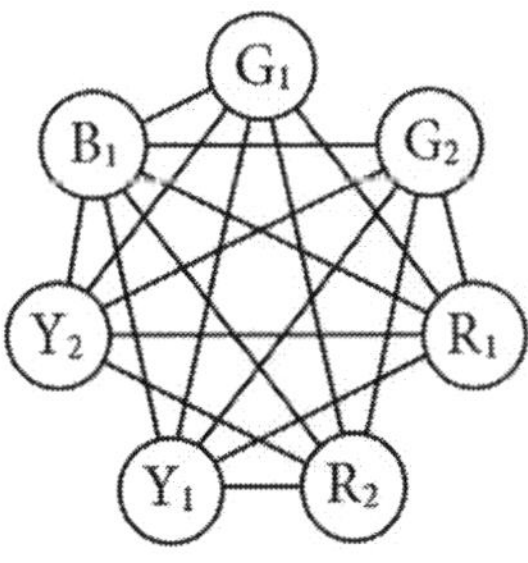

7. An U4CG has $K = L = M = (V - 1)/3$ and $N = 1$ where $(V - 1)$ is evenly divisible by 3 (so V could be 16, 31, 100, or any other integer that is one greater than a multiple of three).

- Verify that $K + L + M + N = V$.
- Show that $E = (V + 2)(V - 1)/3$.

8. An U4CG has $K = L = V/6$ and $M = N = V/3$ where V is evenly divisible by 6.

- Verify that $K + L + M + N = V$.
- Show that $E = 13V^2/36$.

9. A MPG has 24 vertices. None of its vertices has a degree less than 5. What is the maximum possible degree than any single vertex can have in this MPG? What would the degrees of the other vertices be if one vertex has this maximum possible degree?

Now consider an U4CG with 24 vertices with equal numbers of blue, green, red, and yellow vertices. What is the degree of each vertex in this U4CG? Compare this U4CG to the MPG that we asked about earlier in this exercise.

10. A MPG has 100 vertices. None of its vertices has a degree less than 5. What is the maximum possible degree than any single vertex can have in this MPG? What would the degrees of the other vertices be if one vertex has this maximum possible degree?

Now consider an U4CG with 100 vertices with equal numbers of blue, green, red, and yellow vertices. What is the degree of each vertex in this U4CG? Compare this U4CG to the MPG that we asked about earlier in this exercise.

Challenge problem 1: Let's revisit Challenge Problem 1 from Chapter 6 and see if what we now know about U4CG's has any impact on that approach.

- Any U4CG with at least 6 vertices contains a $K_{3,3}$ subgraph.
- $K_{3,3}$ is two-colorable.
- Every U4CG is four-colorable.
- K_5 isn't four-colorable.
- U4CG's don't contain any K_5 subgraphs.
- MPG's don't contain any subgraphs that are subdivisions of K_5 or $K_{3,3}$.

Show that U4CG's don't contain any K_5 subgraphs. Can an U4CG contain a subgraph that is a subdivision of K_5? If so, give an example; if not, explain why.

Can you prove that every MPG can be made by removing edges from a suitable U4CG? If so, you could use this prove the four-color theorem. Alternatively, can you prove that every graph that doesn't contain a subgraph that is a subdivision of K_5 is four-colorable, and if so, can you use this to prove the four-color theorem?

Perhaps you can prove the following alternative: Any graph that doesn't contain a subgraph that is a subdivision of K_5 is a subgraph of a complete multipartite graph with no more than four sets of vertices (the case with four sets of vertices we're referring to as an U4CG), and every complete multipartite graph with no more than four sets of vertices is four-colorable. Since no MPG contains a subgraph that is a subdivision of K_5, it would then follow that all MPG's are four-colorable. If you can't prove this, either explain the flaw in the reasoning or explain why it would be very challenging to prove this.

Note: The answer key doesn't include answers to the challenge problems. These problems are intended to encourage you to think about the ideas. However, you may wish to consider how this problem relates to Chapter 27.

17 Removing Edges

Let's see if we can form MPG's by removing edges from U4CG's. We will draw the U4CG's in the form of closed convex polygons, like those discussed in Problems 5-6 in Chapter 16. In this chapter, we will use K for the color with the most vertices, L for the color with the second most vertices, M for the color with the third most vertices, and N for the color with the fewest vertices. Any U4CG that could be drawn by swapping the values of K and M, or of K and N, or any other combination of K thru N, won't be structurally different.

The two simplest U4CG's have 4 vertices and 5 vertices. These two U4CG's are already MPG's; we don't need to remove any edges from them. These are also the only possible MPG's with 4 and 5 vertices. As we have done in some of the previous chapters, we will use dashed lines to help you visualize which edges could be redrawn outside of the polygon so that you can see that these are indeed MPG's. (Recall that we have defined MPG to be any graph that can be drawn as a maximal planar graph, even if it isn't currently drawn that way.)

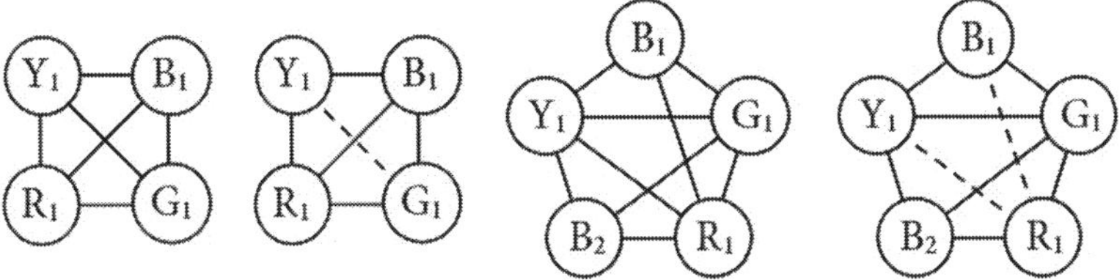

Recall from Chapter 16 that we don't need to consider any U4CG's where one-half of the vertices are the same color. With this in mind, there is only one type of U4CG with 6 vertices: $K = 2$, $L = 2$, $M = 1$, and $N = 1$. (Note that swapping the values of K thru N won't result in an U4CG that is structurally different.) This U4CG has $E = KL + KM + KN + LM + LN + MN = 2(2) + 2(1) + 2(1) + 2(1) + 2(1) + 1(1) = 4 + 2 + 2 + 2 + 2 + 1 = 13$ edges. Since a MPG with $V = 6$ vertices has $E = 3V - 6 = 3(6) - 6 = 18 - 6 = 12$ edges, we must remove 1 edge from this U4CG to form a MPG. There are two possible MPG's with 6 vertices. As shown on the following page, both MPG's can be formed by removing 1 edge from the U4CG.

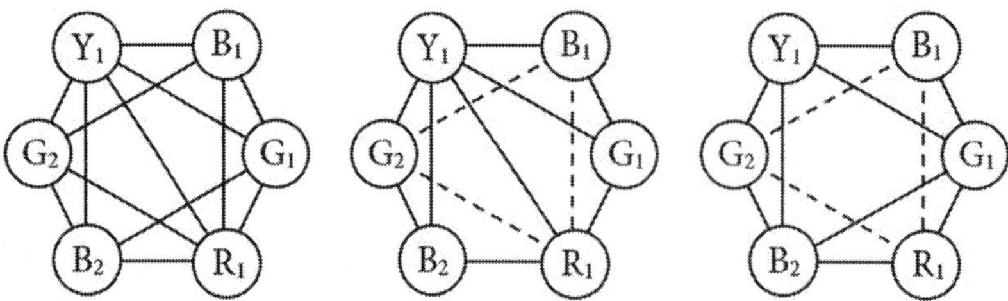

The MPG in the middle was formed by removing edge B_2G_1, while the MPG on the right was formed by removing edge R_1Y_1.

For 7 vertices, we can make an U4CG with $K = 3$, $L = 2$, $M = 1$, and $N = 1$ with $E = KL + KM + KN + LM + LN + MN = 3(2) + 3(1) + 3(1) + 2(1) + 2(1) + 1(1) = 6 + 3 + 3 + 2 + 2 + 1 = 17$ edges. Alternatively, we can make an U4CG with $K = 2$, $L = 2$, $M = 2$, and $N = 1$ with $E = KL + KM + KN + LM + LN + MN = 2(2) + 2(2) + 2(1) + 2(2) + 2(1) + 2(1) = 4 + 4 + 2 + 4 + 2 + 2 = 18$ edges. A MPG with $V = 7$ vertices has $E = 3V - 6 = 3(7) - 6 = 21 - 6 = 15$ edges. We need to remove 2-3 edges from one of these U4CG's to form the different possible MPG's.

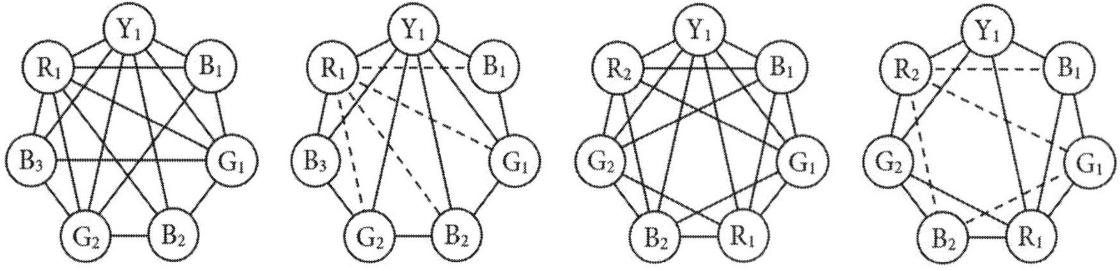

We removed edges B_1G_2 and B_3G_1 from the left U4CG to make one MPG. We removed edges B_1G_2, B_2Y_1, and G_1Y_1 from the other U4CG to make a different MPG on the right. There are 3 additional structurally different MPG's that can be formed from U4CG's with 7 vertices which are not shown here.

As the number of vertices increases, the number of structurally different types of U4CG's grows and the number of structurally different types of MPG's grows much faster. As an example, for $V = 10$, there are 5 structurally different types of U4CG's where no single color takes one-half of the vertices ($K_{4,4,1,1}$, $K_{4,3,2,1}$, $K_{4,2,2,2}$, $K_{3,3,3,1}$, and $K_{3,3,2,2}$) whereas there are 233 structurally different MPG's.

Now let's look at a different way to remove edges from a graph to form a MPG. Instead of starting with an U4CG, we could start with a complete graph. Recall from Chapter 5 that a complete graph with V vertices has $E = V(V - 1)/2$ edges, and recall from Chapter 4 that a MPG with V vertices has $E = 3V - 6$ edges. Subtract these expressions to find out how many edges must be removed from a complete graph to form a MPG:

$$\frac{V(V - 1)}{2} - (3V - 6) = \frac{V^2 - V}{2} - 3V - (-6)$$

$$= \frac{V^2 - V}{2} - 3V + 6 = \frac{V^2 - V}{2} - \frac{6V}{2} + \frac{12}{2}$$

$$= \frac{V^2 - V - 6V + 12}{2} = \frac{V^2 - 7V + 12}{2} = \frac{(V - 3)(V - 4)}{2}$$

We applied the distributive property of algebra in the first steps: $-(X - Y) = -X -(-Y) = -X + Y$. The two minus signs effectively made a plus sign.

To form a MPG from a complete graph, we need to remove $(V - 3)(V - 4)/2$ edges from the complete graph. We can't remove *any* set of $(V - 3)(V - 4)/2$ edges from the complete graph; we must remove edges in such a way that the resulting graph is planar.

Let's look at a few examples. A complete graph with 4 vertices, K_4, already is a MPG. We don't need to remove any edges from K_4 to form a MPG.

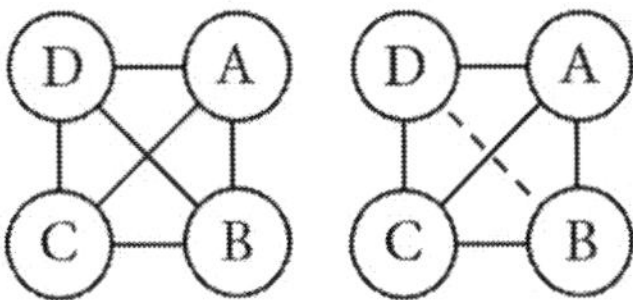

A complete graph with $V = 5$ vertices, K_5, has $E = V(V - 1)/2 = 5(4)/2 = 20/2 = 10$ edges. We need to remove $(V - 3)(V - 4)/2 = 2(1)/2 = 2/2 = 1$ edge from K_5 to form a MPG. In this case, it turns out that it doesn't matter which edge we remove.

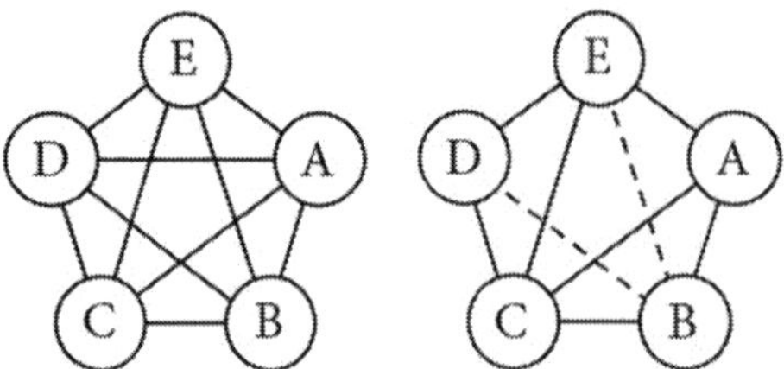

We removed edge AD from K_5 to form the MPG on the right. It doesn't matter which edge is removed from K_5; the result will be structurally equivalent to the MPG shown here.

A complete graph with $V = 6$ vertices, K_6, has $E = V(V - 1)/2 = 6(5)/2 = 30/2 = 15$ edges. We need to remove $(V - 3)(V - 4)/2 = 3(2)/2 = 6/2 = 3$ edges from K_6 to form a MPG. When $V > 5$, it makes a difference which edges are removed, as shown below.

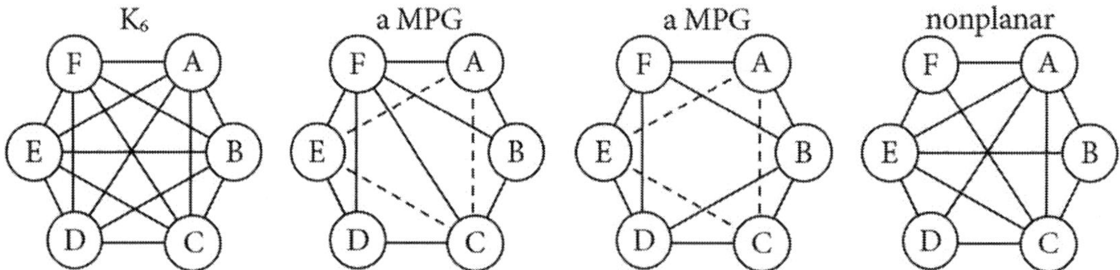

When edges are removed from a complete graph to form a MPG, the edges are removed in a specific way. The removed edges form four (or fewer) complete subgraphs. (By complete subgraph, we mean that the subgraph is itself a complete graph with fewer vertices than the complete graph from which it is removed.) If you could prove that this is always true, you could use this proof to prove the four-color theorem. We will illustrate what we mean about the removed edges forming four complete subgraphs later in this chapter.

First, let's consider a few possible complete subgraphs.

- K_1 is a single vertex with no edges. This may seem sort of silly, but K_1 is important. If we don't remove any edges from a particular vertex, we will associate a removed K_1 from that vertex (even though no edges were removed). Why is this important? This vertex connects to every other vertex, which means that this will be the only vertex of its color. As we'll see, this will be important for the four-color theorem.

- K_2 is a single edge that connects 2 vertices which don't participate in other complete subgraphs. We'll see examples of removed K_2's.

- K_3 is a triangle. It consists of 3 edges and 3 vertices.

- K_4 consists of 6 edges and 4 vertices where each vertex connects to all 3 of the other vertices.

- K_5 consists of 12 edges and 5 vertices where each vertex connects to all 4 of the other vertices.

In general, there may be both active and passive removed edges defined as follows:

- A removed edge is considered **<u>active</u>** if it belongs to a removed complete subgraph.

- A removed edge is considered **<u>passive</u>** if it doesn't belong to a removed complete subgraph.

For example, imagine one removed K_4 that removes all of the edges connecting A, B, C, and D to one another, and a removed K_3 that removes all of the edges connecting E, F, and G to one another. Edges AB, AC, AD, BC, BD, CD, EF, EG, and FG are considered active removed edges. If edge DE is removed in addition to the other edges, then DE is considered a passive edge since it isn't a member of either of the removed subgraphs.

If a single edge is removed from a complete graph, the two vertices it had joined may be the same color. In the graph below, removing edge BE allows B and E to be the same color.

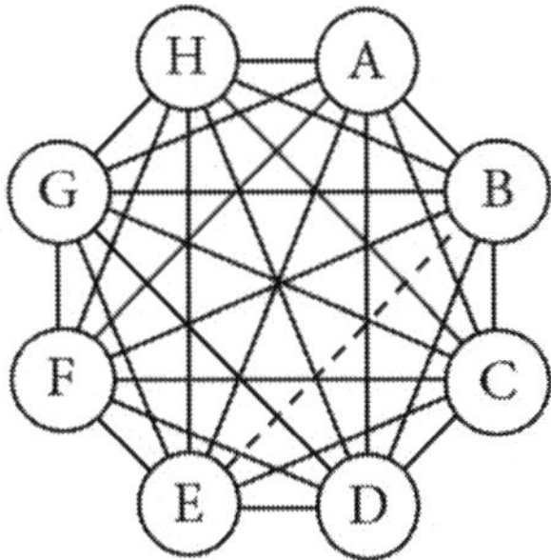

If two edges are removed from a complete graph, the result depends on whether or not these edges join to a common vertex. If the two removed edges don't share a common vertex, like the two cases shown on the left below, then each pair of vertices may be the same color (but they must be two different colors because, for example, A still connects to C and E still connects to G). If the two removed edges do share a common vertex, like the case shown on the right below, then you have a choice: in this case, either A and E could be the same color or B and E could be the same color, but not both (because A still connects to B). Note that it doesn't matter whether or not the two edges cross; it just matters if they join to a common vertex.

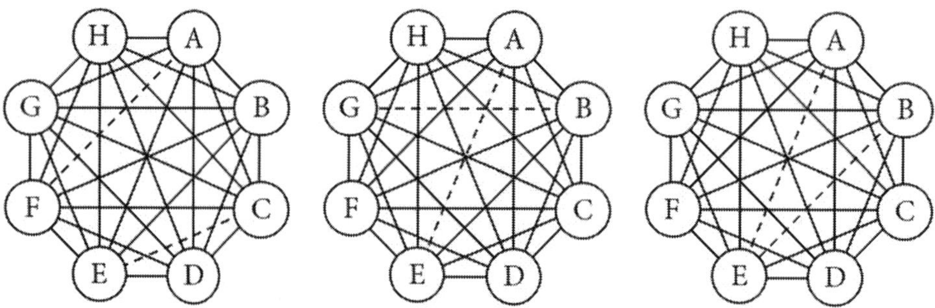

If three edges are removed from a complete graph:

- If they form a triangle, three vertices may be the same color.
- If they don't join to any common vertices, three pairs of vertices may each be the same color (but different from the other pairs).
- If they connect at common vertices to form a single path, the result depends on how many vertices are shared, which we will discuss momentarily.
- If two edges connect at a common vertex, but the third does not join to this pair at a common vertex, two pairs of vertices may each be the same color (but each pair must be different from the other pair). For the two lines that share a common vertex, there is a choice.

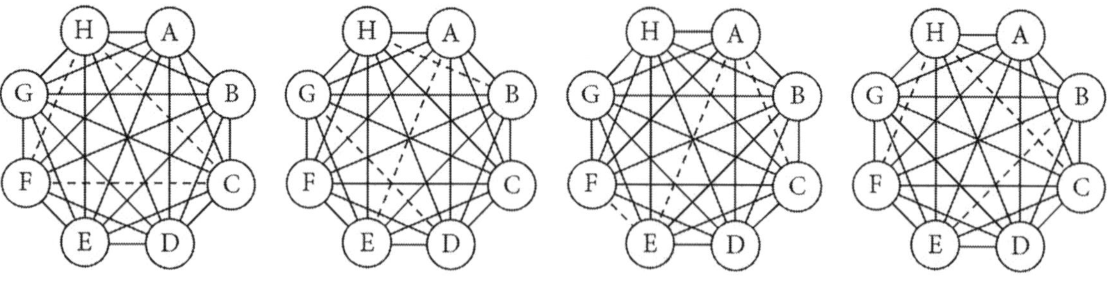

As mentioned previously, when three removed edges share one or more common vertices to form a path, the result depends on how many vertices are shared. In the left example below where two different vertices join the edges, we may color two different pairs the same color. For example, B and D may be red while E and H may be blue. In the right example below where one vertex joins the edges, we may choose one (and only one) pair to be the same color (since the other vertices are still connected by other edges).

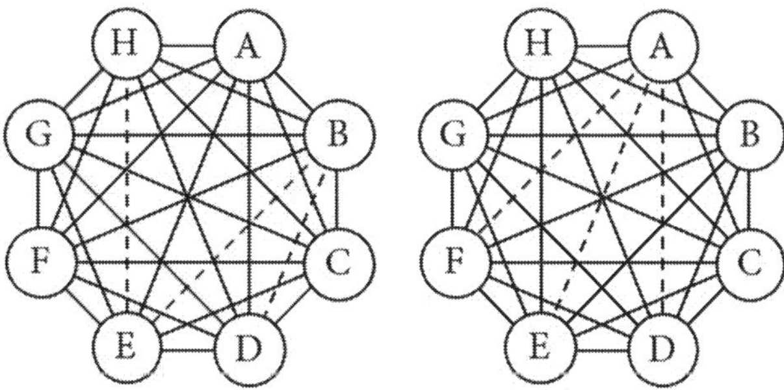

Although a triangle permits three vertices to be the same color, a quadrilateral (a four-sided polygon) does not allow four vertices to be the same color. For example, A and E are still connected in the left graph below, which doesn't allow the opposite corners to be the same color. However, a triangle plus a separate edge allows the three vertices of the triangle to be one color and the edge to be another color. On the other hand, if the edge shares a vertex with the triangle, then at best you can make the triangle's vertices the same color.

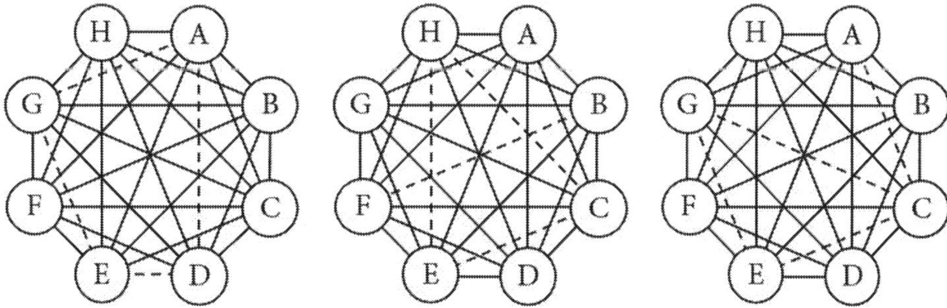

The following example shows how a set of six removed edges may allow four vertices to be the same color. Since every possible edge connecting A, C, D, and G is removed, this allows all four vertices to be the same color. Observe that A, C, D, and G make K_4 (a complete subgraph with 4 vertices; it is a subgraph because it is part of the larger graph K_8).

157

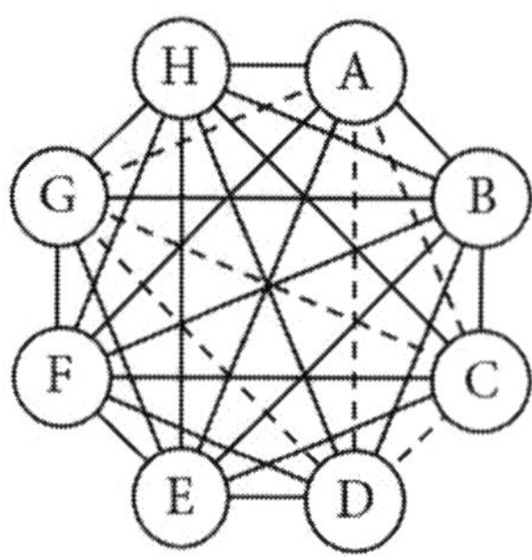

Now we see an important result. If a complete subgraph is removed from a complete graph, all of the vertices of the complete subgraph may be the same color. If two (or more) complete subgraphs are removed from a complete graph, the vertices of each complete subgraph may be the same color provided that there is no sharing of vertices between the removed subgraphs. If there are other removed edges or faces in addition to one or more removed complete subgraphs, each separate complete subgraph may have its vertices colored with the same color, but then any other removed edge or face that shares a vertex with it doesn't cause additional colors to be the same. Here is another way to put it:

- Active edges form removed complete subgraphs and allow a set of vertices to be the same color. The two vertices connected by an active edge may be the same color.

- Passive edges are removed edges that share an edge with a removed complete subgraph (without expanding the complete subgraph into a larger complete subgraph). The two vertices connected by a passive edge can't be the same color.

We will illustrate this with a few examples.

- On the left graph on the following page, B, D, F, and H form a K_4, and C, E, and G form a K_3. The K_3 and K_4 don't share any vertices, which allows the vertices of the K_3 to be one color and the vertices of the K_4 to be another color.

- On the middle graph on the following page, CEG and CFH form two K_3's, but since they share the common vertex C, only one of these K_3's can have its vertices be the same color. We can't use a different color for each K_3 because vertex C can't be two different colors, and we can't make both K_3's the same color because of edges like FG that haven't been removed.

- On the right graph below, CFH and DEG form two K_3's. Since these don't share any vertices, we may color one K_3 red and the other blue, for example. The extra removed edge CD has no additional effect.

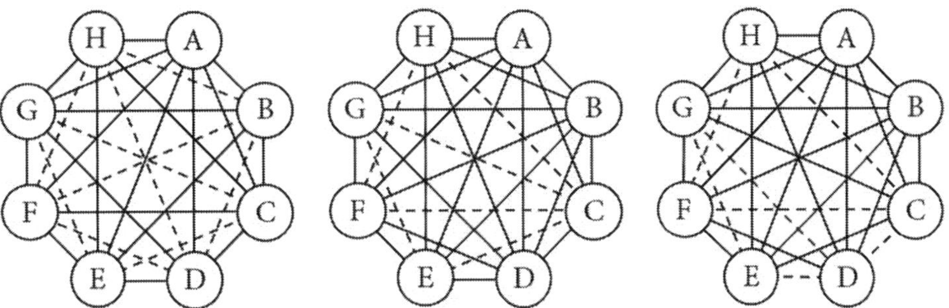

Now we will give some examples of removing edges from a complete graph to form a MPG, and explore how the removed edges form four (or fewer) complete subgraphs. We'll also see the importance of the K_1 and K_2 subgraphs and examples of active and passive edges.

A complete graph with V = 8 vertices, K_8, has E = V(V − 1)/2 = 8(7)/2 = 56/2 = 28 edges. We need to remove (V − 3)(V − 4)/2 = 5(4)/2 = 20/2 = 10 edges from K_8 to form a MPG.

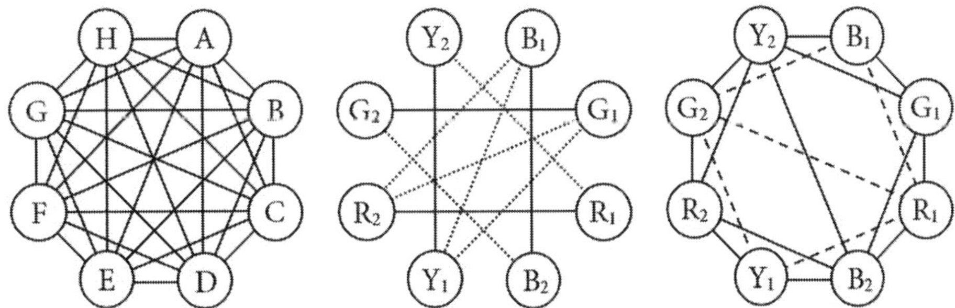

In the example above, the middle graph shows the 10 edges that are removed from the K_8 on the left. These 10 removed edges form 4 K_2's: active edges AD, HE, BG, and CF, shown as solid lines. (Recall that K_2 is a single edge joining two vertices.) The other 6 edges that are removed are passive: AF, AE, BF, BE, CH, and DH. These passive edges connect vertices from different K_2's. Each removed K_2 allows one pair of vertices to be the same color: AD allows A and D to both be blue, HE allows E and H to both be yellow, BG allows B and G to both be green, and CF allows C and F to both be red. The right graph shows the MPG that is formed once these 10 edges are removed from the K_8. Note how the coloring of the four removed complete subgraphs matches the coloring of the MPG.

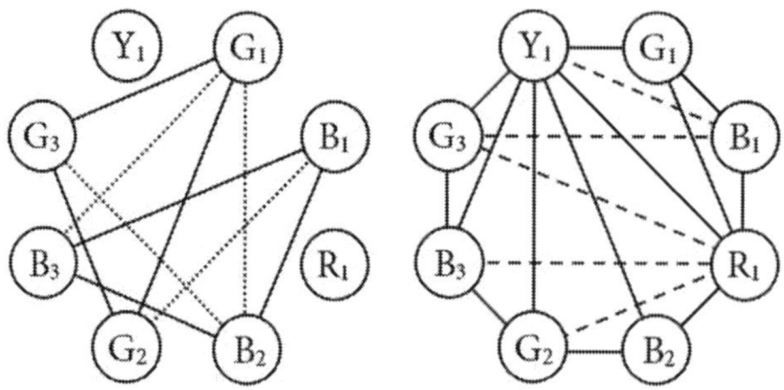

In the example above, the left graph shows a different set of 10 edges that can be removed from a K_8. This forms the MPG on the right. In this example, the removed edges form 2 K_3's and 2 K_1's. The 6 active removed edges form the two removed K_3's: triangles $B_1B_2B_3$ and $G_1G_2G_3$. The other 4 removed edges are passive: B_1G_2, B_2G_1, B_2G_3, and B_3G_1. These passive edges connect vertices from different K_3's. Two of the vertices don't have any edges removed from them: R_1 and Y_1. We associate the two K_1's with these vertices. Each of the 4 complete subgraphs allows a set of vertices to be the same color: each K_3 allows 3 vertices to be the same color (one is blue, the other is green), and each K_1 requires a vertex to be the only vertex of a given color (one is red, the other is yellow).

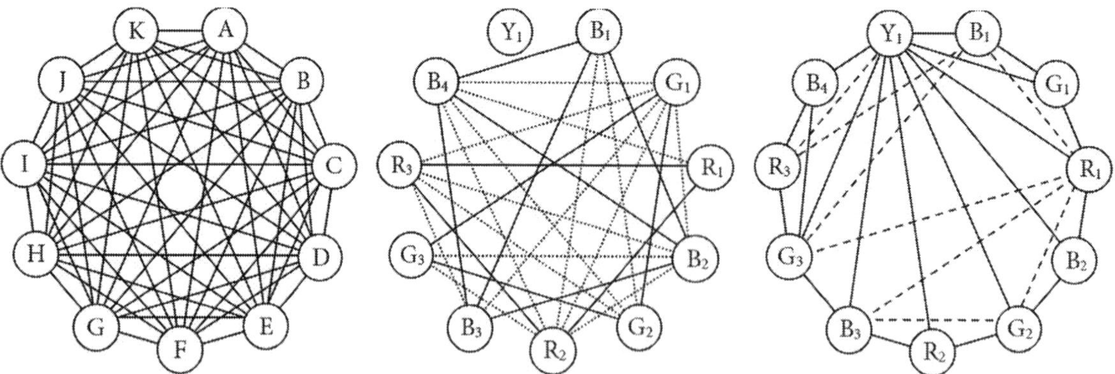

A complete graph with $V = 11$ vertices, K_{11}, has $E = V(V - 1)/2 = 11(10)/2 = 110/2 = 55$ edges. We need to remove $(V - 3)(V - 4)/2 = 8(7)/2 = 56/2 = 28$ edges from K_{11} to form a

MPG. One way to do this is shown on the previous page. In this example, the removed edges form 1 K_4, 2 K_3's, and 1 K_1. The removed K_4 is ADGJ (which has 6 active edges: AD, AG, AJ, DG, DJ, and GJ). The removed K_4 allows A, D, G, and J to all be blue. The removed K_3's are BEH and CFI (which together have 6 more active edges: BE, BH, EH, CF, CI, and FI). The K_3's allow B, E, and H to all be green and C, F, and I to all be red. No edges are removed from vertex K. Therefore, a K_1 subgraph is associated with vertex K. Vertex K is connected to all 10 of the other vertices, so it must be its own color: yellow. The 12 active edges forming these 4 removed complete subgraphs are shown as solid lines in the middle graph. There are 16 passive edges shown as dotted lines in the middle graph. The right graph shows the MPG formed by removing these 28 edges from the K_{11}.

Chapter 17 Exercises

1. Verify that the left graph below meets our definition of an U4CG.

- Determine K, L, M, N, V, and E for this U4CG.

- How many edges would a MPG with this many vertices have?

- How many edges need to be removed from the U4CG to form a MPG?

- Draw a MPG on the right that can be formed by removing this number of edges from the U4CG. Draw the corresponding removed edges in the middle graph.

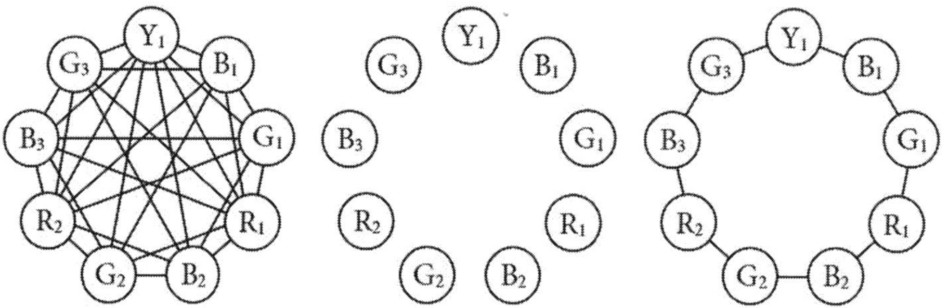

2. The right graph below is a MPG. Determine V and E for this MPG.

- Color the MPG with four different colors so that two vertices connected by an edge don't have the same color.

- Once the MPG has been properly four-colored, how many edges need to be added to the MPG in order to form an U4CG?

- Draw an U4CG on the left that can be formed by adding this number of edges to the MPG. Draw the corresponding removed edges in the middle graph.

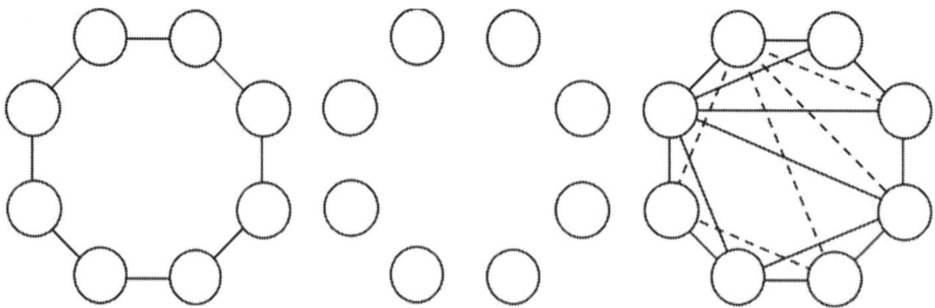

3. The left graph below is a complete graph. Determine V and E for this complete graph.

- How many edges would a MPG with this many vertices have?
- How many edges need to be removed from the complete graph to form a MPG?
- Draw a MPG on the right that can be formed by removing this number of edges from the complete graph so that the active removed edges form a K_3, a K_2, and two K_1's.
- How many active removed edges and how many passive removed edges are there?
- Draw the corresponding removed edges in the middle graph.
- Color the MPG with four colors according to the removed complete subgraphs.

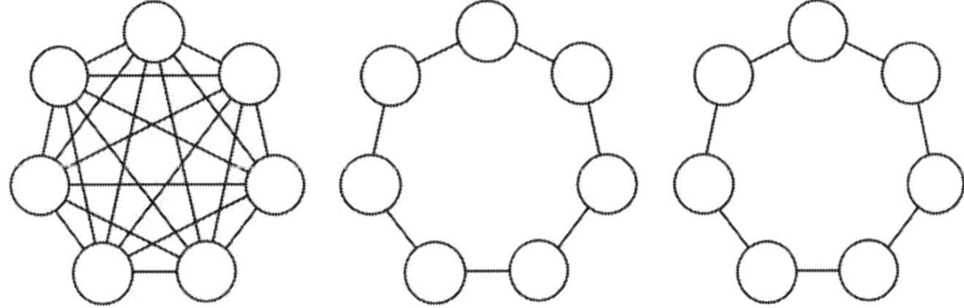

4. The left graph below is a complete graph. Determine V and E for this complete graph.

- The right graph below is a MPG. Determine V and E for this MPG.
- How many edges need to be removed from the complete graph to form this MPG?
- Color the MPG with four different colors so that two vertices connected by an edge don't have the same color.
- Draw the corresponding removed edges in the middle graph. Identify four removed complete subgraphs that match the coloring of the MPG.

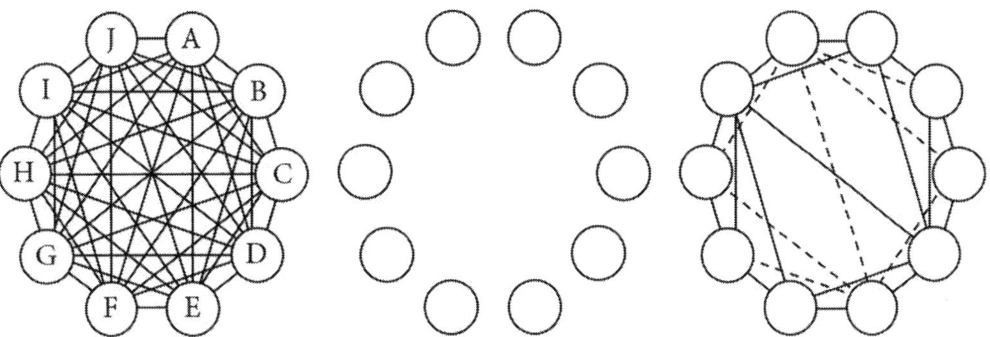

5. List the possible sets of values of K, L, M, and N (where these are ordered from largest to smallest, though it's possible for two or more values to be equal) for an U4CG with V = 11, where no single color has more than one-half of the vertices.

- For each case, determine how many edges the U4CG has.
- How many edges would a MPG with V = 11 have?
- For each case, how many edges must be removed from the U4CG to form a MPG?

6. In order to form a MPG, how many edges need to be removed from a complete graph? Work out the answers separately for V = 4, 5, 6, 7, 8, 9, 10, 11, and 12. What is the pattern?

7. What are the possible sets of 4 complete subgraphs that can be removed from K_7 to form a MPG?

8. What are the possible sets of 4 complete subgraphs that can be removed from K_{12} to form a MPG?

9. Is it possible for a MPG with 7 vertices to be three-colorable? If this is possible, draw and color an example, and identify the 3 complete subgraphs that need to be removed from K_7 plus the passive edges that need to be removed in order to make the MPG. If it isn't possible, show or explain why.

10. How many edges need to be removed from K_9 to form an U4CG? Explain why there is more than one answer. Give the answers for each possible case. (However, only work out cases where no single color has at least one-half of the vertices.)

11. Show that the formula $E = KL + KM + KN + LM + LN + MN$ from Chapter 16 for the number of edges in an U4CG with K blue vertices, L green vertices, M red vertices, and N yellow vertices may alternatively be expressed as:

$$E = \frac{K(V - K) + L(V - L) + M(V - M) + N(V - N)}{2}$$

After showing this in general, show that both formulas agree for the specific case of $K = 7$, $L = 5$, $M = 3$, and $N = 1$. Also, provide a conceptual interpretation of the above formula.

Challenge problem 1: Can any possible MPG be formed by removing edges from a suitable U4CG? Do one of the following:

- Prove this and use this proof to prove the four-color theorem. Unlike the challenge problem from Chapter 16, here you're not confined to answering in terms of K_5 or $K_{3,3}$ subgraphs. You may use any means necessary (except, for example, such fallacies as circular reasoning).

- Explain why it would be impossible or very difficult to prove the four-color theorem with this approach.

Challenge problem 2: Can any possible MPG be formed by removing edges from a complete graph with the same number of vertices? Do one of the following:

- Prove this and use this proof to prove the four-color theorem.

- Explain why it would be impossible or very difficult to prove the four-color theorem with this approach.

Note: The answer key doesn't include answers to the challenge problems. These problems are intended to encourage you to think about the ideas. However, you may wish to consider how these problems relate to Chapter 27.

Challenge problem 3: Remove K_5, K_4, K_3, and K_2 subgraphs from K_{14} to form a MPG. Draw the complete graph, the MPG, and the removed edges. Identify both the active and passive removed edges. Four-color the MPG according to the four removed complete subgraphs.

Now draw an U4CG that corresponds to the four-coloring of the MPG. Which edges need to be removed from K_{14} to form this U4CG? Which edges need to be removed from this U4CG to form this MPG? Do you see any relationship between the removed K_5, K_4, K_3, and K_2 subgraphs and this U4CG? (We're not suggesting that there is or isn't one. It's a *question*.)

Challenge problem 4: How many edges can you remove from K_{12} without removing a K_3 subgraph, if you're allowed to remove any edge from the K_{12}, including those from the closed convex polygon, and including the removal of edges that wouldn't make it possible to form a MPG. Just to be clear, we're saying that you may remove *any* edges that you wish from a K_{12} in any order, provided that you don't make a K_3 subgraph.

Now generalize your answer to the case of K_V, where V is unknown. State your answer as a simplified formula involving V.

List several restrictions that could be placed on this problem if your goal is to make a MPG that you could use in a possible proof of the four-color theorem.

Note: The answer key doesn't include answers to the challenge problems. These problems are intended to encourage you to think about the ideas.

18 Vertex Splitting

We will provide a method for building *any* MPG by starting with the K_4 graph shown below (and will attempt to four-color the graph along the way). A K_4 graph is a complete graph (recall Chapter 5) with 4 vertices. In the K_4 graph, every vertex connects to all three of the other vertices. It has the fewest vertices of any complete graph that is a MPG (as we saw in Chapter 5, the next step up, K_5, is neither a MPG nor four-colorable). Note that K_4 is four-colorable with each vertex being a different color.

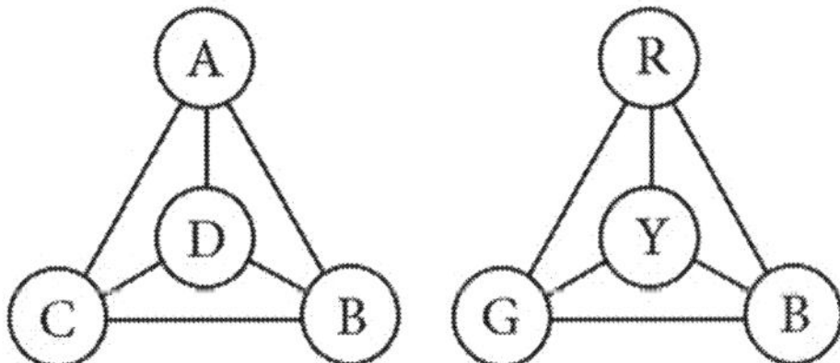

By always starting with the K_4 graph above, we can build *any* MPG adding one vertex at a time by dividing one of the vertices into two vertices and pushing the two vertices apart. We will refer to this as **vertex splitting**.

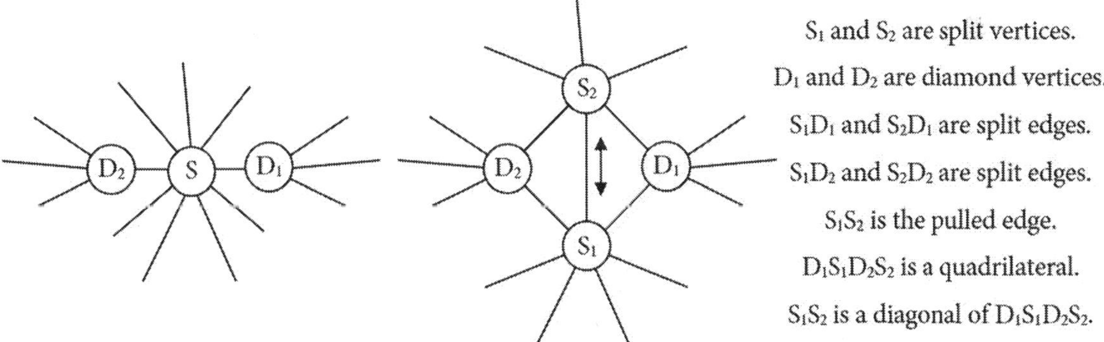

In the figure above, vertex S in the left graph splits into two vertices S_1 and S_2 in the right graph. Each diagram is zoomed in to illustrate the vertex splitting. We will refer to S_1 and S_2 as **split vertices**.

Since the new graph has one additional vertex compared to the original MPG, in order for the new graph to also be a MPG, it must have three new edges according to Euler's formula

(Chapter 4): $E = 3V - 6$. For example, when $V = 8$ we get $E = 18$ and when $V = 7$ we get $E = 15$, showing that a MPG with 8 vertices has 3 more edges than a MPG with 7 vertices. If we increase V by 1 for a MPG, E will always increase by 3. In the previous diagram, there are three new edges:

- We will refer to S_1S_2 as the **pulled edge**. The pulled edge joins the two split vertices. The pulled edge grows as S_1 and S_2 are pulled apart.
- Edge D_1S in the original MPG is split into two edges (D_1S_1 and D_1S_2) in the new MPG. We will refer to D_1S_1 and D_1S_2 as one pair of **split edges**.
- Edge D_2S in the original MPG is split into two edges (D_2S_1 and D_2S_2) in the new MPG. We will refer to D_2S_1 and D_2S_2 as a second pair of **split edges**.

We will refer to D_1 and D_2 as diamond vertices. The split vertices and diamond vertices form quadrilateral $D_1S_1D_2S_2$ with the pulled edge (S_1S_2) as a diagonal of the quadrilateral.

In the example on the previous page, vertex S in the original MPG had degree 9, whereas the split vertices S_1 and S_2 have degree 7 and 6, respectively. In general, the degrees of the split vertices will add up to 4 more than the degree of the original vertex. Here, you can see that $7 + 6 = 13$ is equal to $4 + 9 = 13$. Why? Two of the degrees are from the pulled edge and two are from the split edges. Note that D_1 and D_2 also gain one degree each, so that the sum of the degrees for the new MPG is 6 more than for the original MPG (which is consistent with having 3 more edges).

In general, the original vertex S connects to multiple vertices in the original graph. Of these, D_1 and D_2 are special in that the edges connecting D_1 and D_2 to S get split as S splits into S_1 and S_2 (forming the split edges D_1S_1, D_1S_2, D_2S_1, and D_2S_2 out of D_1S and D_2S).

Let's look at some examples of vertex splitting. These will illustrate different ways to build MPG's by splitting vertices. We will also consider how to color the new MGP in accordance with the four-color theorem.

The simplest case of vertex splitting occurs when one of the split vertices has degree three. We will use S_2 for the vertex with degree three. The diagonal vertices each connect to S_1 and S_2; these were colored B and R in the original diagram (and S was originally colored G, but this isn't shown). The dashed lines show the two split edges (SB split into S_1B and S_2B, and SR split into S_1R and S_2R; of course, there is more than one B and R, but we mean the ones connected to dashed lines) and the pulled edge (S_1S_2). We will refer to this case as VS1 (for Vertex Splitting Type 1). In VS1, one of the split vertices (we labeled this S_2) is degree 3.

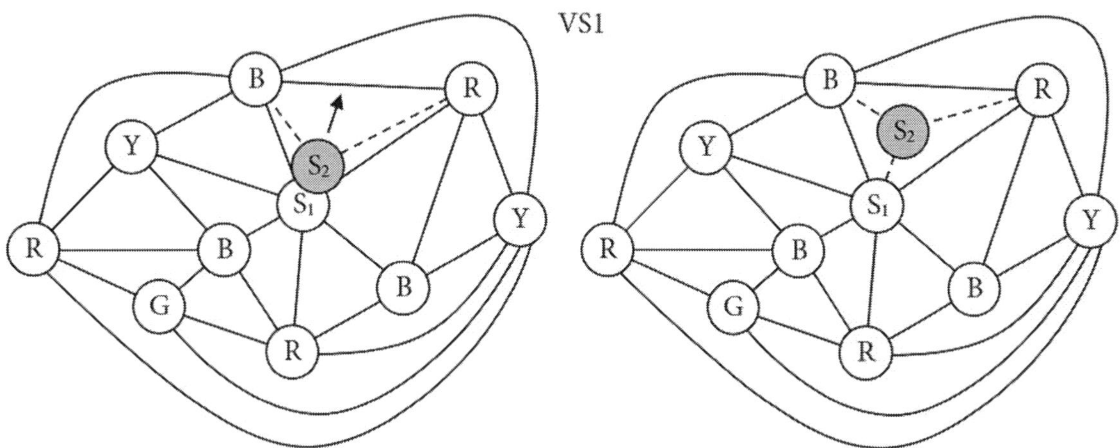

Vertex Splitting Type 1 (VS1) pushes one of the split vertices into a triangular face. Note that an alternative way to achieve VS1 would be to add a new vertex to any face (including the infinite area outside) and add edges to connect it to the vertices of that face.

Coloring the new MPG is trivial in the case of VS1. One of the split vertices only connects to three other vertices (S_2 only connects to S_1, B, and R), so that vertex (S_2) simply takes whichever color is left over while the other vertex (S_1) is colored the same as S was in the original MPG. In the example above:

- S was originally G. (You can tell that G works for S before it was split into two.)
- One split vertex (S_2) has degree three. We will color this vertex last.
- The other split vertex (S_1) will be G (the same color S had been).
- Since S_2 connects to B, R, and G, S_2 will be Y, as illustrated on the following page.

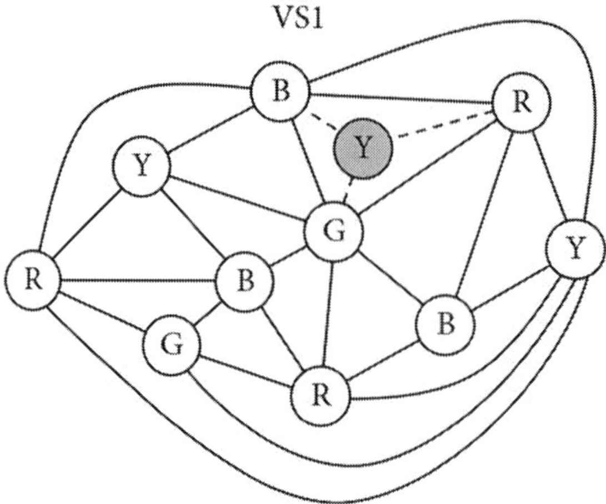

The second simplest case of vertex splitting occurs when one of the split vertices has degree four. We will use S_2 for the vertex with degree four. The diagonal vertices each connect to S_1 and S_2; these were each colored B in the original diagram (and S was originally colored G, but this isn't shown). Although the original MPG is the same as in the previous example, note that the diagonal vertices are different in this example. The dashed lines show the two split edges (each SB split into S_1B and S_2B) and the pulled edge (S_1S_2). We will refer to this case as VS2 (for Vertex Splitting Type 2). In VS2, one of the split vertices (we labeled this S_2) is degree 4.

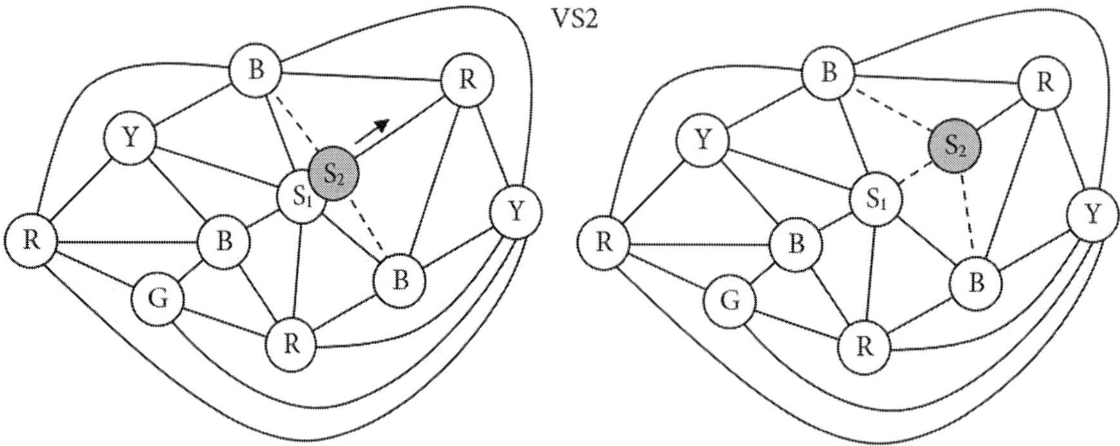

Vertex Splitting Type 2 (VS2) pushes one of the split vertices along an edge from the other split vertex. Note that an alternative way to achieve VS2 would be to **subdivide** (it may help to review Chapter 6) one of the edges from S and then add two edges (these are the diagonal edges) to this new vertex to triangulate the graph. In the previous example, subdivide SR, rename S as S_1, call the new vertex S_2, and add the pair of edges S_2B. An alternative way to think of VS2 is to imagine pushing one of the split vertices along an edge (whereas we may think of VS1 as pushing one of the split vertices into a face).

Since VS2 results in a vertex of degree four, we may apply Kempe's argument (discussed in Chapter 7) to color the new MPG.

- One of the split vertices (S_2) has degree four. This vertex is connected to the other split vertex (S_1), the two diagonal vertices, and one another vertex (labeled A below).

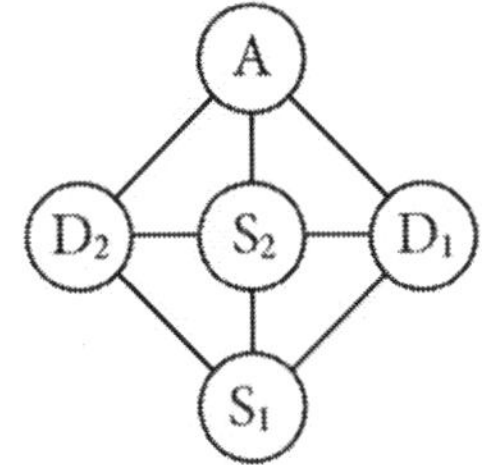

- If the diagonal vertices D_1 and D_2 happen to be the same color, we may color S_1 the same as S had been and S_2 will be whichever color is unused. For example, in the VS2 example shown on the previous page, D_1 and D_2 are both B, A is R, and S had been G (this isn't shown, but you can tell that G works for S), such that S_2 is Y.

- If D_1 and D_2 are different colors, we may either make D_1 and D_2 the same color or make S_1 and A the same color using Kempe's argument (discussed in Chapter 7).

  o If D_1 and D_2 each lie on separated sections of D_1-D_2 Kempe chains, reverse the colors of one of these sections so that D_1 and D_2 have the same color. Now color S_1 the same as S had been and S_2 will be whichever color is unused.

  o Otherwise, S_1 and A must lie on separated sections of S-A Kempe chains (where S refers to the color of the original vertex before the split). Reverse the colors of one of these sections so that S_1 and A have the same color, and S_2 will be whichever color is unused. (As discussed in Chapter 7, if D_1 and D_2 don't like on separated sections of Kempe chains, then S_1 and A must.)

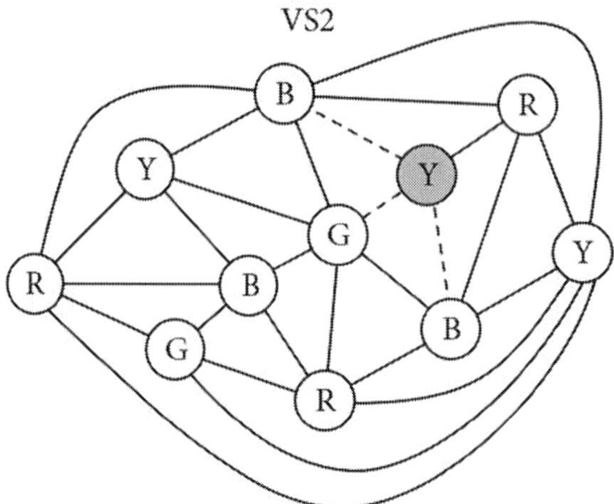

The figure above shows how to color the previous example of a VS2. This turned out to be the trivial case where the two diagonals both have the same color (B). We colored S_1 the same as S had been (G). In this example, A is R. The remaining color for S_2 is Y.

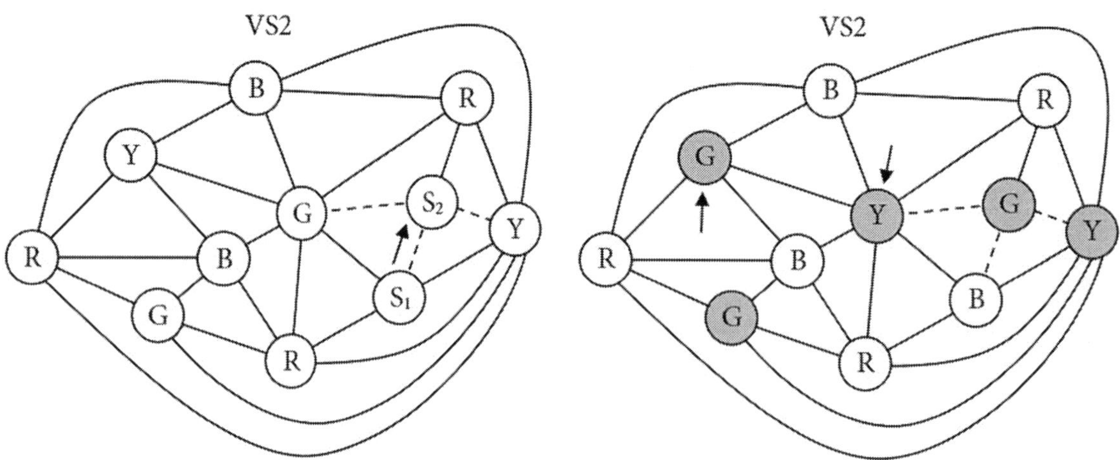

A different example of a VS2 is shown above (starting from the same original MPG as the previous example). In this case, S was originally B (this isn't shown, but you can tell that B works for S) and A is R. Here the diagonal vertices are different colors: G and Y. In this case, the two diagonal vertices lie on separated sections of G-Y chains (note that the Y on the right connects to a G far to the left, but that the two sections of G-Y chains don't meet in the left figure). We reversed the colors of the G-Y section on the left so that both diagonal vertices are now Y. We colored S_1 the same as S had been (B), and S_2 is the unused color

(G) from its degree four vertex. Note that S_1 and A (colored B and R, respectively) in the previous figure lie on the same section of a B-R chain, whereas D_1 and D_2 lie on separated sections of G-Y chains. In general, either S_1 and A will lie on separated sections of Kempe chains or D_1 and D_2 will (as discussed in Chapter 7).

In the more general case, the two split vertices (S_1 and S_2) each have degrees of at least five. We will refer to this as VS3 (for Vertex Splitting Type 3).

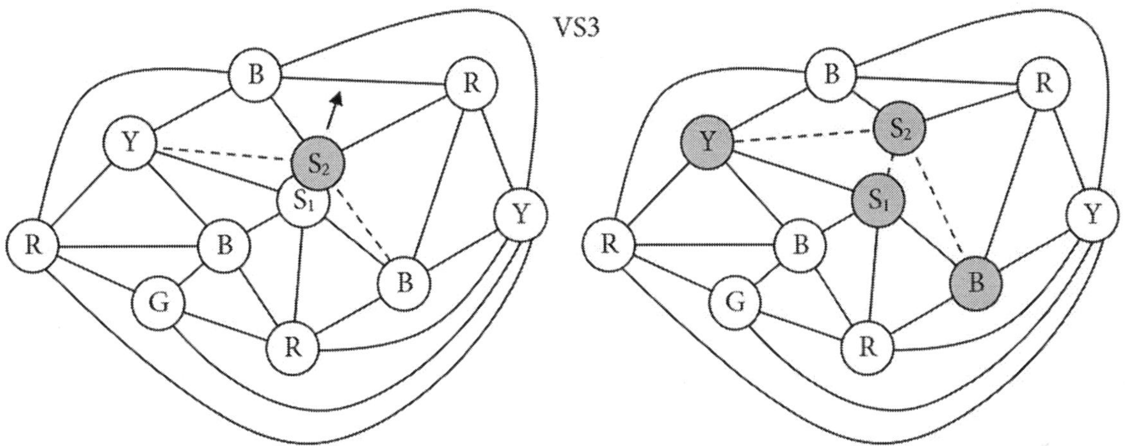

The figure above shows an example of VS3. The diagonal vertices each connect to S_1 and S_2; these were colored B and Y in the original diagram (and S was originally colored G, but this isn't shown). The dashed lines show the two split edges (SB split into S_1B and S_2B, and SY split into S_1Y and S_2Y; of course, there is more than one B and Y, but we mean the ones connected to dashed lines) and the pulled edge (S_1S_2). The shaded vertices show the quadrilateral (BS_1YS_2) formed by the split vertices and diagonal vertices.

Note that in the new MPG, neither S_1 nor S_2 connects to any other vertices that are the same color that S had been in the original MPG. We will be able to color S_1 or S_2 with the color of S, but we will need to color the other split vertex a different color. The figure above shows the simplest possible case of VS3, where each split vertex has degree five. We will argue that we can always split vertices such that at least one split vertex has degree five.

If you visualize the different cases of vertex splitting (VS1, VS2, and VS3) in reverse, such that the two split vertices are pushed along the pulled edge to merge together back into S, this should remind you of contracting edges to form minors for Wagner's theorem (Chapter 6). Just as any graph may be reduced to smaller and smaller minors by contracting edges, any MPG may be expanded into a larger MPG through vertex splitting.

Recall from Chapter 4 that the average degree of the vertices of any MPG is always below 6. This means that every MPG must have at least one vertex with a degree equal to 5 or less. We will use this to show that we can build any MPG beginning with K_4 by splitting vertices one at a time so that **at least one of the two split vertices always has a degree equal to 5 or less**.

Why? It's easier to visualize this backwards. Imagine starting with the complete MPG and merging one pair of vertices at a time in the reverse of VS1, VS2, or VS3. Each time, three edges contract, making a new MPG with one less vertex. Since these merged vertices always result in a new MPG, once there are only 4 vertices left, the graph will be K_4. This process of merging vertices is the reverse of vertex splitting that we've been discussing (with vertex splitting, we start with K_4 and split vertices one at a time to build MPG's).

(Note the distinction between contracting an edge, which was discussed in Chapter 6, and merging vertices, which we are defining to be the reverse of VS1, VS2, or VS3. Merging vertices always contracts 3 edges at a time so that the new graph is always a MPG. That's how we know that when only 4 vertices are left, it will be K_4.)

The order in which we merge the vertices is completely arbitrary. Once only 4 vertices are left, the result will always be K_4, regardless of the order in which pairs of vertices are merged. At least one vertex in every MPG always has a degree equal to 5 or less. This means that we can always choose to merge a pair of vertices where one of the vertices has a degree equal to 5 or less. Since the new graph is always a MPG, after merging one pair of vertices, the new graph will always have at least one vertex with a degree equal to 5 or less.

This means that we can begin with K_4 and split vertices according to VS1, VS2, and VS3 in order to make any MPG without ever splitting two vertices where neither vertex has a degree under 6. Given any MPG, first visualize how to merge its vertices one pair at a time so that at least one vertex of each pair always has a degree equal to 5 or less, and then carry out the process in reverse in order to build the MPG starting with K_4 by splitting vertices.

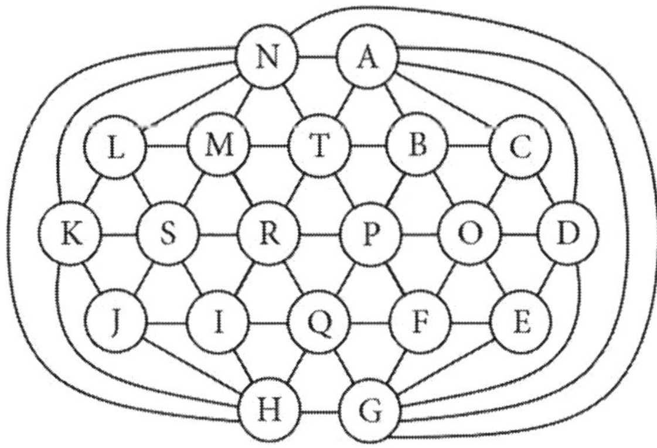

The MPG above has V = 20 vertices and E = 3V − 6 = 3(20) − 6 = 60 − 6 = 54 edges. The degrees of its vertices add up to 108 (twice the number of edges), so that the average degree of its vertices equals 108/20 = 5.4, which is below 6. Here is a breakdown:

- G and N are degree 7.
- A, H, O, P, Q, R, S, and T are degree 6.
- B, D, F, I, K, and M are degree 5.
- C, E, J, and L are degree 4.

Vertex G is degree 7. It connects to vertices D, E, and F, which have degrees equal to 4-5. Similarly, vertex N is degree 7, and it connects to vertices K, L, and M, which have degrees equal to 4-5. We can start out by merging higher degree vertices with lower degree vertices. On the following page, we merged the following pairs of vertices:

- E and G (contracting edges ED, EF, and EG)

- L and N (contracting edges LM, LN, and LS)

- A and C (contracting edges AC, BC, and CO)

- H and J (contracting edges HJ, IJ, and JS)

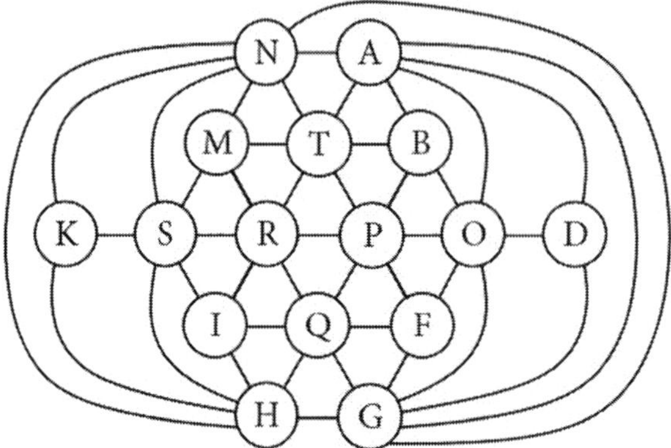

Now we have two vertices with degree 3: D and K. Let's merge D with O and K with S.

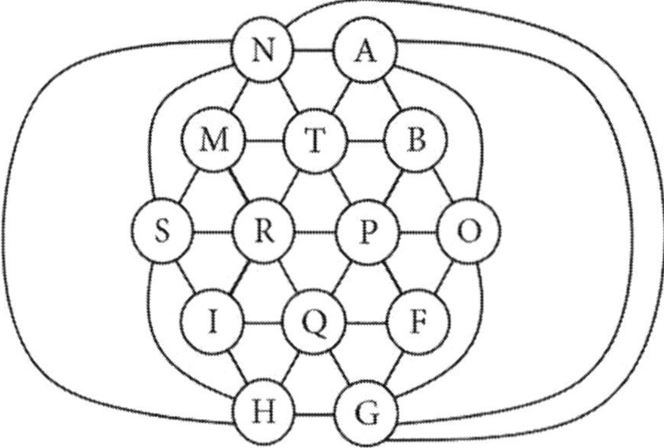

Now G and N are each degree 6 instead of 7, and A and H are degree 5 instead of 6. We can continue merging pairs of vertices like this without ever having a pair where both merged vertices have degrees above 5. If you're not yet convinced, try continuing this process, and also explore a variety of other graphs until you are.

Now the problem of coloring VS3 is more manageable. We can require one of the two split vertices to have a degree equal to 5. (If one of the split vertices has degree 3, that's VS1, and if one of the split vertices has degree 4, that's VS2.) We no longer need to be concerned about VS3's where both vertices have degree 6 or higher. We can always arrange it so that one of the two split vertices has degree 5. The other split vertex may have a higher degree. The case of VS3 with one vertex with degree 5 is illustrated below, where the original vertex before splitting had color 4.

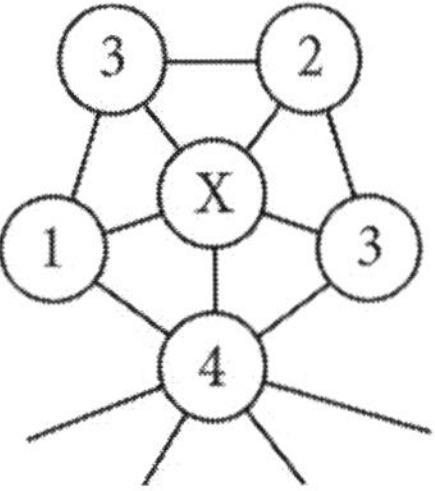

We colored the higher degree split vertex (at the bottom) the same as the original vertex (color 4) and labeled the split vertex with degree 5 with an X. The diagram above is zoomed in, focused on split vertex X and its neighbors. Other edges and vertices are not shown.

It seems that we have returned back to square one: recoloring the graph to accommodate the vertex labeled X appears to be the same as the vertex with degree 5 which Kempe had originally attempted to recolor without success. However, there is a difference. Instead of choosing a random vertex with degree 5 in a MPG, as Kempe did, we are building a MPG from a MPG which had one less vertex. Perhaps there is a way to split vertices in such a way (or in such an order) as to avoid Kempe's problem. We'll let you contemplate this in one of the challenge problems at the end of the chapter.

Chapter 18 Exercises

1. Show that whether you apply VS1 or VS2 to K_4, regardless of which vertex it is applied to, either way the resulting graph is structurally equivalent to the only MPG with 5 vertices, which is shown below on the right.

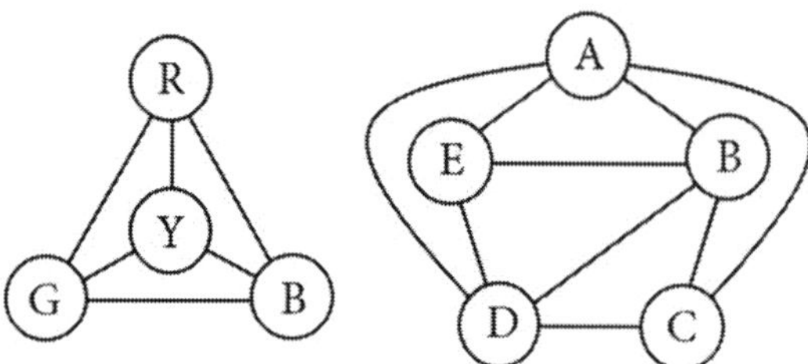

2. Show how the MPG below on the left can be transformed into the MPG below on the right through vertex splitting.

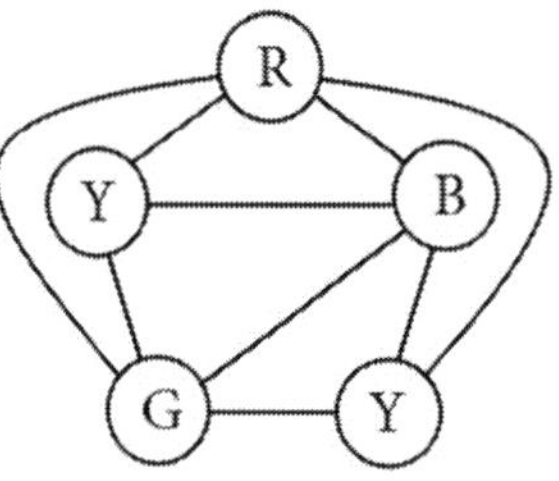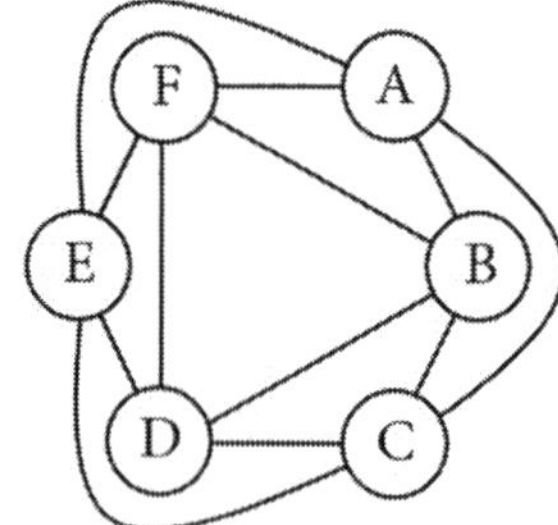

3. Shown below on the left, a four-colored MPG had one of its vertices split. One of the new vertices, labeled X, spoils the original four-coloring. Use a Kempe chain to recolor the new MPG on the right.

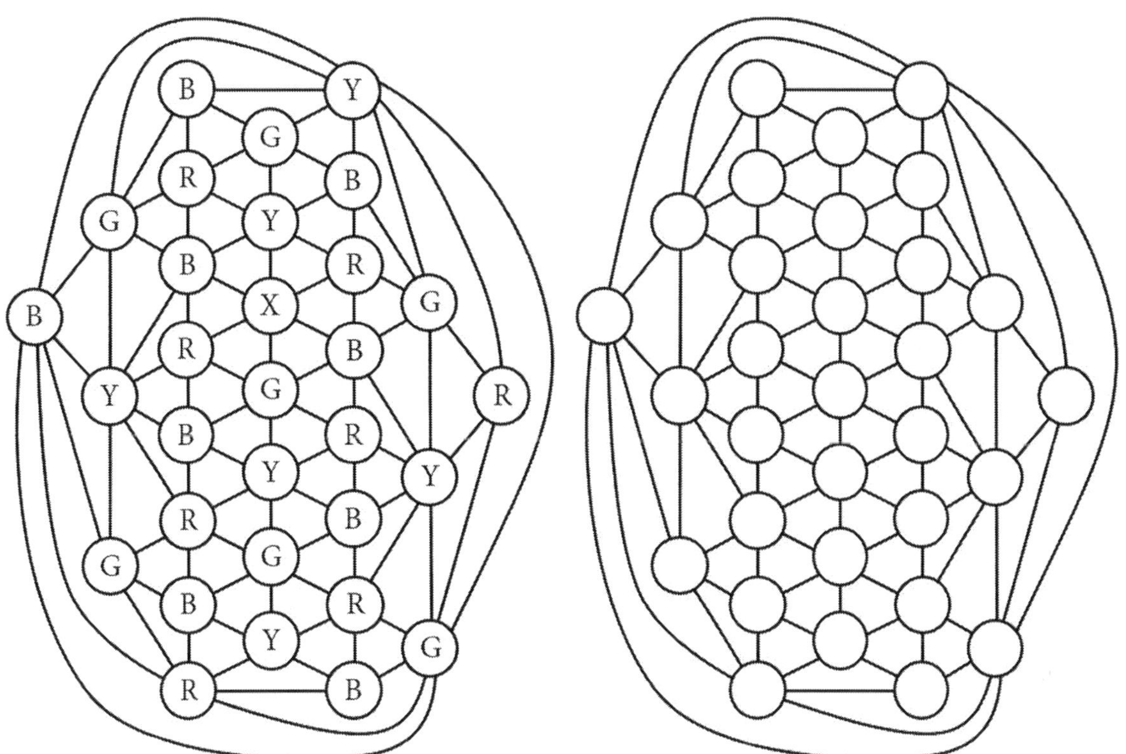

4. Show how to apply a combination of VS1, VS2, or VS3 to transform the left MPG below into the right MPG. (You're not "required" to use more than one type of vertex splitting.)

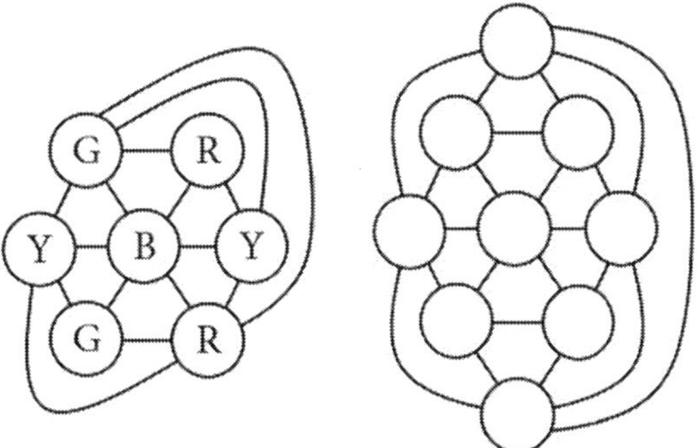

5. Starting with K₄ (where every vertex has degree 3), show how the icosahedral MPG (where every vertex has degree 5) can be built by splitting vertices one at a time. Do this in such a way that at least one vertex of any VS3 always has degree 5.

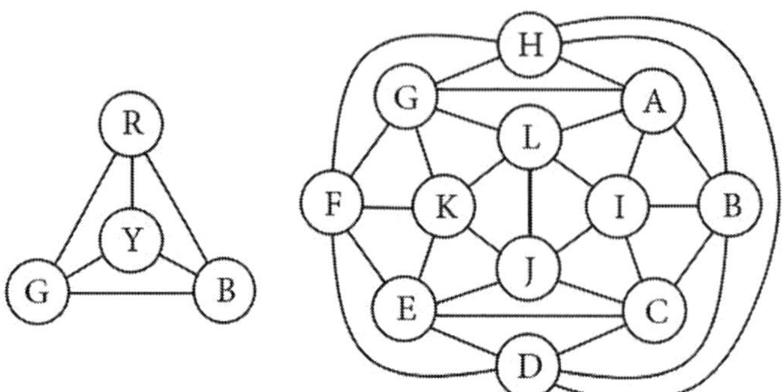

6. The vertex splitting shown below isn't VS1, VS2, or VS3. Explain why. Also, explain why we don't need to consider cases like this if our goal is to prove the four-color theorem.

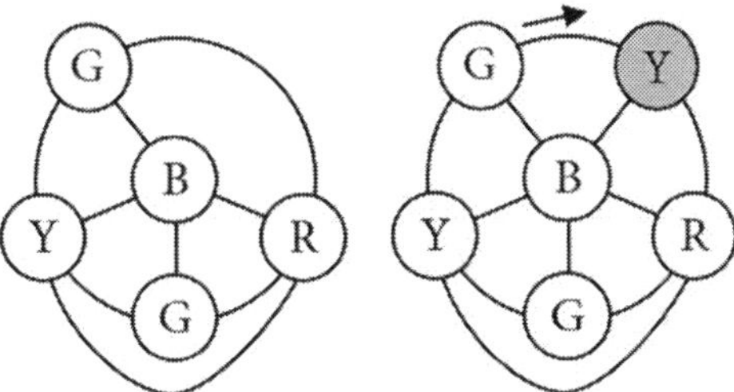

Challenge problem 1: The diagrams below are portions of graphs; <u>**many vertices and edges of the complete MPG's are not shown**</u>. In each case of VS3 below, the original vertex before splitting had been colored blue. Either S_1 or S_2 may be blue, but both can't be blue.

- How does the structure of the left graph affect the idea of reversing the colors of a section of a Kempe chain that involves B in an attempt to color S_1 and S_2?

- What features of the right graph might complicate the idea of using G-R, G-Y, and R-Y color reversals of sections of Kempe chains (one reversal at a time) in order to move one color (such as G) completely away from either S_1 or S_2?

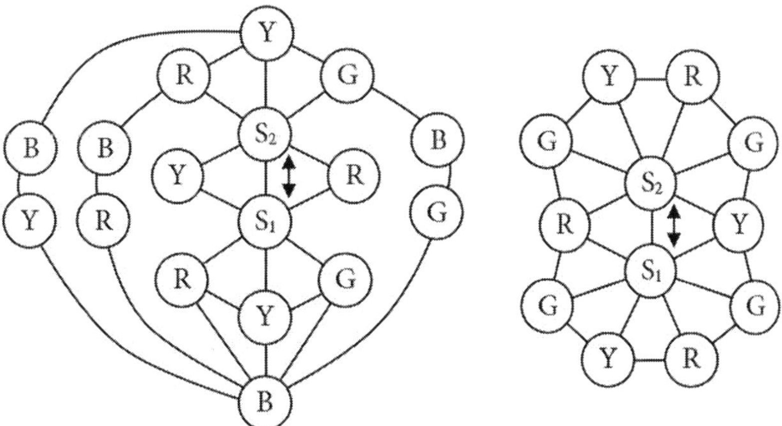

Note: The answer key doesn't include answers to the challenge problems. These problems are intended to encourage you to think about the ideas.

Challenge problem 2: We showed that we can make any MPG where VS3's always have at least one vertex with degree 5. Such a case is shown below. The original vertex had color 4. One of the split vertices has been colored 4; the other is labeled X.

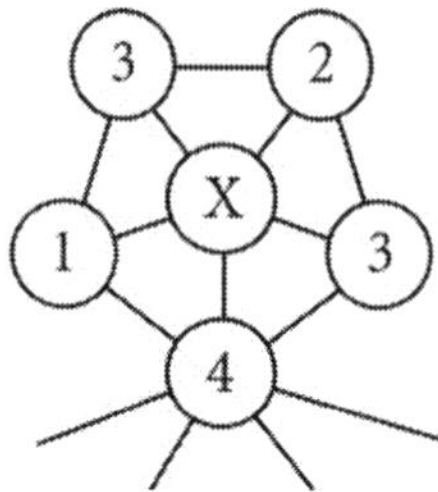

List and describe the various ways in which this VS3 and the process by which it was made is distinguished from Kempe's problem of coloring a random vertex with degree 5. Can you think of any ways to exploit any of these differences that might lead to a proof of the four-color theorem, or to exploit the process itself (for example, through a wise choice of which vertices to split in which order)?

Does every MPG always have at least one edge that connects two vertices with degree five or an edge which connects a vertex with degree five to a vertex with degree six (or an edge where one vertex has a degree less than five)? Either prove this or provide a counterexample. If this is true, can we always arrange it so that in the case of VS3, vertex 4 in the diagram above has degree five or six? If so, can you exploit this to help prove the four-color theorem? Explain.

Note: The answer key doesn't include answers to the challenge problems. These problems are intended to encourage you to think about the ideas.

Challenge problem 3: Let's go a step beyond vertex splitting. Instead of making a pair of new vertices connected by an edge, let's make three new vertices that form a triangle. This is equivalent to splitting a vertex and then splitting one of the new vertices, except that we will wait until both splits are performed before we recolor the graph. Show that each time we do this, we obtain a new MPG with 2 new vertices, 6 new edges, and 4 new faces. Can we always choose to split vertices in such a way that 2 of the 3 vertices to be colored always have degree 5 or less? If not, explain why, and discuss the "best" that can be done in terms of the degrees of the vertices. If yes, can we use the idea behind Challenge Problem 2 from Chapter 7 to prove the four-color theorem? (This may depend upon the answer to Challenge Problem 2 from Chapter 7.) If not, explain why. If so, use this idea to prove the four-color theorem.

The diagram below shows an example. The original vertex had color 4; we assigned color 4 to the higher degree split vertex. Are X and Y always four-colorable? Note that the colors 1, 2, and 3 will generally not be the same as shown below.

Another point to consider: Are there any MPG's with an even number of vertices that we can't make with this idea starting with K_4? Are there any MPG's with an odd number of vertices that we can't make with this idea starting with a MPG with 5 vertices?

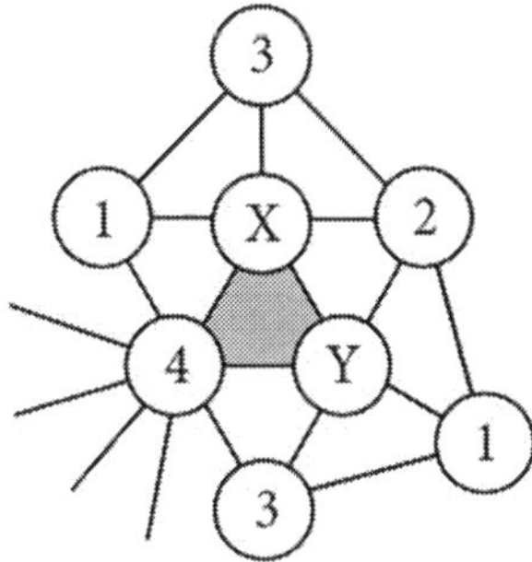

Note: The answer key doesn't include answers to the challenge problems. These problems are intended to encourage you to think about the ideas.

19 Quadrilateral Switching

Since a MPG is triangulated, every edge in a MPG is shared by two triangles. (Recall that we use the word "triangle" in a loose sense, allowing for an edge to be curved instead of a straight line.) The two triangles that share an edge form a quadrilateral, like the one below. This diagram shows two adjacent faces of a much larger MPG. Edge AC is the diagonal of quadrilateral ABCD, which divides the quadrilateral into triangles ABC and ACD.

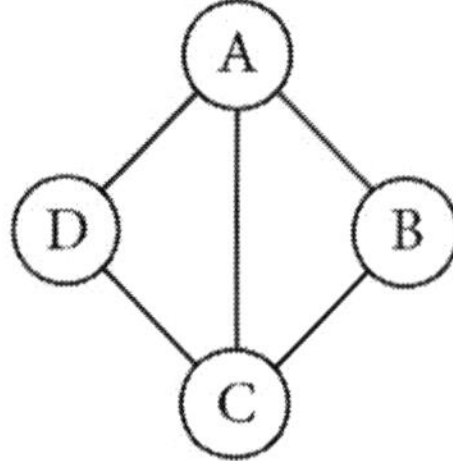

There are two possible ways that such a quadrilateral may be colored:

- B and D may turn out to be the same color.
- B and D may turn out to be different colors.

Either way, A is different from C, and B and D are each different from A and C. The example below colored A red and C green. On the left, B and D are the same color (both blue). On the right, B and D are different colors (one yellow, the other blue).

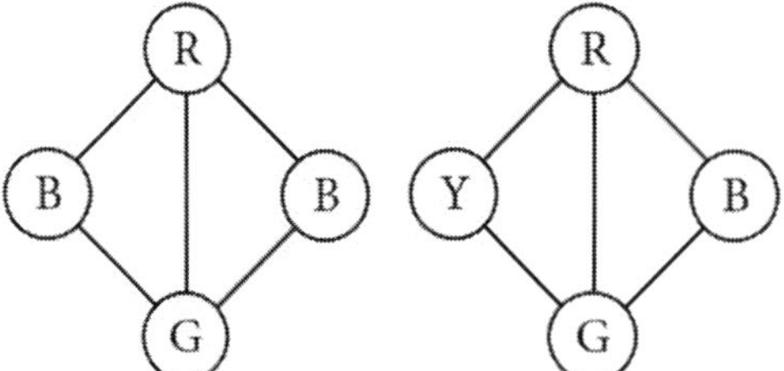

Imagine turning the diagonal edge AC ninety degrees so that it connects B and D instead of A and C. Doing so, we're treating the diagonal edge like a switch. Since the turned edge is the diagonal of a quadrilateral, we will refer to this as a **quadrilateral switch**. There are two different types of quadrilateral switches corresponding to the two different ways that such a quadrilateral may be colored (illustrated by the examples above):

- When two of the quadrilateral's vertices are the same color before turning the switch, we will refer to this as a **color-changing** quadrilateral switch. Since the two vertices can't be the same color after the switch is turned, the MPG will need to be recolored. That's why we refer to this as a color-changing switch.

- When all four vertices of the quadrilateral are different colors, we will refer to this as a **color-preserving** quadrilateral switch. Since all four vertices are different colors, no recoloring is necessary when the switch is turned. If the original MPG had been four-colored, the same coloring will work for the new MPG that is created by turning a color-preserving switch.

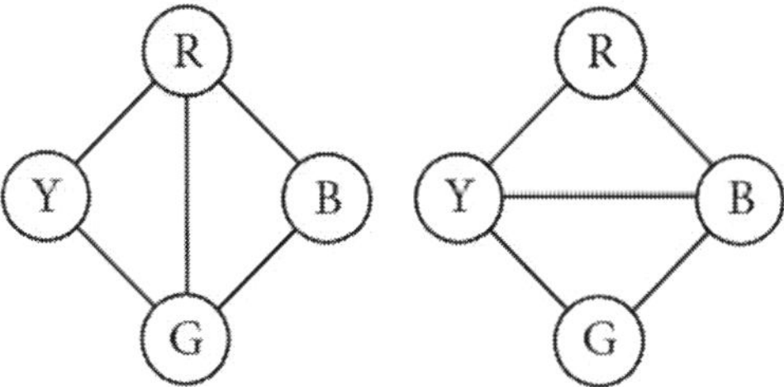

The diagram above illustrates the turning of a color-preserving quadrilateral switch.

Before you turn a quadrilateral switch, you need to make sure that turning the switch doesn't create a double edge. For example, if B and Y had already been connected by an "outside" edge in the above diagram (in which case the diagram above is actually part of a K_4), then turning the quadrilateral switch would have resulted in two edges connecting B and Y. We need to remember to watch out for this possibility before turning quadrilateral switches.

Let's consider some examples of quadrilateral switches in MPG's. A MPG must have at least 5 vertices to have a quadrilateral switch. (The only MPG with 4 vertices is K_4; turning any diagonal of a quadrilateral in K_4 results in a double edge.) However, it will take at least 6 vertices before anything significant happens. With 5 vertices, turning a quadrilateral switch results in a MPG that is structurally equivalent to the original MPG. To see this, try looking for quadrilateral switches in the MPG on the next page.

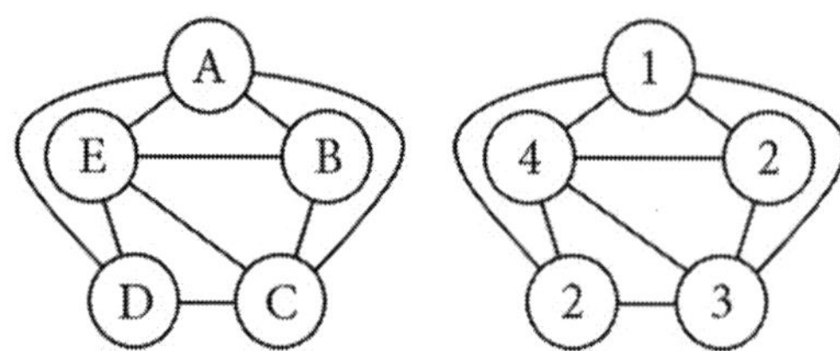

Which edges in the pentahedral graph shown above function as quadrilateral switches?

- There are only color-changing quadrilateral switches. For example, CE can turn to BD, as illustrated below. Other examples include AC to BD (bear in mind that the infinite area outside counts as a face; in this case, it is ACD) and AE to BD (since AE is the diagonal of quadrilateral ABED).

- The following turns would create double edges: AB to CE, AD to CE, BC to AE, BE to AC, CD to AE, and DE to AC.

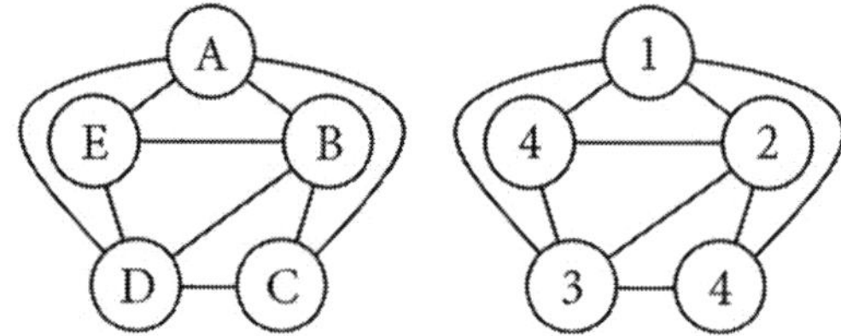

Now consider the MPG below, which has 6 vertices.

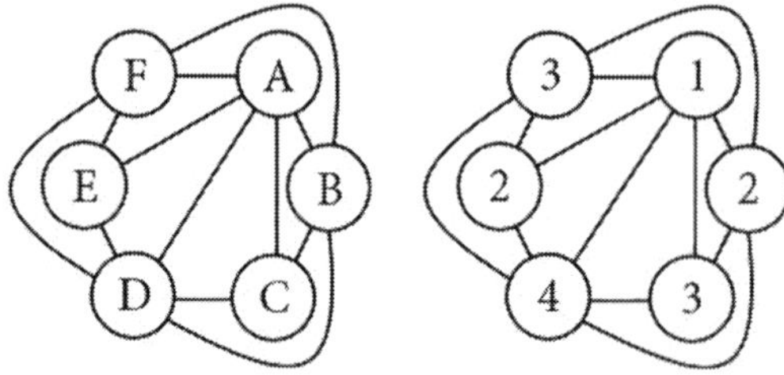

This MPG has an example of a color-preserving switch. When we turn AD to CE, not only does the same coloring work for the new MPG, but the new MPG is structurally different from the original MPG.

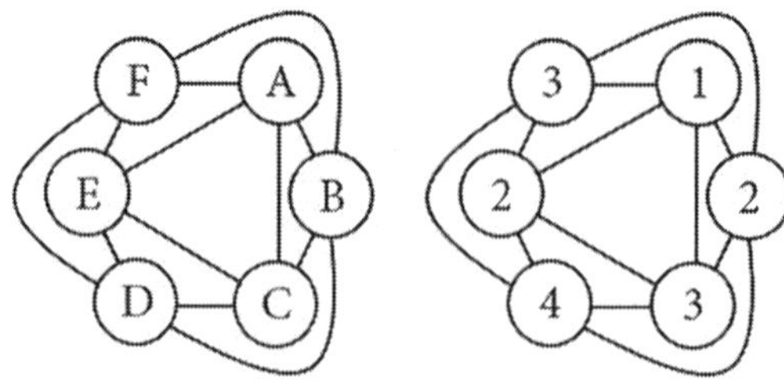

Take a moment to appreciate what just happened. There are two structurally different MPG's with 6 vertices.

- One MPG has two vertices with degree 5, two vertices with degree 4, and two vertices with degree 3. This MPG is four-colorable a single way (once three colors are set).
- The other MPG has six vertices with degree 4. This MPG has less restrictive coloring; it can be four-colored multiple ways (once three colors are set).
- After four-coloring the restrictive MPG, by simply turning one color-preserving quadrilateral switch, we made the other MPG and it was already four-colored.

Note that the reverse would not necessarily have worked. If we had first colored the MPG with less restrictive coloring, it may have had a color-changing switch instead of a color-preserving switch. The less restrictive MPG can be four-colored multiple ways (it is even *three*-colorable). Most of those ways wouldn't have worked for the more restrictive MPG.

Consider the MPG below, which has 7 vertices and maximally restrictive coloring.

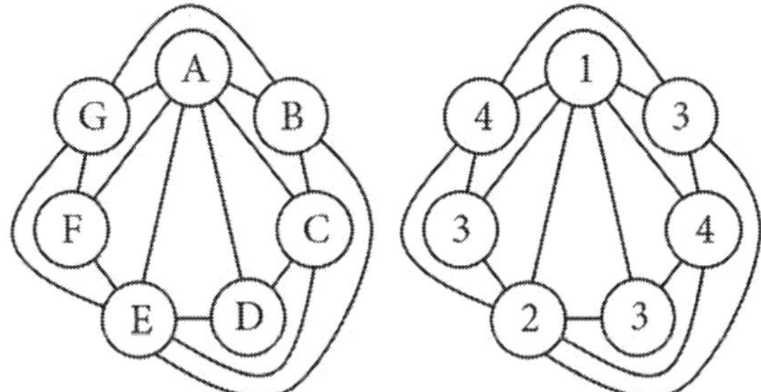

Even though the coloring of this MPG was maximally restrictive, this time there are color-changing quadrilateral switches. For example, we turned AE to DF in the diagram below.

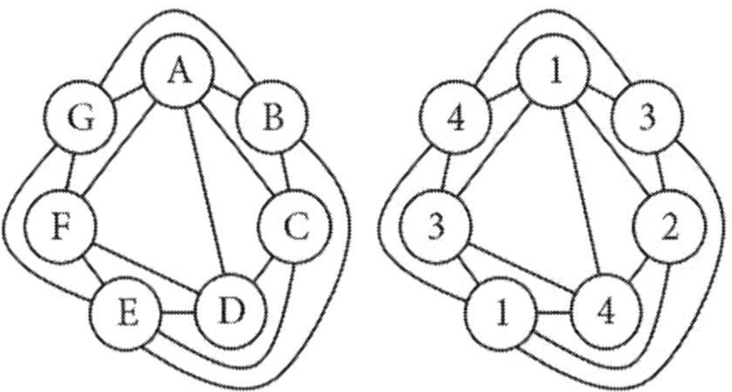

Thus, we see that going from a more restrictive MPG to a less restrictive MPG (with regard to coloring) can result in a color-changing quadrilateral switch; it doesn't always result in color-preserving quadrilateral switches.

Below is a different MPG with 7 vertices.

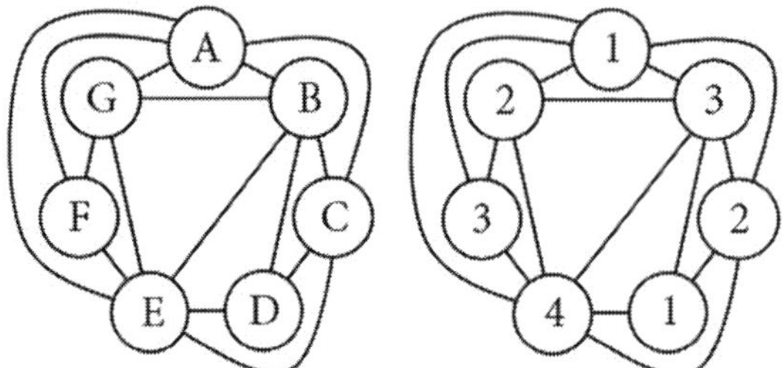

This time we'll turn a color-preserving quadrilateral switch. Turn BE to DG.

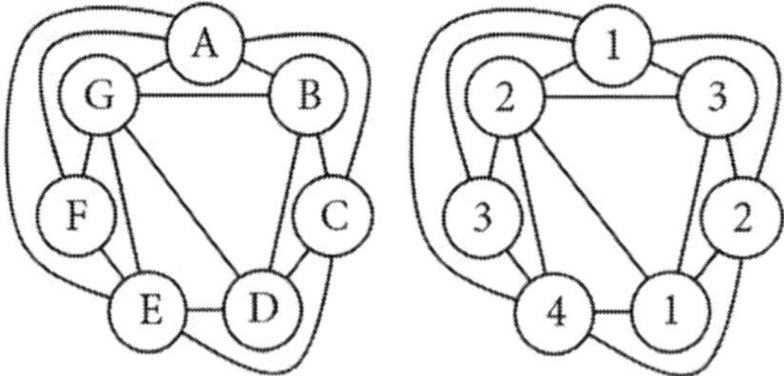

The MPG below has 8 vertices and illustrates something new. In the left diagram, we turned a color-preserving quadrilateral switch from 1-4 to 2-3. In the middle diagram, we turned a different color-preserving quadrilateral switch from 1-3 to 2-4. This second switch hadn't been possible until the first switch was changed. As some quadrilateral switches are turned, they may open up opportunities for other switches to turn which had not previously been available.

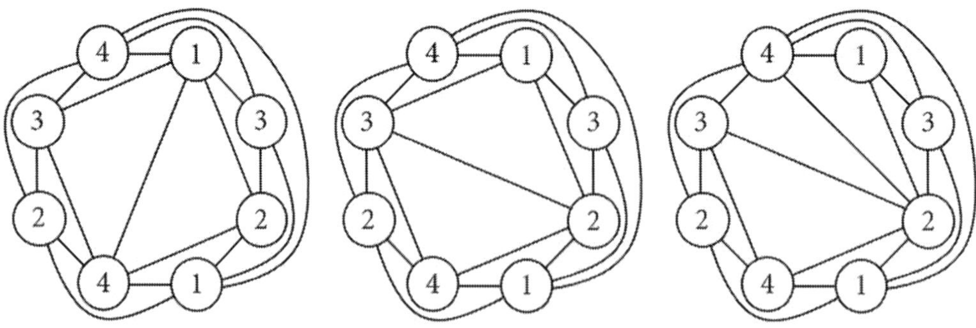

190

Let's consider the possible quadrilateral switches in a vertex with degree five. In any four-colorable MPG, a vertex with degree five will always have two pairs of neighbors of the same color plus one neighbor with a lone color. In the example below, A and D are both color 1, B and E are both color 2, and C is the only neighbor of F with color 3.

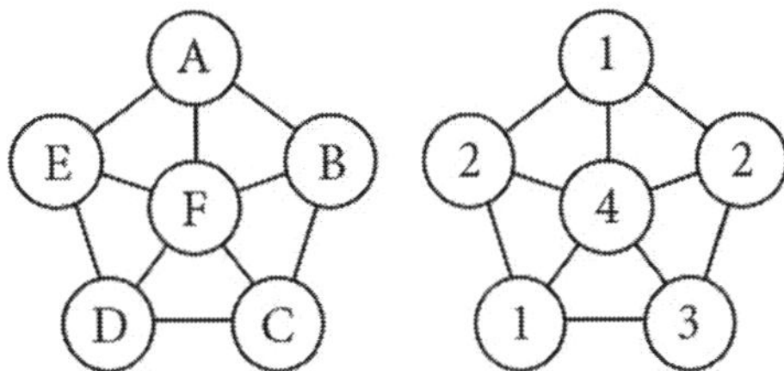

This means that three of the radial edges of any vertex with degree five are color-preserving quadrilateral switches. In the example above, the three radial color-preserving quadrilateral switches are:

- BF to AC
- CF to BD
- DF to CE

We don't have to worry about a vertex with degree five being three-colored; it's easy to show that this isn't possible. If you're wondering about the possibility of it being five-colored (in the event that the four-color theorem turned out to be false, for example), in that case, we get an additional radial color-preserving quadrilateral switch. In the example below, these are BF to AC, CF to BD, DF to CE, and EF to AD.

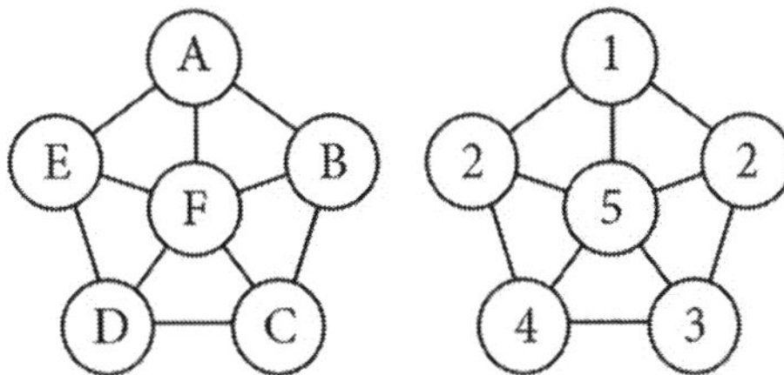

Let's return to the question of four-coloring a MPG. In Chapter 12, we showed that we don't need to consider MPG's with K_4 subgraphs. If we can prove the four-color theorem for all MPG's without K_4 subgraphs, it will follow that the four-color theorem is also true for MPG's with K_4 subgraphs.

Every vertex with degree five in every MPG without a K_4 subgraph has three radial color-preserving quadrilateral switches, as illustrated in a previous example. (This is true even if the graph isn't four-colorable; we noted in the case of five-coloring that there would be an extra radial color-preserving quadrilateral switch; the three that we're discussing would still apply.)

Why does it matter if there are any K_4 subgraphs? If there aren't any K_4 subgraphs, we don't have to worry about creating a double edge when we turn a quadrilateral switch. (However, if you turn multiple quadrilateral switches, in that case you need to check that two of the switches don't combine to form a double edge.) For example, consider turning BF to AC. If there is already an edge connecting AC outside of the pentagon, then ABCF would be a K_4 subgraph. If we know that there aren't any K_4 subgraphs in the MPG, then we don't need to worry about this.

In the example below, we turned two of the radial color-preserving quadrilateral switches. What is special about these two color-preserving quadrilateral switches?

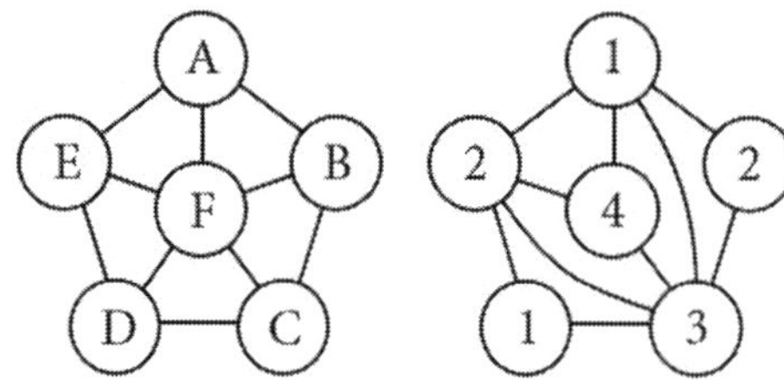

We turned quadrilateral switch BF to AC and quadrilateral switch DF to CE. Both are color-preserving switches. As a result, vertex F is now degree three, triangle ACE is a ST (Chapter 12), and ACEF is a K_4 subgraph (when before there hadn't been any K_4 subgraphs).

We must be careful with color-preserving quadrilateral switches.

- Before coloring a MPG, we don't know which quadrilateral switches will be color-preserving and which will be color-changing.

- After turning a color-preserving quadrilateral switch, if any vertices are recolored, the switches may no longer be color-preserving. If you turn quadrilateral switches *before* you color a graph, they might become color-changing switches when you turn the switches back to their original positions.

We can make all of the possible MPG's for a given number of vertices by turning quadrilateral switches in every possible combination (and ignoring any graphs with double edges). For example, consider the MPG below. By turning the right combination of switches, we may connect an edge to any pair of vertices.

For example, consider vertices H and K in the example below. Can you find a way to turn quadrilateral switches such that H will connect to K?

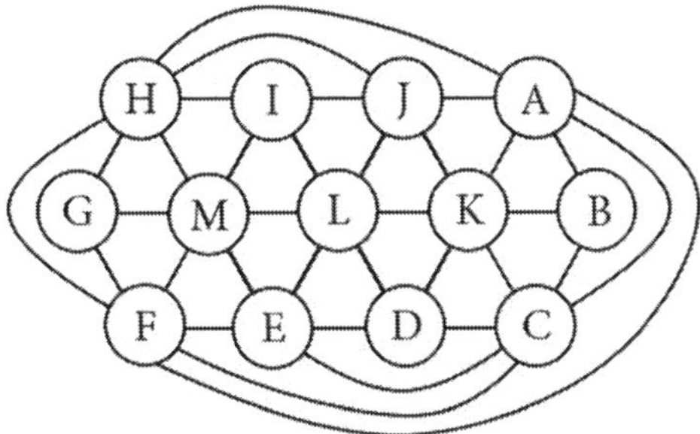

Here is one way to do it. First turn JL to IK. Now turn IL to KM. Finally, turn IM to HK.

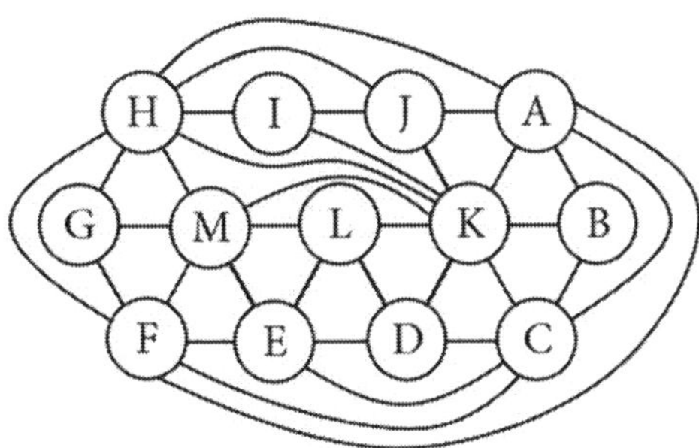

Note that some of these quadrilateral switches may be color-changing switches.

The first exercises of this chapter explore this concept further.

Chapter 19 Exercises

1. There are 5 structurally different MPG's with $V = 7$ vertices. One is shown below. Make all 4 other structurally different MPG's by turning one quadrilateral switch at a time (without making any double edges).

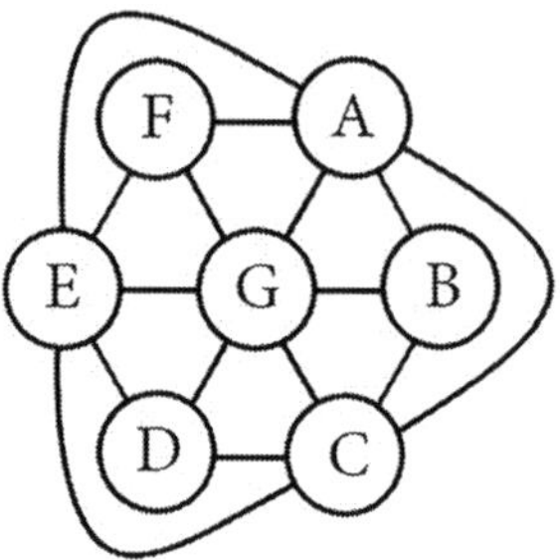

2. This problem is similar to the first problem in that the MPG below also has 7 vertices. The layout of the vertices is different, but so is the question. This time, let's pay attention to the colors of the vertices. Of the 5 structurally different MPG's with 7 vertices, how many can you make using the same coloring by turning only **color-preserving** quadrilateral switches? If there are any that can't be made using the same coloring, describe the problem.

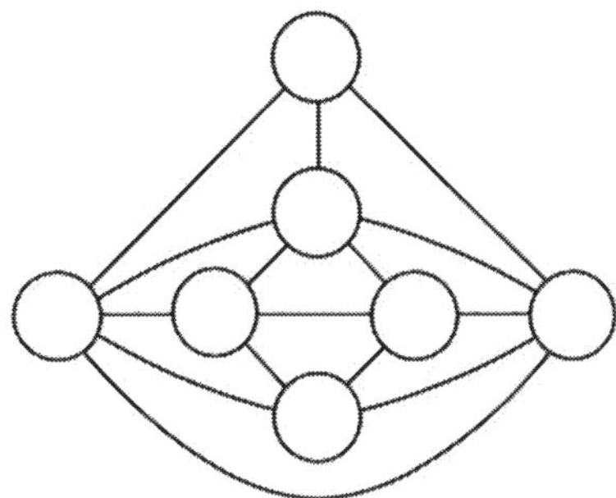

3. In the MPG below, turn quadrilateral switches (without creating any double edges) such that an edge will connect vertex C to vertex H.

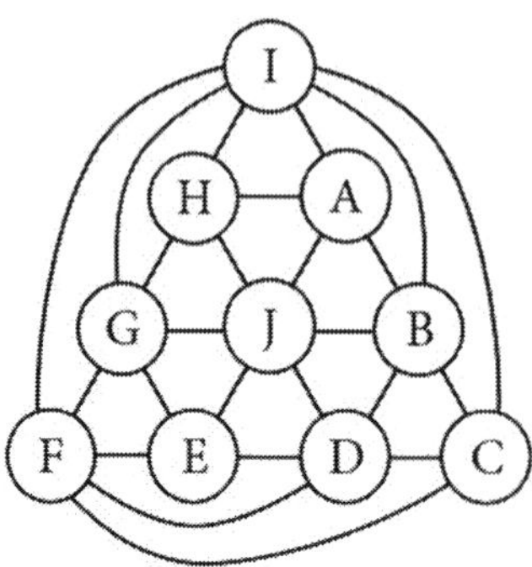

4. Each time a quadrilateral switch is turned (it doesn't matter which kind), what happens to the degree numbers of the vertices of the MPG? Which vertices are affected?

5. What is the minimum number of quadrilateral switches that would have to be turned in order to transform the first MPG described below into the second MPG described below?

- a MPG with 2 vertices with degree 8, 5 vertices with degree 4, and 2 vertices with degree 3?
- a MPG with 6 vertices with degree 5 and 3 vertices with degree 4.

Challenge problem 1: We showed that two radial quadrilateral switches of any vertex with degree five can always be turned in such a way as to make the vertex degree three (with a ST), like the example below. Can we use this idea to prove the four-color theorem? We may not know which quadrilateral switches can be turned in this way, but we do know that they are there. When these quadrilateral switches are turned, they appear to greatly simplify the vertex. It is now trivially four-colorable and may even be removed using the ST. Either show what the problem is with this approach or use this approach to prove the four-color theorem.

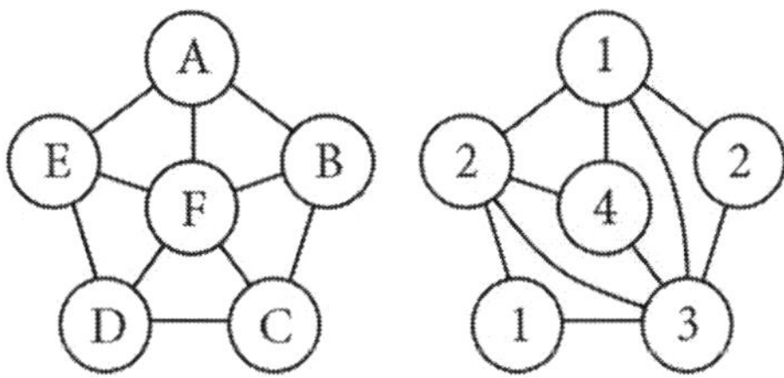

Challenge problem 2: Suppose that a MPG is four-colorable with K blue vertices, L green vertices, M red vertices, and N yellow vertices. We can draw a trivially four-colorable MPG with the same number of vertices of these colors by nesting together K_4's. (For a couple of examples of this, see Exercise 5 in Chapter 12.) We can then turn color-preserving quadrilateral switches to transform the trivially four-colorable MPG into another MPG. Can we make *any* MPG this way? Even if we don't know how many vertices of each color there will be, we can try every possible combination and eventually make the MPG… or can we? Can we use this approach to prove the four-color theorem? Either prove the four-color theorem this way or explain what the problem is with this approach.

Note: The answer key doesn't include answers to the challenge problems. These problems are intended to encourage you to think about the ideas.

Challenge problem 3: In Chapter 18, we showed that we can split vertices to make a MPG that is four-colorable except for one of the split vertices, which could be labeled X; the other split vertex could have the same color as the original vertex. Can you think of a way to apply the concept of quadrilateral switches to solve the problem of coloring the last vertex X? An alternative might be to use quadrilateral switches to transform the MPG into a different form before splitting vertices. Either explain the difficulty with carrying out these ideas or show how they could be used to prove the four-color theorem.

Note: The answer key doesn't include answers to the challenge problems. These problems are intended to encourage you to think about the ideas.

20 Kirchhoff's Rules

Kirchhoff has two general rules (which are conservation laws) for analyzing any electric circuit. Let's see if a similar concept applies to four-colored graphs.

Kirchhoff's junction rule expresses conservation of charge. For a circuit with resistors and batteries, it is expressed in this form: the sum of the currents entering a junction equals the sum of the currents exiting the junction. (How does that express conservation of charge? Current is the rate of flow of charge. Kirchhoff's junction rule states that the rate at which charge enters a junction equals the rate that charge leaves a junction, such that no charge is created or destroyed in the process.)

Kirchhoff's loop rule expresses conservation of energy. For a circuit with resistors and batteries, it is expressed in this form: the sum of the potential differences around any closed loop is zero. (How does that express conservation of energy? The potential difference across a resistor or battery is the work per unit charge needed to move a positive test charge across the circuit element, and energy is the ability to do work.)

A four-colored graph doesn't have any circuit elements, but we can treat the edges as if they were a sort of circuit element as follows. We will let the numbers 1-4 represent the colors blue, green, red, and yellow. Every edge connects two colors, which are two of the numbers 1 thru 4. We will define the **color difference** to be the value that we get when we subtract these two numbers. Specifically, if we travel along an edge from vertex X to vertex Y, we will subtract the color number for the first vertex from the color number of the second vertex: $Y - X$. Note that color difference may be positive or negative; along a single edge, the color difference can't be zero. For example, going from a vertex colored 1 to a vertex colored 4, the color difference is $4 - 1 = 3$. Going in the opposite direction, that is from a vertex colored 4 to a vertex colored 1, the color difference is $1 - 4 = -3$. For color difference, as we've defined it, direction matters.

In terms of color differences, Kirchhoff's loop rule takes the following form: the sum of the color differences traveling along any closed loop is zero. For example, consider the graph below, which has already been four-colored. The vertices shaded gray form one of many possible closed loops. (Recall that when a set of vertices in a graph forms a closed chain, it is referred to as a **circuit**.)

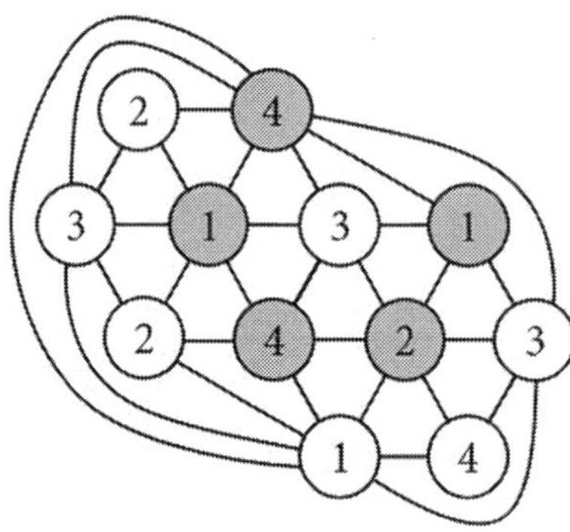

Let's apply Kirchhoff's loop rule to the closed loop shaded gray above:

- Let's begin at the vertex colored 2 and travel clockwise.
- Going from 2 to 4, the color difference is $4 - 2 = 2$.
- Going from 4 to 1, the color difference is $1 - 4 = -3$.
- Going from 1 to 4, the color difference is $4 - 1 = 3$.
- Going from 4 to 1, the color difference is $1 - 4 = -3$.
- Going from 1 to 2, the color difference is $2 - 1 = 1$.
- The sum of these color differences is: $2 + (-3) + 3 + (-3) + 1 = 6 - 6 = 0$.

Note that we have only defined one sort of circuit element with a property called color difference. If you want to think about three properties (like resistance, current, and voltage, or like charge, capacitance, and voltage), you will need to develop more than just color difference.

One way to introduce the concept of charge to a graph is to literally sprinkle charges on the vertices, edges, or faces. (Placing a positive charge at each vertex and inside each face and a negative charge along each edge, where the charges are equal apart from the signs, one can prove that $Vq + Fq = Eq + 2q$, which reduces to Euler's formula, $V + F = E + 2$. One way to do this is to show that all of the charges cancel except for the charges at two of the vertices. Challenge problem 2 of this chapter asks you to do this.)

Another example that applies the concept of charge to graph theory is the process of discharging used in Appel and Haken's computer-assisted proof of the four-color theorem [14] (Appel and Haken's proof is mentioned briefly in Chapter 28, but to learn about the discharging process, you'll need to read their paper).

(If you wish to learn more about Kirchhoff's rules for electric circuits, consult a standard introductory physics textbook. The second volume typically covers electricity and magnetism with some chapters on circuits. The author's *Essential Physics Study Guide Workbooks*, Volume 2, include a chapter on Kirchhoff's rules.)

Chapter 20 Exercises

1. Four-color the graph shown below and show that the color differences satisfy Kirchhoff's loop rule for each of the following loops:

- starting at G, traverse along N-F-G.
- starting at K, traverse along A-C-G-K.
- starting at K, traverse along O-G-N-C-M-K.
- starting at A, traverse B-C-N-G-O-K-L-A.
- starting at J, traverse along K-L-A-B-C-D-E-F-G-H-I-J.

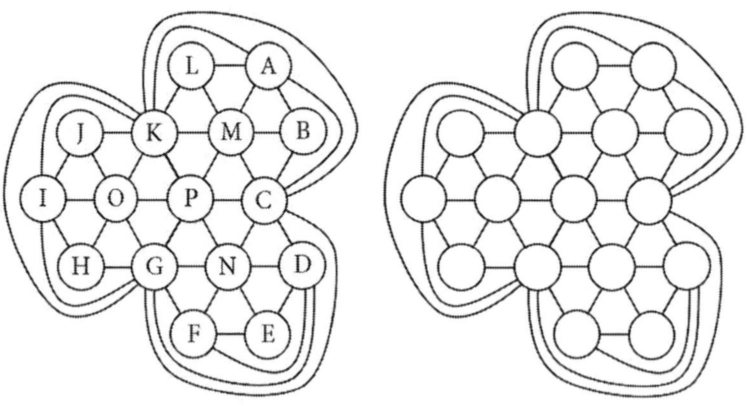

Challenge problem 1: Devise a useful way of applying Kirchhoff's junction and loop rules to a planar graph that is relevant to the four-color theorem. Can you use the junction and loop rules to setup a system of equation with colors as the unknowns? Either setup an example where you can, or explain the difficulty with doing this.

Challenge problem 2: Use the concept of charge to derive Euler's formula (Chapter 4).

Note: The answer key doesn't include answers to the challenge problems. These problems are intended to encourage you to think about the ideas.

21 Building Blocks

Any MPG can be built and four-colored using the basic building blocks shown below. These building blocks are the triangular faces. The numbers in the corners represent the colors of the vertices. When the faces are put together to form a MPG, the colors at the corners must match up. Since none of the building blocks have two corners with the same number, this prevents us from building a MPG where two vertices with the same color share an edge. (Note that if you proceed to build a MPG with these building blocks and the last piece would require a building block with two corners of the same number – unlike those shown below – this means that you need to revise the building blocks that you already have in place.)

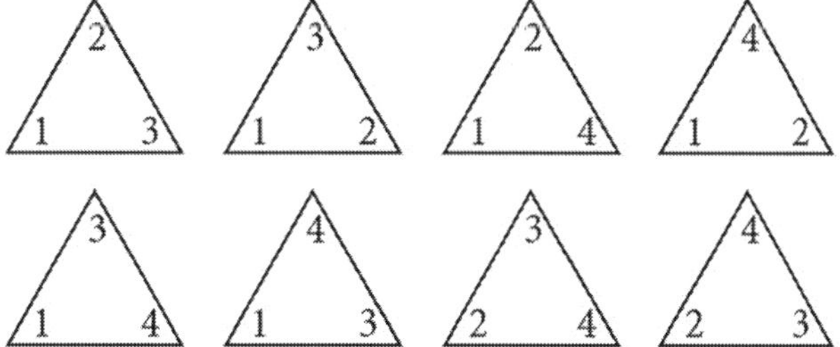

The following example shows a MPG that was built and four-colored by putting the above building blocks together. In this MPG, C and E are color 1, B and F are color 2, G is color 3, and A and D are color 4. Note that building blocks ABC, CDF, DEF, and ACF have curved edges; sometimes, a building block needs to be reshaped to build the MPG. What's important is that each building block is triangular (it has 3 edges; whether they are straight, curved, or bent doesn't matter) and that each corner has a different color.

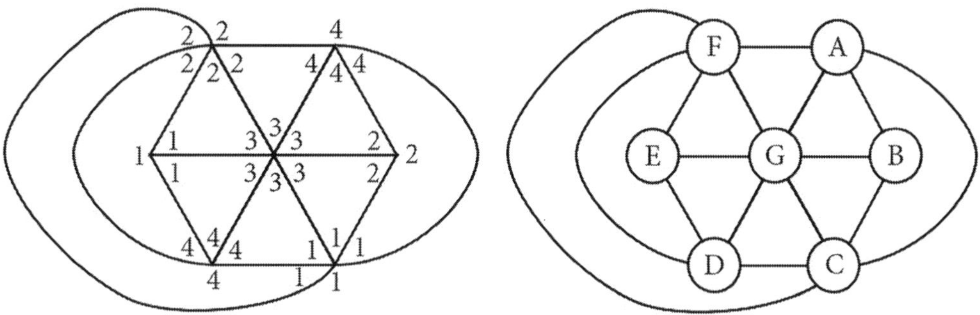

These building blocks provide a way of thinking about the four-color theorem in terms of faces rather than vertices. (Similarly, the color differences of the previous chapter show how to think about the four-color theorem in terms of edges rather than vertices. Imagine that each edge is bi-colored: put one number at the end of each edge. Then instead of coloring vertices, we just need to make sure that the edges are joined together so that the same number meets at every point where the ends of edges meet.)

Thinking about how the faces are put together to form a MPG and how the faces affect the coloring can be helpful because each face involves three vertices which can only be put together in eight different arrangements.

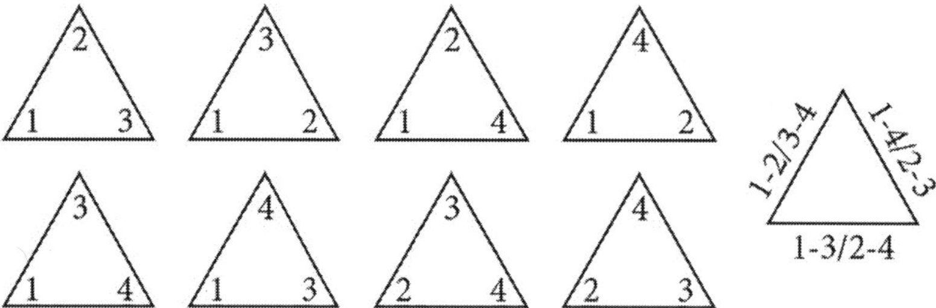

In fact, we can go a step further. All eight building blocks have the *same* structure. All eight building blocks have three different types of edges:

- One edge has a pair of colors from either a 1-2 or a 3-4 Kempe chain.
- A second edge has a pair of colors from either a 1-3 or a 2-4 Kempe chain.
- A third edge has a pair of colors from either a 1-4 or a 2-3 Kempe chain.

The figure above shows that there are four combinations of Kempe chains on each face:

- The two left faces on the top row have a 1-2 edge, a 1-3 edge, and a 2-3 edge.
- The two left faces on the bottom row have a 3-4 edge, a 1-3 edge, and a 1-4 edge.
- The two right faces on the top row have a 1-2 edge, a 2-4 edge, and a 1-4 edge.
- The two right faces on the bottom row have a 3-4 edge, a 2-4 edge, and a 2-3 edge.

Recall from Chapter 7 that any MPG may be divided three ways. It may be divided into 1-2 and 3-4 Kempe chains, it may be divided into 1-3 and 2-4 Kempe chains, or it may be divided into 1-4 and 2-3 Kempe chains. These three divisions are inherent in each building block.

We can illustrate that each edge of a building block has three different types of edges – one edge associated with each way of dividing the MPG into Kempe chains – by drawing faces with three visually different types of edges. In the figure below, we used thin solid lines for 1-2/3-4 chains, thick solid lines for 1-3/2-4 chains, and dashed lines for 1-4/2-3 chains.

The MPG below was drawn using faces like the one above. The three sets of edges make it easy to identify the three ways of dividing the MPG into Kempe chains. Every face has one of each type of edge.

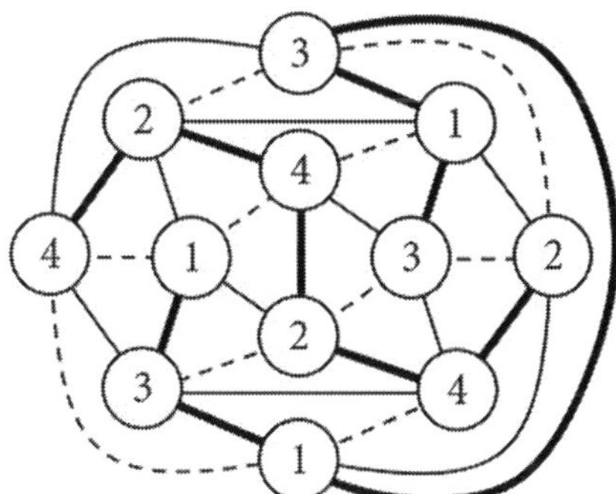

A MPG with V vertices has $E = 3V - 6 = 3(V - 2)$ edges (Chapter 4). This shows that the edges can be divided into 3 sets of $(V - 2)$ edges. The MPG above has $V = 12$ vertices and $E = 3V - 6 = 3(12) - 6 = 30$ edges. The edges have been divided into three sets of $V - 2 = 12 - 2 = 10$ edges: there are 10 thin solid edges, 10 thick solid edges, and 10 dashed edges. Each face has one edge of each type. These three types of edges divide the four-coloring of the MPG into three sets of Kempe chains: 1-2 and 3-4 Kempe chains, 1-3 and 2-4 Kempe chains, and 1-4 and 2-3 Kempe chains.

Let's look at how these basic building blocks with three types of edges (one for each possible division of Kempe chains) can be put together to form vertices with specific degrees in a MPG. Every vertex with degree three, for example, has the following structure; it is K_4.

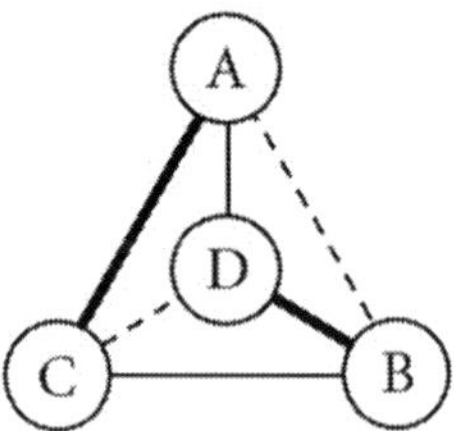

Vertices with degree four fall into two classes (note that these are just the vertices of MPG's below; these are not complete MPG's):

- In one class (shown at the left below), one type of edge forms a closed loop around the vertex. The other two types of edges alternate inside.

- In the other class (shown at the right below), two pairs of edges form a closed loop around the vertex. All three types of edges are used inside.

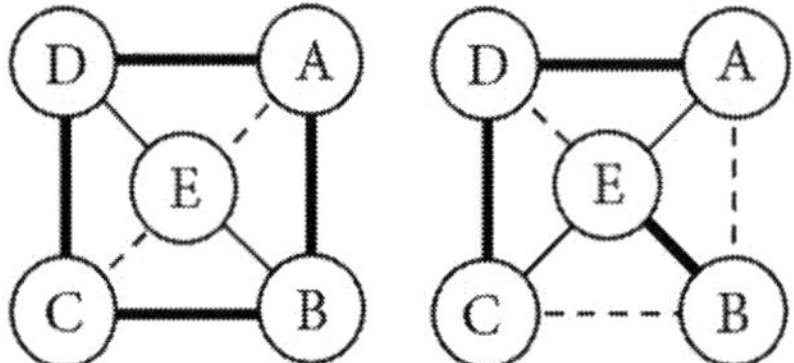

All vertices with degree five (in a MPG) have the same general structure. Two pairs of edges alternate inside along with one edge of the third type, and the third type of edge forms three consecutive edges outside.

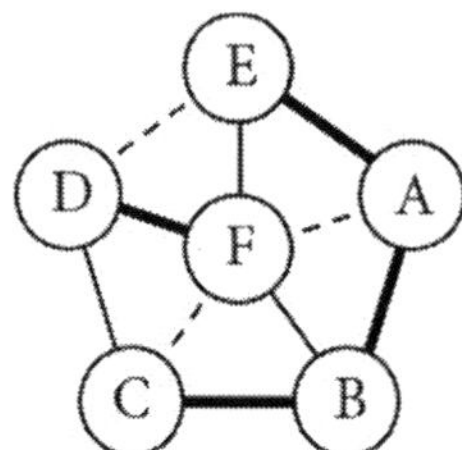

Four general types of vertices with degree six are shown below.

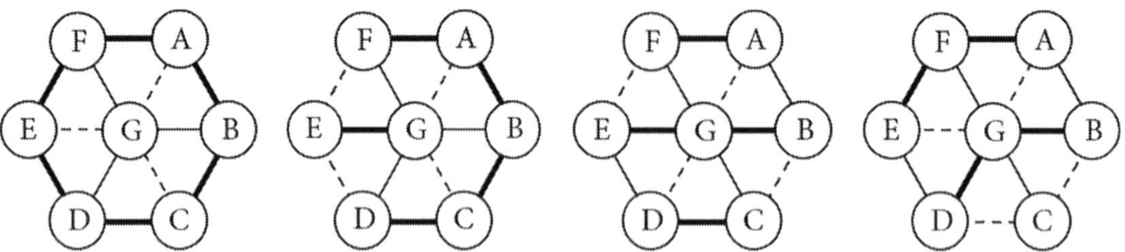

Three general types of vertices with degree seven are shown on the following page.

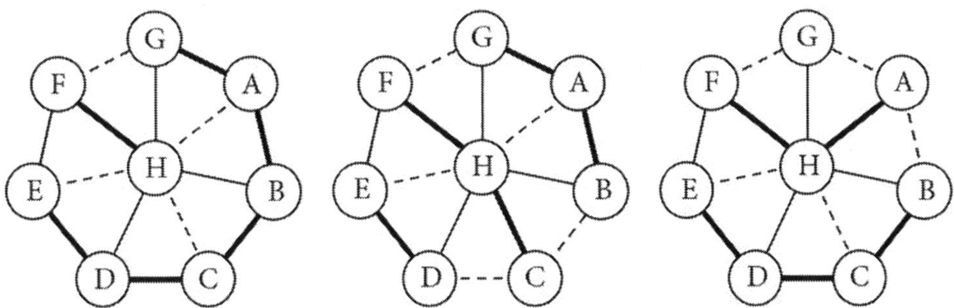

We will continue to explore the role that these building blocks have in the four-coloring of MPG's in the next two chapters.

Chapter 21 Exercises

1. Use the basic building blocks from this chapter to build and four-color the MPG below.

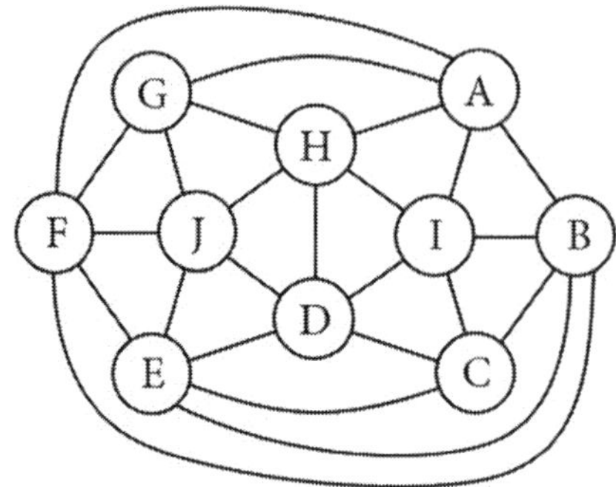

2. Redraw the MPG above using three styles of edges. For example, you might draw solid edges for 1-2 and 3-4 Kempe chains, a series of plus signs (+++) or double lines (or whatever is easy to draw and identify) for 1-3 and 2-4 Kempe chains, and dashed edges for 1-4 and 2-3 Kempe chains. One alternative is to draw each type of edge with a different color (but where the color of the edge isn't to be confused with the colors of the vertices).

Challenge problem 1: Prove that four-coloring a MPG by putting tri-colored faces (which we have referred to as building blocks) together is equivalent to coloring vertices to four-color a MPG. Prove that four-coloring a MPG by putting bi-colored edges (as described in the parenthetical note on the second page of this chapter) together is equivalent to coloring vertices to four-color a MPG.

Note: The answer key doesn't include answers to the challenge problems. These problems are intended to encourage you to think about the ideas.

22 Four-Coloring by Pairing Faces

Following is a simple and efficient approach to four-coloring a MPG.

- Identify a single quadrilateral consisting of two triangles that share a common edge. To help make our quadrilaterals clear, we'll write a pair of numbers in the two faces of each quadrilateral. In the example below, the faces of a quadrilateral are labeled 1.

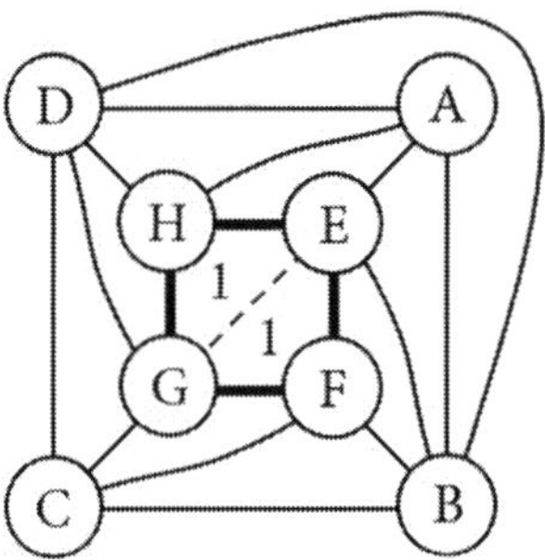

- Identify another quadrilateral that shares an edge with the first quadrilateral.
- Continue to identify quadrilaterals to form a cluster of quadrilaterals. Don't create any **island** faces like the shaded face shown below on the left, where quadrilaterals AJIE, FGKJ, and GHIK made it impossible for face IJK to be part of a quadrilateral.

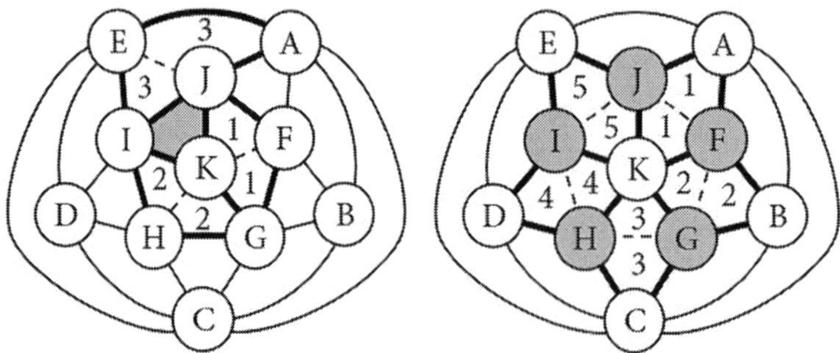

- Be careful not to let an odd number of diagonals of quadrilaterals form a closed loop, like dashed loop FGHIJ shown above on the right. Why? An odd closed loop can't be colored B-G or R-Y like a Kempe chain can. (An *even* closed loop is allowed.)
- Your goal is to pair every face of the MPG into quadrilaterals. The diagonals of the quadrilaterals form sections of Kempe chains. Color one chain B-G and color the other chain R-Y. Ensure that no vertices of the same color are connected by an edge.

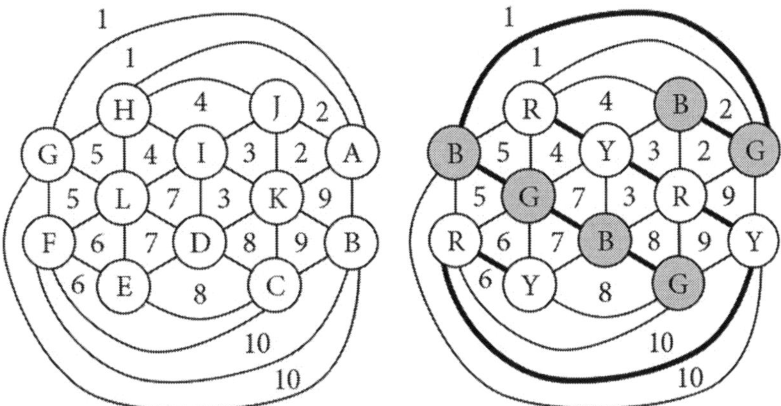

The diagram above illustrates our four-coloring method. The colors are B, G, R, and Y.

- The left figure above shows how we numbered pairs of faces. Each numbered pair shares one edge. Each numbered pair forms a quadrilateral.

- The thick lines in the right figure above show the edge that is shared by each pair. The thick lines are the diagonals of the quadrilaterals.

- Observe that no single face has two thick edges. This occurs automatically since we only highlighted edges shared by a pair of faces labeled with the same number.

- We avoided island faces. (If we hadn't, we wouldn't have been able to complete the pairing process.)

- The thick lines form two separated paths. Neither path makes a loop. It's important to ensure that the thick lines don't form a closed loop with an odd number of edges in the loop (like the pentagon shown on the previous page). Coloring Kempe chains only works if we avoid such odd loops. In our example above, we avoided this by avoiding loops all together. However, an even loop would be okay. (The right figure on the previous page shows a closed loop with an odd number of diagonals. Try to color pentagon F-G-H-I-J with a R-Y chain and you will see that F and J would have the same color, which isn't allowed).

- We colored one of the two separated paths B-G (shaded gray) and colored the other separated path R-Y. (Sometimes, we get three or more separated paths. In that case, each path is still colored B-G or R-Y.)

As you pair faces together, if you accidentally create an island face or a closed loop with an odd number of highlighted edges, you must do some backtracking to fix the problem. This happens when you pair faces at random. If instead you plan ahead as you pair faces, you can avoid much potential backtracking, making the four-coloring method more efficient.

The diagram below shows two ways that an island face can form. In the left figure, all three faces that share an edge with the central face have been paired, making it impossible to pair the central face. In the right figure, imagine that all of the faces in a MPG have been paired except for these last four faces. It's impossible to make two pairs of faces in this configuration because any pairing would involve the central face and that face can only be part of one pair. Island faces either need to be prevented by thinking ahead or solved by backtracking.

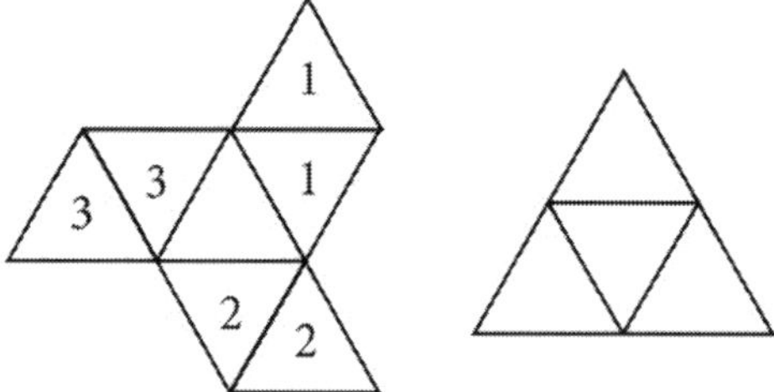

The simplest way for an odd number of highlighted edges to form a closed loop is when all of the faces of a vertex with an odd degree number are paired like the figure below. We don't need to worry about a vertex with degree three. As discussed in Chapter 11, any vertex with degree three may simply be removed from the MPG along with its connecting edges. Once the MPG is four-colored, it is a trivial matter to add the vertex back in and color it.

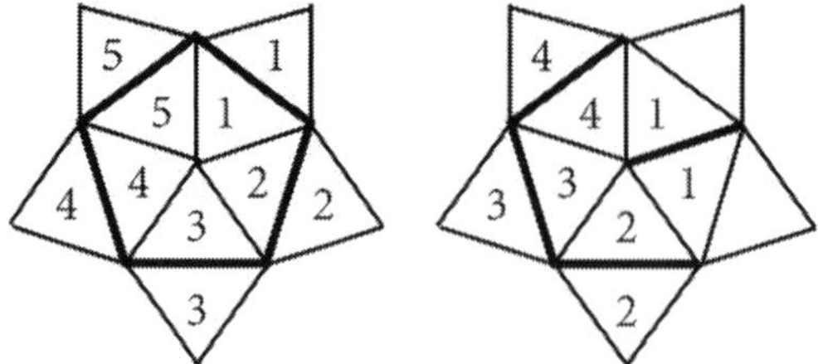

Situations like the one shown above on the left can be avoided by ensuring that at least one quadrilateral is formed from adjacent faces in every vertex with an odd degree. For example, the figure on the right avoids this problem (but we still need to watch out for larger loops).

Another way for an odd number of highlighted edges to form a closed path is when a graph contains a ST. (Recall from Chapter 12 that ST stands for "separating triangle.") This is easy to prevent. As discussed in Chapter 12, the ST allows us to divide the MPG into two smaller graphs. We can color the two graphs separately, and then recolor one graph such that the vertices of the ST match in both graphs.

Note that our four-coloring method greatly simplifies the four-coloring process. Amazingly, you don't even need to look at the vertices to color them. All you need to do is pair the faces together in a way that avoids island faces and odd loops. After that, the vertices of the MPG practically color themselves; a B-G or R-Y Kempe chain is assigned to each separated section of diagonal edges (as you will be able to explore in the exercises for this chapter).

Will this four-coloring method work for *every* possible MPG? Is it always possible to pair all of the faces into quadrilaterals? Is it always possible to avoid an island face? Is it always possible to avoid an odd loop? The answer to all of these questions is "Yes!" if the four-color theorem is true. We will demonstrate this in the following chapter.

We will learn how to improve upon our face-pairing four-coloring technique in Chapter 26 (and Chapter 24 will discuss how to recolor the MPG if you run into island faces or odd loops of diagonal edges).

Chapter 22 Exercises

1. Use our four-coloring method to number the faces and color the MPG below.

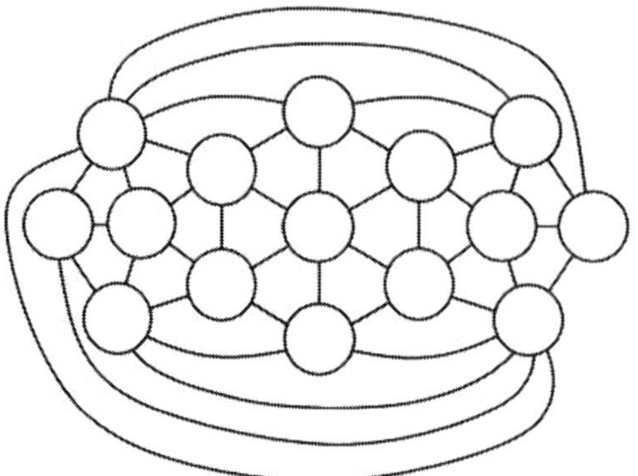

2. Use our four-coloring method to number the faces and color the MPG below.

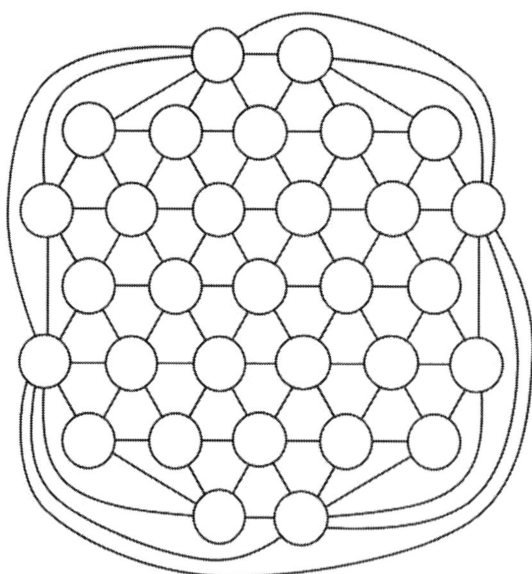

Challenge problem 1: Consider the MPG below, which has been four-colored using our four-coloring method. The B-G Kempe chain features a closed loop with four diagonals of quadrilaterals. Either show that we could have prevented a closed loop or explain why it couldn't have been prevented.

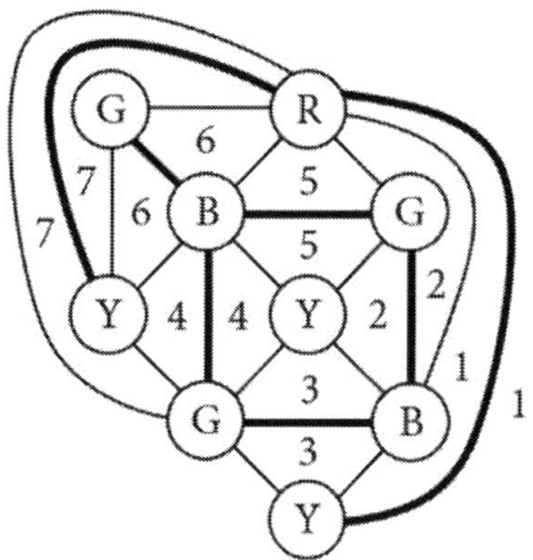

What is special about the vertex colored Y in the center of the square? Could this feature have been prevented?

The thick solid lines above correspond to separated sections of B-G and R-Y Kempe chains. Look at the B-R and G-Y Kempe chains, and also the B-Y and G-R Kempe chains. Discuss whether or not there are similar features in these chains.

Challenge problem 2: Our four-coloring method applies to MPG's. Since some nonplanar graphs aren't four-colorable, our four-coloring method doesn't apply to *all* graphs. Which features of some types of nonplanar graphs pose a problem for our four-coloring method?

Note: The answer key doesn't include answers to the challenge problems. These problems are intended to encourage you to think about the ideas.

Challenge problem 3: Consider the map below. (Note that this is a *map*, not a *graph*.)

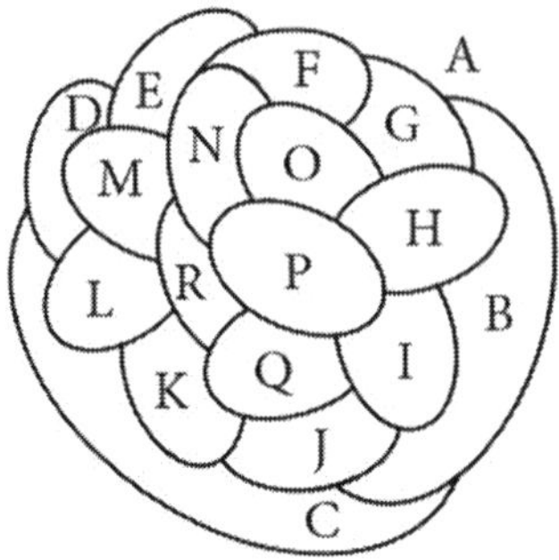

Our method for four-coloring a graph involves pairing faces together. In order to apply our four-coloring method to a map (instead of a graph), what should be paired together? Apply our four-coloring method to the map above. Once the map is four-colored, draw a colored graph corresponding to the map. Is the corresponding graph a MPG?

Challenge problem 4: A graph can have an island face. What would be analogous to this for a map? Draw an example.

A graph can have an odd number of diagonals of quadrilaterals in a closed loop. Draw an example of the same concept applied to a map.

Note: The answer key doesn't include answers to the challenge problems. These problems are intended to encourage you to think about the ideas.

Challenge problem 5: Prove that it is always possible to pair the faces so that each pair of faces shares one edge. (This is equivalent to showing that an island face can always be avoided.)

Prove that it is always possible to pair the faces (with each pair sharing one edge) in such a way as to avoid having an odd number of highlighted edges form a closed loop.

Prove that if our face-pairing four-coloring method is applied in such a way that there are no island faces or odd loops of highlighted edges, then the MPG will be properly four-colored.

If you can't prove one or both of these, show or explain what makes the proof difficult. If you can prove all three, you can use these proofs to prove the four-color theorem.

Challenge problem 6: We mentioned that our four-coloring method can be optimized by planning ahead to some extent in order to avoid island regions and odd loops. Develop a strategy for pairing faces that helps to avoid island regions and odd loops automatically so that little to no planning is needed to prevent backtracking.

Note: The answer key doesn't include answers to the challenge problems. These problems are intended to encourage you to think about the ideas. However, when you finish working on these problems, you should read Chapter 23.

23 The Three-Edges Theorem

The conventional approach to the four-color theorem focuses on how to color the vertices using no more than four different colors so that no two vertices that share an edge have the same color. We will show how the four-coloring of a MPG can be broken down into simpler terms by focusing on the faces and edges, using ideas from Chapters 21-22.

Recall from Chapter 4 that $E = 3V - 6$ and $F = 2V - 4$ for a MPG. We can factor out $(V - 2)$ from each of these equations to write $E = 3(V - 2)$ and $F = 2(V - 2)$. This means that a MPG with V vertices has $E = 3(V - 2)$ edges and $F = 2(V - 2)$ faces.

We will break down a MPG into simpler terms that interpret these formulas for E and F.
- Any MPG can be separated into $V - 2$ quadrilaterals, where each quadrilateral consists of two faces. These $V - 2$ quadrilaterals have $V - 2$ diagonals. These $V - 2$ diagonals form separated sections of Kempe chains, which we may color B-G and R-Y. Our four-coloring method from Chapter 22 applied this principle.
- Each face of a MPG has three edges where one edge is part of a B-G or R-Y Kempe chain, another edge is part of a B-R or G-Y Kempe chain, and the last edge is part of a B-Y or G-R Kempe chain. We discussed this idea in Chapter 21.
- Any MPG can be separated into three sets of $V - 2$ edges: one for each of the three divisions of Kempe chains described in the previous bullet point.

The MPG that follows illustrates this break down.
- There are $V = 8$ vertices (A thru H).
- The are two sets of $V - 2 = 8 - 2 = 6$ faces (two pairs of numbers 1-6). These pairs form $V - 2 = 6$ quadrilaterals. Their diagonals (thick lines) form B-G and R-Y chains.
- There are three sets of $V - 2 = 6$ edges (solid lines, thick solid lines, and dashed lines).

- The $V - 2 = 6$ solid lines form one B-R chain and one G-Y chain.

- The $V - 2 = 6$ thick solid lines form one B-G chain and two sections of R-Y chains.

- The $V - 2 = 6$ dashed lines form one B-Y chain and one G-R chain.

- Every face has one solid edge, one thick solid edge, and one dashed edge.

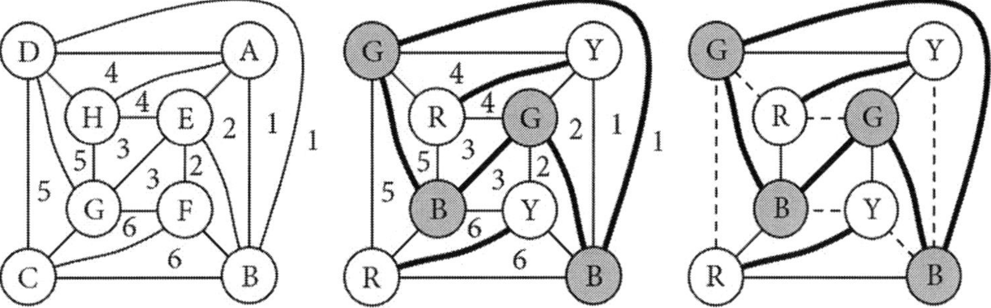

What makes this approach simpler?

- We don't need to think about coloring vertices. This comes about naturally once the edges are formed into separated sections of Kempe chains.

- We don't need to think about finding three sets of edges (one for each division of Kempe chains). These come about naturally once the faces are paired.

- We just need to focus on pairing faces into quadrilaterals, as we did in Chapter 22. Once this is done, four-coloring the MPG is trivial. The diagonals of these quadrilaterals form separated sections of Kempe chains, which we can color B-G and R-Y.

- Pairing faces into quadrilaterals is generally simpler than trying to four-color all of the vertices of a MPG with numerous vertices.

In addition to simplifying the four-coloring process, our approach also offers a seemingly simpler means to attempt to **prove** the four-color theorem **by hand**. (As of the publication of this book, the only known and accepted proof of the four-color theorem involves use of the computer to perform numerous calculations [Ref. 14].)

We could use our approach to prove the four-color theorem if we could prove each of the following points:

- Prove that it is always possible to pair all of the faces of a MPG into quadrilaterals (meaning that it is always possible to avoid an island face).
- Prove that it is always possible to avoid having an odd number of diagonal edges of quadrilaterals form a closed loop.
- Prove that if the above conditions are met, the MPG will be properly four-colored.

You should recognize the first two points from the four-coloring method in Chapter 22. Proving these three points would prove that our four-coloring method from Chapter 22 works for every possible MPG, which would prove that every MPG is four-colorable, which would prove the four-color theorem.

We will see that these three points can be condensed down to a single condition, which we will call the **three-edges theorem**.

- Prove that every possible MPG with V vertices can be constructed with three sets of V – 2 edges where each face has exactly one edge from each set of edges.

You should recognize this idea from Chapter 21. The three sets of V – 2 edges correspond to the three divisions of Kempe chains (B-G/R-Y, B-R/G-Y, and B-Y/G-R).

If the three-edges theorem is true, it follows that the four-color theorem is true. We will demonstrate this by showing that if the three-edges theorem is true, all three points regarding our face-pairing four-coloring method will be satisfied (since if these three points are satisfied, every possible MPG can be four-colored).

According to the three-edges theorem, we can draw any MPG with V – 2 solid edges, V – 2 thick solid edges, and V – 2 dashed edges, where every face has exactly one solid edge, thick solid edge, and dashed edge, like the right graph on the previous page.

There are 2(V – 2) faces on any MPG, and each face has one of each type of edge. Let's focus on one set of V – 2 edges for a moment, such as the thick solid edges.

Every face has one thick solid edge. Each thick solid edge is shared by two faces, and that is the only thick solid edge that each of those faces has. Each thick solid edge is the diagonal of a quadrilateral consisting of one pair of faces. Therefore, the V – 2 thick solid edges divide the MPG into V – 2 quadrilaterals, as shown below. The thick solid edges are the diagonals of these quadrilaterals. This shows that if the three-edges theorem is true, then all of the faces of any MPG can always be paired into quadrilaterals.

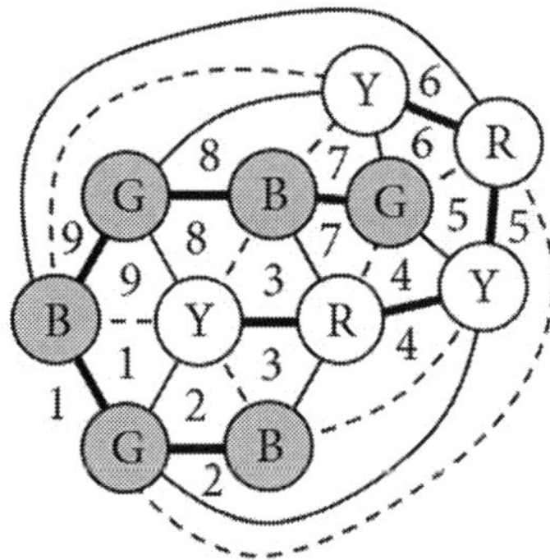

For the second point, consider a simple closed path with N edges of the same kind (such as thick solid edges) on a MPG for which the edges have been divided in accordance with the three-edges theorem. The diagram shows a couple of potential closed paths. The complete MPG has many vertices and edges that aren't shown; the diagram is focused on the closed path. We will show that N must be an even number like the second figure below, and not an odd number like the left figure below. Note that the path is *simple* (defined in Chapter 1).

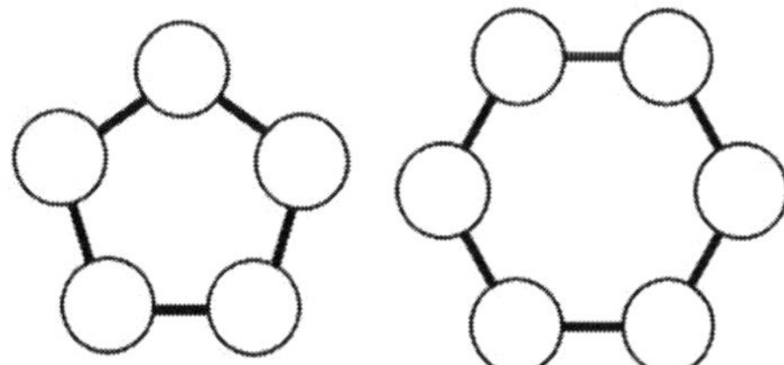

To do this, we'll make a new MPG as follows.

- Keep the closed path and all of the vertices and edges inside of the closed path from the original MPG.
- Remove all of the vertices and edges outside of the closed path from the original MPG.
- Duplicate the inside vertices and edges, drawing them outside of the closed path.

The example below will help to visualize what we mean.

- The original MPG (top left) has V = 16 vertices and E = 3V – 6 = 3(16) – 6 = 42 edges.

- The edges of the original MPG have been divided into three sets of V – 2 = 14 edges such that every face has exactly one solid, one thick solid, and one dashed edge.

- The original MPG has two closed paths of 6 thick solid edges. Focus on path AIDEJH.

- On the top right, we removed the vertices and edges outside of closed path AIDEJH.

- On the bottom, we drew duplicates of the inside vertices and edges on the outside.

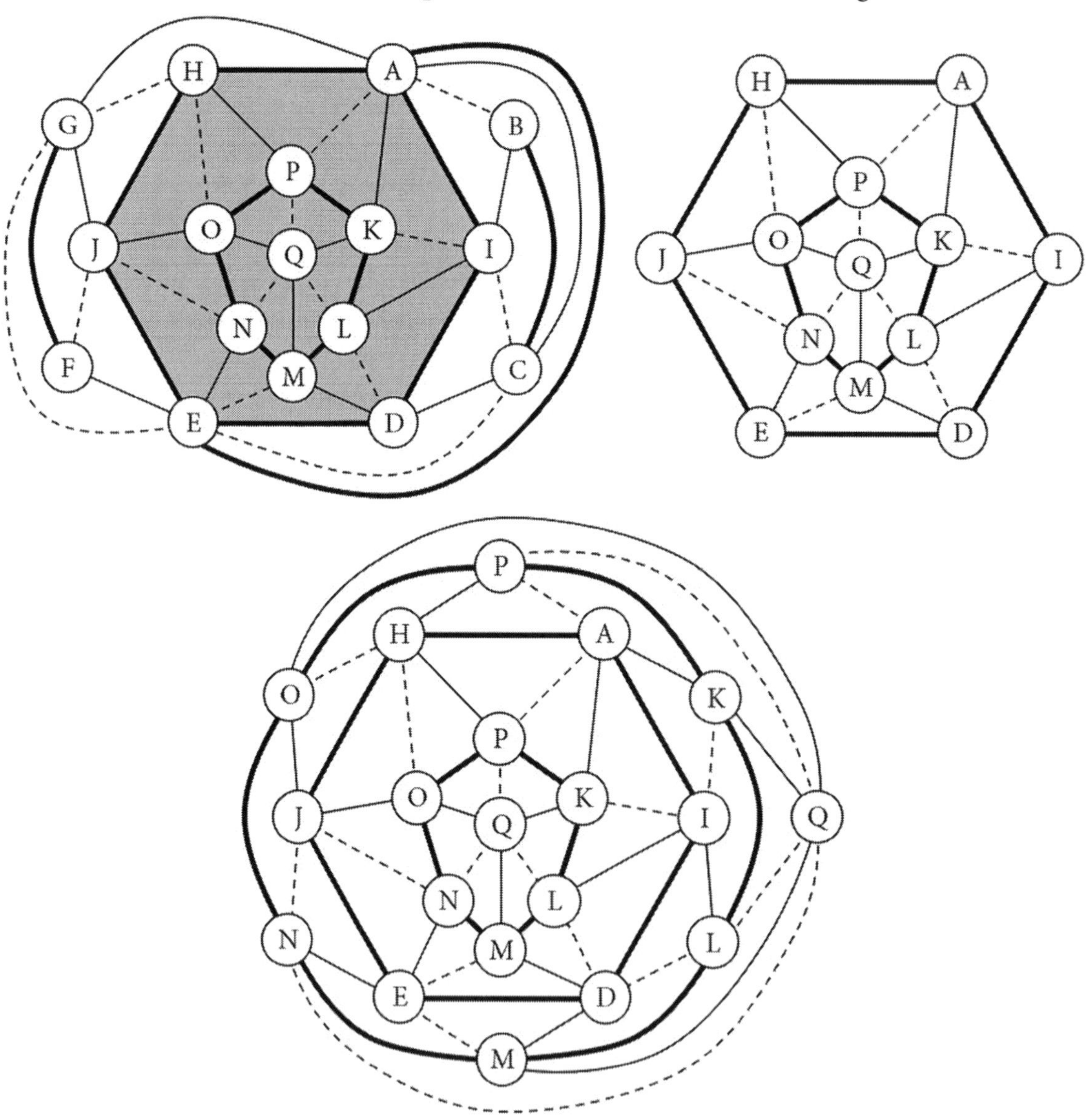

Observe that the new graph is a MPG; the regions inside and outside of closed path AIDEJH are each triangulated. The portion of the MPG outside of the closed path has the same set of edges connecting the same pairs of vertices as the inside portion; that's what we mean by duplicating the inside vertices and edges on the outside. Note that the new MPG has its edges divided in a way that satisfies the three-edges theorem; this occurs automatically since the inside portion satisfied the three-edges theorem.

If the original MPG has a simple closed path of one type of edge (like a thick solid edge), we can always make another MPG in this manner.

Suppose that the closed path consisted of N edges (and thus N vertices) and that there are M vertices inside of the closed path. The new MPG has:

- $V = 2M + N$ vertices (since there are N vertices on the path, M inside, and M outside).
- $E = 3V - 6 = 3(2M + N) - 6 = 6M + 3N - 6 = 3(2M + N - 2)$ edges.
- There are $2M + N - 2$ solid edges, $2M + N - 2$ thick solid edges, and $2M + N - 2$ dashed edges.
- Since there are N thick solid edges on the closed path itself, there must be $M - 1$ thick solid edges inside of the path and $M - 1$ thick solid edges outside the path.
- There aren't any solid edges or dashed edges on the path itself, so to figure out how many solid edges (or dashed edges) are inside and outside of the path, we just need to divide $2M + N - 2$ by two. This will only be possible if N is an even number.

This proves that if the three-edges theorem is true, then any simple closed path consisting of the same type of edges (such as thick solid edges) must have an even number of edges. This means that any MPG that has its edges divided in a way that satisfies the three-edges theorem will never have a closed loop with an odd number of the same type of edges. (What if the closed path isn't a *simple* closed figure, but has the shape of a figure eight, for example? We simply break it up into simple closed figures. For a figure eight, this would mean that each closed loop of the figure eight must have an even number of edges.)

For the third point, consider a MPG for which the edges have been divided in accordance with the three-edges theorem.

- The vertices connected by the V – 2 thick solid edges form Kempe chains. Each separate chain may be colored B-G or R-Y. We don't have to worry about three vertices from the same chain being mutually connected (with all three vertices connected to each of the other two vertices, forming a triangle with three edges) because no single face on the MPG has two thick solid edges. We also don't have to worry about any closed loops with an odd number of thick solid edges (since we already showed that this can't happen if the three-edges theorem is satisfied).

- Similarly, the vertices connected by the V – 2 solid edges may be colored B-R and G-Y, and the vertices connected by the V – 2 dashed edges may be colored B-Y and G-R. The other notes from the previous bullet point also apply to these sets of edges.

- Only the solid edges join B to R or join G to Y, only the thick solid edges join B to G or R to Y, and only the dashed edges join B to Y or G to R.

Only four colors are used: B, G, R, and Y. No edges join vertices of the same color. We know this because no face has two of the same kind of edge, which prevents the same Kempe chain (like B-G) from appearing on two edges of the same triangle, and because there aren't any closed loops with an odd number of the same kind of edge. (A Kempe chain is two-colored, and an odd closed loop doesn't permit two-coloring; see Chapter 28, Sec. 28.10.)

Regarding the third point, we are basically saying that if the three-edges theorem is true, then each face has one solid edge, one thick solid edge, and one dashed edge, so that every face is one of the basic building blocks from Chapter 21. Each edge of a building block corresponds to one of the three types of Kempe chains. If the three-edges theorem is true, then every possible MPG can be built from these basic building blocks. By the very nature of these building blocks, any MPG that is built from them ("properly" so that every face has exactly one of each type of edge) will be four-colorable.

We have now shown that if the three-edges theorem is true, then all three points of the face-pairing four-coloring method from Chapter 22 can be satisfied. This means that if the three-edges theorem is true, then every MPG is four-colorable. Therefore, if the three-edges theorem is true, then the four-color theorem must also be true.

It follows from the three-edges theorem that any MPG can be built by doing the following:

- Draw V vertices.
- Draw V − 2 thick solid edges in at least two sets of connected vertices without letting an odd number of thick solid edges form a closed loop.
- Add V − 2 solid edges and V − 2 dashed edges to triangulate the graph such that every face has exactly one solid edge, thick solid edge, and dashed edge. (Also check that neither of these new edges forms any odd closed loops.)

For V = 4 vertices, the V − 2 = 2 thick solid edges connect two pairs of vertices. Once we add 2 dashed edges and 2 solid edges, the result is K_4.

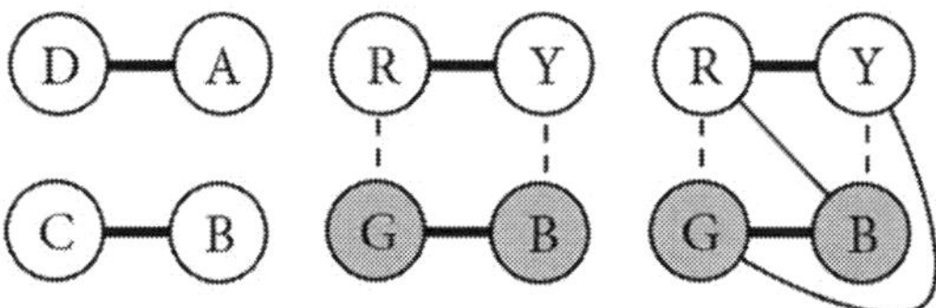

For V = 5 vertices, there are V − 2 = 3 thick solid edges. We used two of these edges to form one chain with 3 vertices and another chain has 2 vertices. After adding 3 dashed edges and 3 solid edges, the result is the pentahedral graph (Chapter 8).

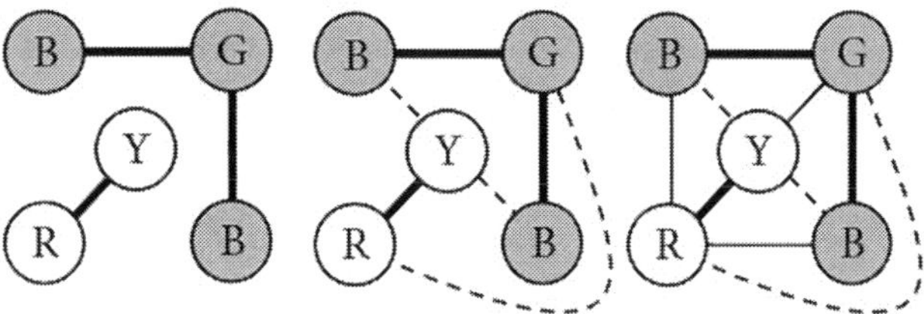

When there are more than five vertices, the choices that you make (such as which vertices to connect with thick solid edges) when you draw a MPG with V vertices according to the above instructions result in MPG's with different structures.

For example, for V = 6 vertices, there are V − 2 = 4 thick solid edges. We can separate these 4 thick solid edges into two groups of two like the top figures below, into three edges plus one edge like the middle figures below, or into one closed loop like the bottom figures below. In top and middle examples, the MPG has two vertices with degree five, two vertices with degree four, and two vertices with degree three. In the bottom example, the MPG has six vertices with degree four.

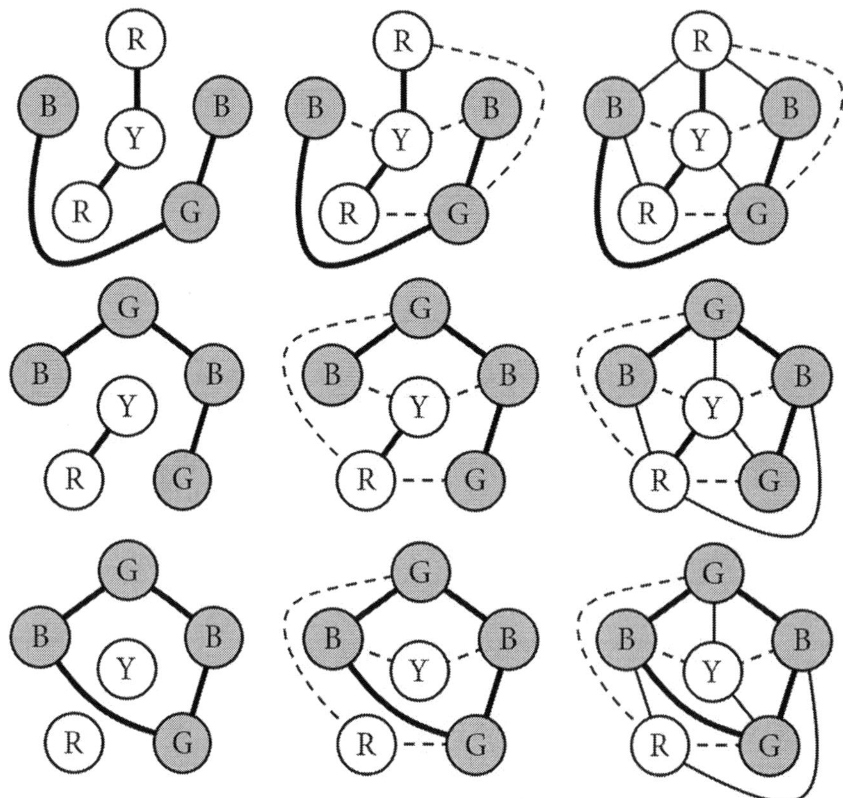

Note that the three-edges theorem applies to all MPG's, but not to PG's that aren't MPG's, whereas the four-color theorem applies to all PG's, including all MPG's. It should be easy to see why the three-edges theorem doesn't apply to PG's that aren't MPG's; we'll save this for one of the end-of-chapter exercises.

Chapter 23 Exercises

1. Color the MPG below and redraw its edges so that it simultaneously satisfies the four-color theorem and the three-edges theorem. (A simple way to make three kinds of edges by hand is to draw ++++'s and >>>>'s over some of the existing edges.)

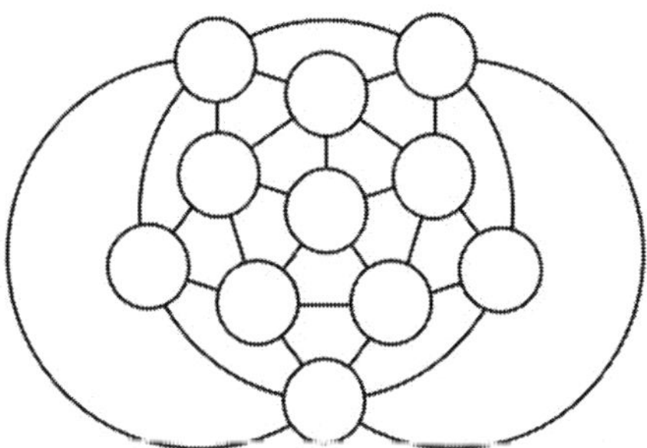

2. Color the MPG below and redraw its edges so that it simultaneously satisfies the four-color theorem and the three-edges theorem.

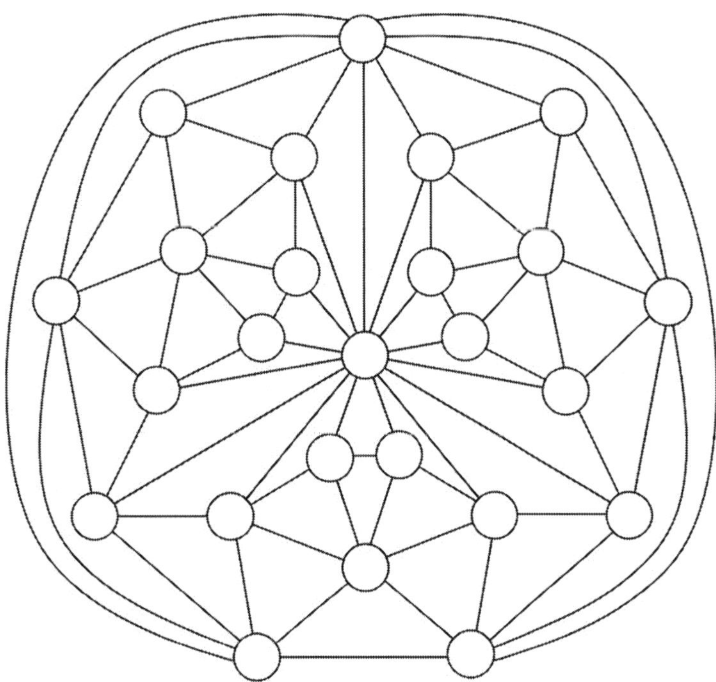

3. Ultimately, what is the origin of the number 2 that is subtracted from V in the formulas $E = 3(V - 2)$ and $F = 2(V - 2)$?

4. If a MPG has 32 vertices, how many solid edges, thick solid edges, and dashed edges does it have according to the three-edges theorem? How many edges and faces does it have in total?

5. A MPG has 17 solid edges, 17 thick solid edges, and 17 dashed edges. How many vertices and faces does it have?

6. A MPG contains a subgraph where the outer border of the subgraph contains 16 edges (and thus 16 vertices). There are another 24 vertices inside of the subgraph's border. All 16 edges along the outer border are **dashed edges**. Determine the total number of vertices, edges, and faces for the subgraph.

The subgraph is removed from the MPG. The inside vertices and edges are duplicated outside of the subgraph (just like we discussed in this chapter). Determine each of the following:

- the total number of vertices, edges, and faces for the new MPG
- the total number of solid edges, thick solid edges, and dashed edges for the new MPG
- the total number of solid edges, thick solid edges, and dashed edges in each of the three regions (inside of the subgraph, along the border of the subgraph, and outside of the subgraph) such that the edges of the new MPG satisfy the three-edges theorem

7. A MPG contains a subgraph where the outer border of the subgraph contains 11 edges (and thus 11 vertices). There are another 14 vertices inside of the subgraph's border. Determine the total number of vertices, edges, and faces for the subgraph.

The subgraph is removed from the MPG. The inside vertices and edges are duplicated outside of the subgraph (just like we discussed in this chapter). Determine each of the following:

- the total number of vertices, edges, and faces for the new MPG
- the total number of solid edges, thick solid edges, and dashed edges for the new MPG

Show that the 11 edges along the outer border of the subgraph can't all be dashed edges (unlike Problem 6) by determining how many solid edges, thick solid edges, and dashed edges there would be in each of the three regions (inside of the subgraph, along the border of the subgraph, and outside of the subgraph) if the edges of the new MPG satisfy the three-edges theorem.

8. We showed that if the three-edges theorem is true, a MPG will never have a closed loop with an odd number of the same type of edges. Show visually that it would be impossible to make each closed loop below **<u>with the same type of edge along the border of the loop</u>** (such that each face has one of each of three different types of edges). Show this *visually* (without counting edges and without duplicating the inside vertices and edges). Note that these closed loops consist of an odd number of edges along the border.

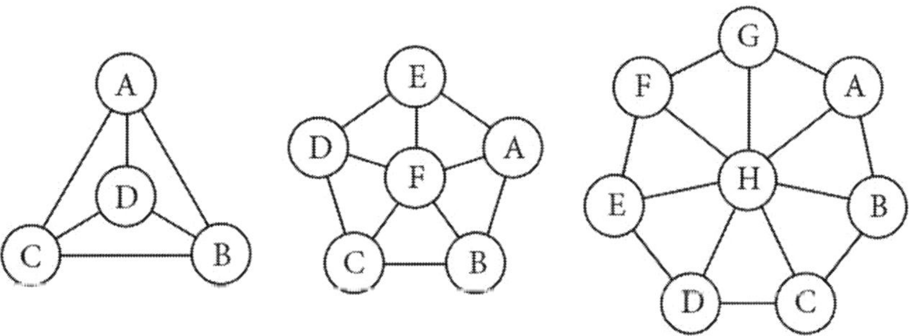

9. Consider the closed loop ACDF in the MPG below. Redraw loop ACDF and all of the vertices and edges within it, but not those outside of it. Form a new MPG by "duplicating" the vertices and edges inside of ACDF and drawing this duplicate set outside of ACDF (like we discussed in this chapter). Assign three types of edges to the new MPG in such a way that closed loop ACDF has just one type of edge along it and the new MPG satisfies the three-edges theorem. (Note that this is an even loop.) Four-color the new MPG. Can these same edges be used for the original MPG (except, of course, for the vertices and edges lying outside of ACDF)? If so, show this. If not, show or explain why.

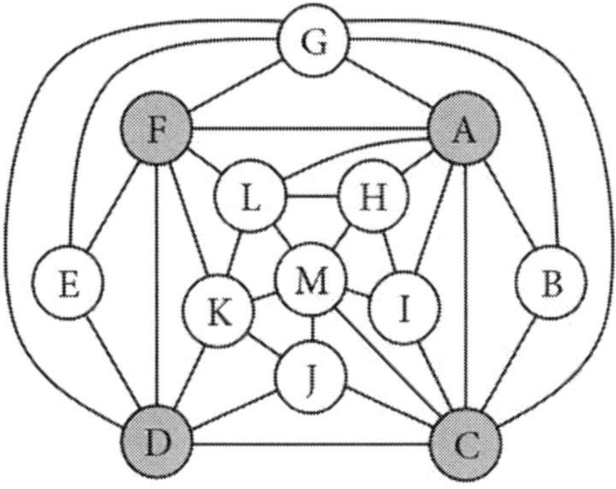

10. Consider the closed loop AHIEF in the MPG below. Redraw loop AHIEF and all of the vertices and edges within it, but not those outside of it. Form a new MPG by "duplicating" the vertices and edges inside of AHIEF and drawing this duplicate set outside of AHIEF (like we discussed in this chapter). Show visually that it isn't possible to assign three types of edges to the new MPG in such a way that closed loop AHIEF will have just one type of edge along it and the new MPG will satisfy the three-edges theorem. (Note that this is an odd loop.) Does this mean that the MPG below (or the new MPG formed from the duplicated edges) disproves the three-edges theorem? Explain.

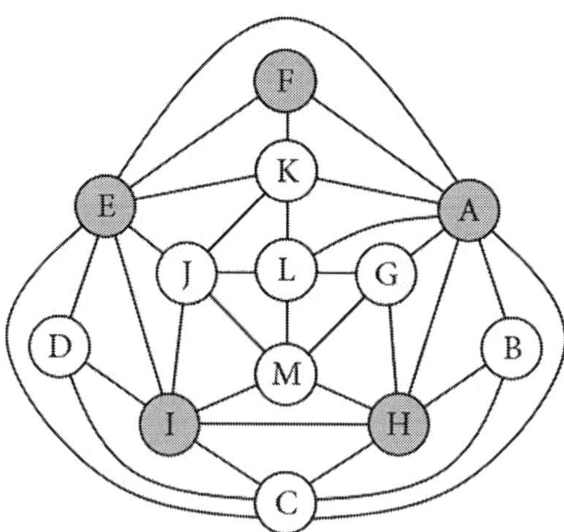

11. Why doesn't the three-edges theorem apply to PG's that aren't MPG's?

12. Build five structurally different MPG's for the sets of 7 vertices below using the method for building MPG's with three different types of edges discussed at the end of the chapter.

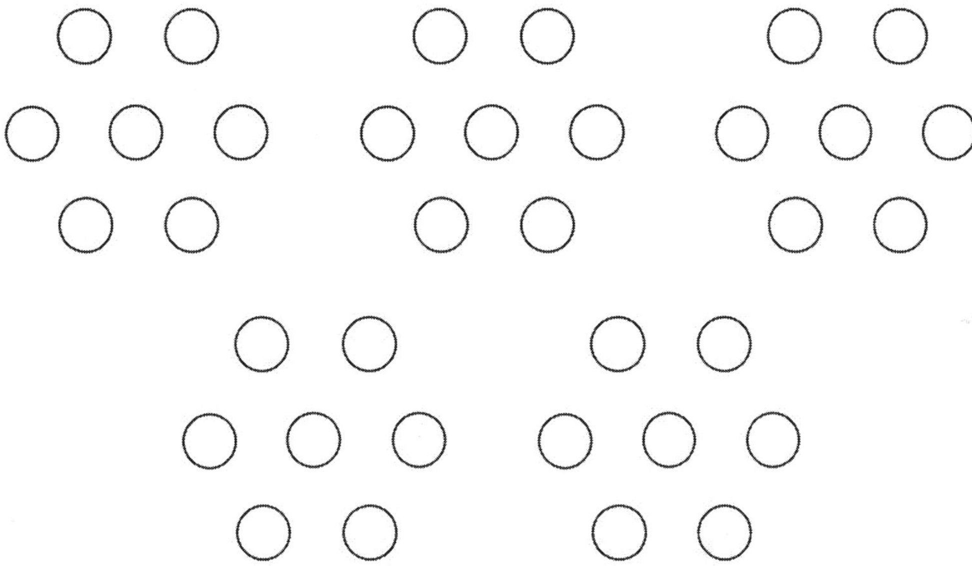

Challenge problem 1: In this chapter, we argued that if the three-edges theorem is true, then the four-color theorem must also be true. Does the converse argument also work? That is, if the four-color theorem is true, must the three-edges theorem also be true?

Challenge problem 2: Every face of a maximal planar graph has three edges (where an edge may be curved or bent). Can every maximal planar graph be made by adding edges to a planar graph where every face is a quadrilateral? Explain. If so, would it be possible to prove the three-edges theorem (or the four-color theorem) by first proving that every planar graph where every face is a quadrilateral is four-colorable? Either prove this or describe why not (or identify any obstacles that would make this approach very challenging).

Note: The answer key doesn't include answers to the challenge problems. These problems are intended to encourage you to think about the ideas.

Challenge problem 3: In this chapter, we argued that if the three-edges theorem is true, then it is always possible to pair the faces of a MPG into quadrilaterals. We also argued that if the three-edges theorem is true, it is always possible to pair the faces of a MPG into quadrilaterals in such a way as to avoid having an odd number of diagonals of quadrilaterals form a closed loop. Does the converse of each separate argument necessarily hold? If not, do the converse of the two arguments together necessarily hold? If not, why not? Is there another condition that would make it work? If so (for either case, separate or together), show or explain why.

Challenge problem 4: According to Problem 11, the three-edges theorem doesn't apply to a PG that isn't a MPG. Consider a possible modification of the three-edges theorem: Instead of stating that every face must not have two of the same type of edge, suppose that we revise this to state that adjacent edges in the same face must not be the same type of edge. With this modification, would the three-edges theorem apply to any PG's that aren't MPG's? Explain. Would the three-edges theorem apply to all PG's? If so, show or explain why. If not, draw a counterexample. If the edges of a MPG satisfy the three-edges theorem and an edge is removed, would the new PG (that is no longer a MPG) necessarily still satisfy the three-edges theorem (that is, the modified version that we are discussing in this problem) without having to change the edges? If so, show or explain why. If not, draw a counterexample.

Note: The answer key doesn't include answers to the challenge problems. These problems are intended to encourage you to think about the ideas.

24 A Recoloring Technique

The main challenge with attempting to prove the four-color theorem can be summarized as follows. If a MPG is four-colored and a small change is made to the structure of the MPG, the new MPG can require extensive recoloring. The challenge lies in developing a general recoloring method that applies to every possible scenario. Here are a couple of examples:

- Kempe developed a method for recoloring a vertex with degree five, but Heawood later showed that Kempe's method wasn't foolproof (Chapter 7).
- Any MPG can be made by splitting vertices (Chapter 18) on a smaller graph that is already four-colored, but the most general case can require extensive recoloring.
- A color-changing quadrilateral switch (Chapter 19) can require extensive recoloring.

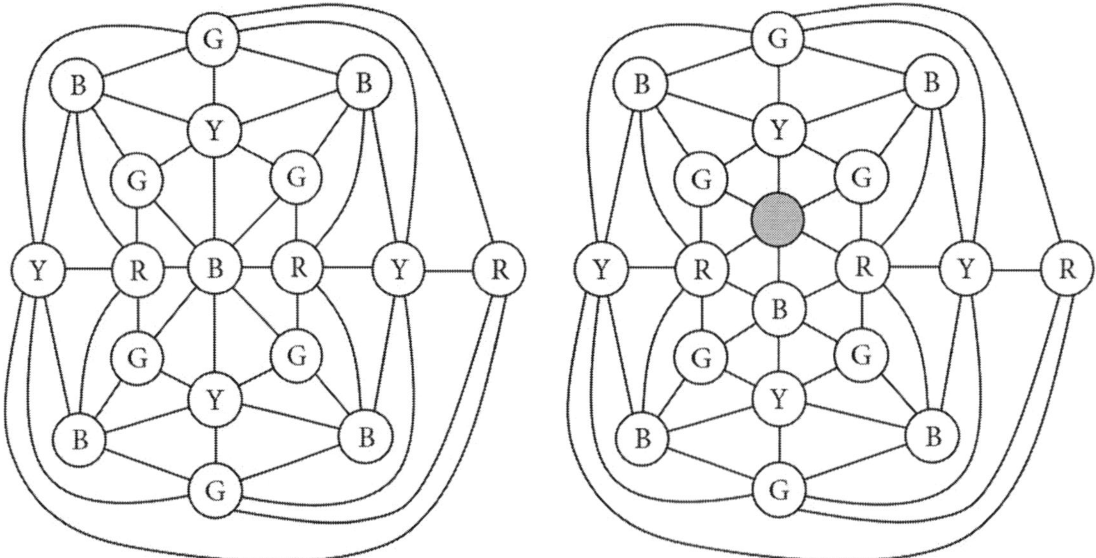

As an example, we added one vertex and three edges to the MPG on the left. Although the original MPG was four-colored, we need to change some colors in the right MPG in order for it to be four-colored. If we can develop a general method for recoloring a MPG that works for all possible scenarios, we could use this method to prove the four-color theorem. When we focus on coloring vertices, this proves to be quite challenging. An alternative is to focus on three types of edges. Our recoloring method is based on this alternative.

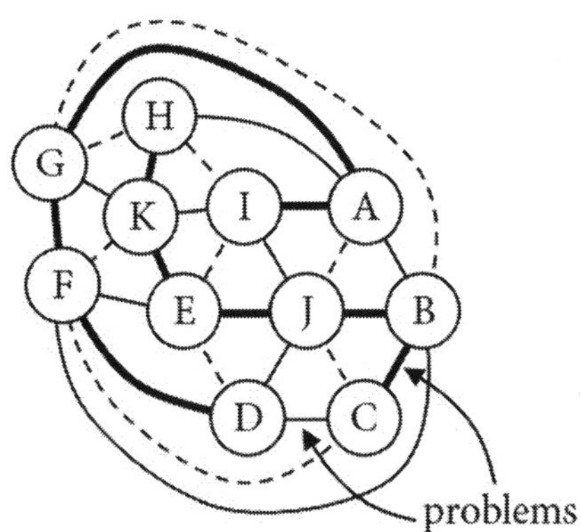

The MPG shown above has V = 11 vertices and E = 3(V − 2) = 3(9) = 27 edges. If you try to assign 9 thick solid edges, 9 solid edges, and 9 dashed edges, your first attempt might not work out. For example, consider the attempted edge assignment above. Face CDJ has two solid edges but no thick solid edges, while face BCJ has two thick solid edges but no solid edges (where we are considering "solid" and "thick solid" to be different types).

We need to trade one solid edge (DJ) for one thick solid edge (BJ). In this example, observe that we can travel one face at a time from face CDJ to BCJ by jumping across alternate solid and thick solid edges:

- Start at face CDJ.
- Jump across edge DJ (solid) to face DEJ.
- Jump across edge EJ (thick solid) to face EIJ.
- Jump across edge IJ (solid) to face AIJ.
- Jump across edge AI (thick solid) to face AHI.
- Jump across edge AH (solid) to face AGH.
- Jump across edge AG (thick solid) to face ABG.
- Jump across edge AB (solid) to face ABJ.
- Jump across edge BJ (thick solid) to face BCJ.

These alternate solid and thick solid edges allow us to create a domino effect. We will change edge DJ from solid to thick solid, and then we will switch solid edges to thick solid edges and vice-versa as we jump alternate edges until we reach face BCJ.

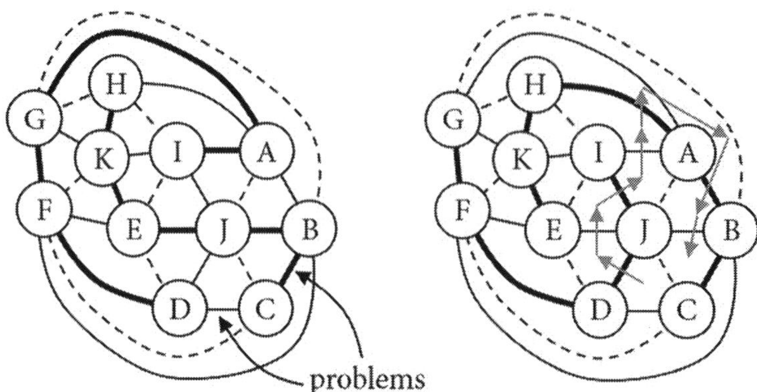

The diagram above shows how we solved this problem. The gray arrows show that the path of leaping over alternate solid and thick solid edges is part of a closed loop.

Why does this work?

- All but two of the faces originally had one edge of each kind. We'll refer to these as **three-edge faces**.
- Every three-edge face has one solid edge and one thick solid edge (and obviously one dashed edge, too).
- If we swap the solid and thick solid edges in a three-edge face, it will still be a three-edge face.
- Each three-edge face shares a solid edge with one face and a thick solid edge with a different face.
- When we leap over a solid edge to enter a three-edge face, we next leap over a thick solid edge to exit the three-edge face. Similarly, when we leap over a thick solid edge to enter a three-edge face, we next leap over a solid edge to exit the three-edge face.
- If the starting face and finishing face are connected in the same loop, switching the solid and thick solid edges in all of the connecting faces creates a domino effect with that carries through to the finishing face.
- We could similarly switch solid and dashed edges, or thick solid and dashed edges.
- This allows us to effectively trade one edge of one face with one edge of another face, provided that they are part of a loop formed by leaping over these edges.

Swapping edges works when the MPG already has the same number of each type of edge. If the MPG doesn't already have the same number of each type of edge, reassign edges until the MPG does have the same number of each type of edge. For example, suppose that a MPG has 8 thick solid edges, 7 solid edges, and 6 dashed edges. If we change one of the thick solid edges into a dashed edge, there will now be 7 thick solid edges, 7 solid edges, and 7 dashed edges. Once the numbers of each type of edge are equal, you are ready to swap edges.

What if the two problem faces don't lie on the same loop of alternating edges? For example, suppose that one face has two solid edges and another face has two dashed edges, but leaping over alternate solid and dashed edges doesn't take you from one face to the other. This can happen if there is a closed loop of thick solid edges separating the problem faces, like the example below. The solution to this problem is "break open" the closed loop by swapping thick solid edges with solid or dashed edges (or both). Once the closed loop opens, we will be able to swap solid and dashed edges.

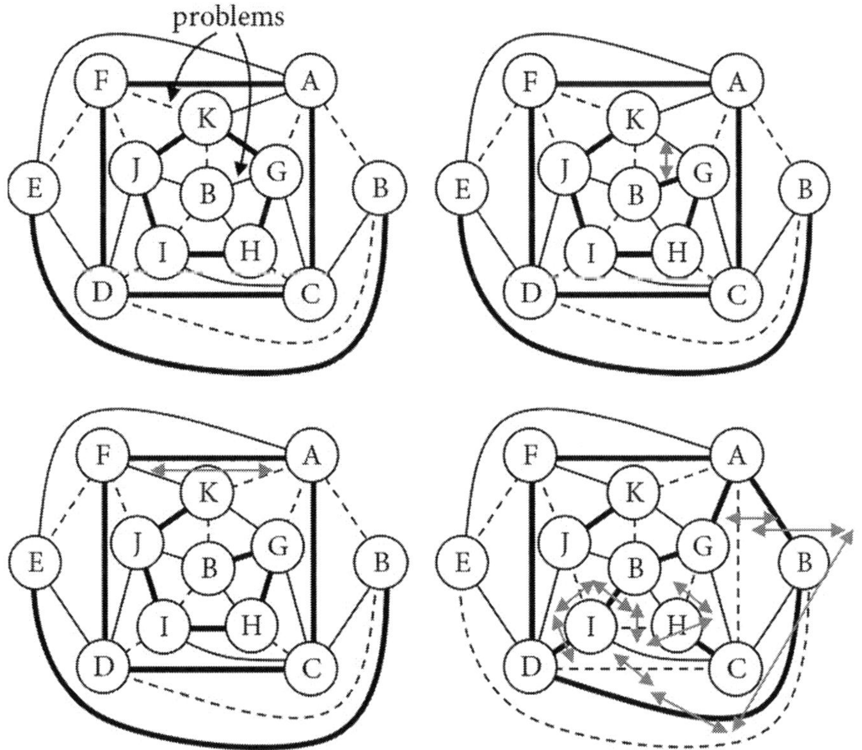

In the example on the previous page, face BGH has two solid edges while face FJK has two dashed edges. Faces BGH and FJK are separated by the thick solid edges of pentagon GHIJK. Swapping solid edges and dashed edges won't help us travel from a face inside the pentagon to a face outside the pentagon because the thick solid edges form a closed barrier.

We solved this problem as follows:

- First we broke open the pentagon of thick solid edges by swapping solid edge BG with thick solid edge GK (top right).
- Next we swapped a dashed edge from problem face FJK with solid edge AK (bottom left). This corrects the solid edges.
- Finally, we swapped thick solid edges with dashed edges (bottom right).

When swapping edges, you need to think about what you're doing. For example, in some cases it is possible to leap over alternate edges in such a way that you travel around a loop without any progress. The top center and top right figures on the following page show an example of this.

In the example on the next page, the problem faces (FGI and BCH) have two dashed edges or two thick solid edges. Simply swapping dashed and thick solid edges isn't adequate here because the problem faces lie on separate loops (which is why the swap shown on the top center and right doesn't work). The problem faces are separated by a closed loop of solid edges (ACH). We solved this problem as follows:

- We first swapped one thick edge and one dashed edge (left middle).
- Next we swapped a solid edge and a dashed edge (center). This breaks open closed loop ACH in the sense that the solid edges no longer form a closed loop.
- We swapped another thick edge and dashed edge (middle right).
- Finally, we swapped dashed and solid edges (bottom).

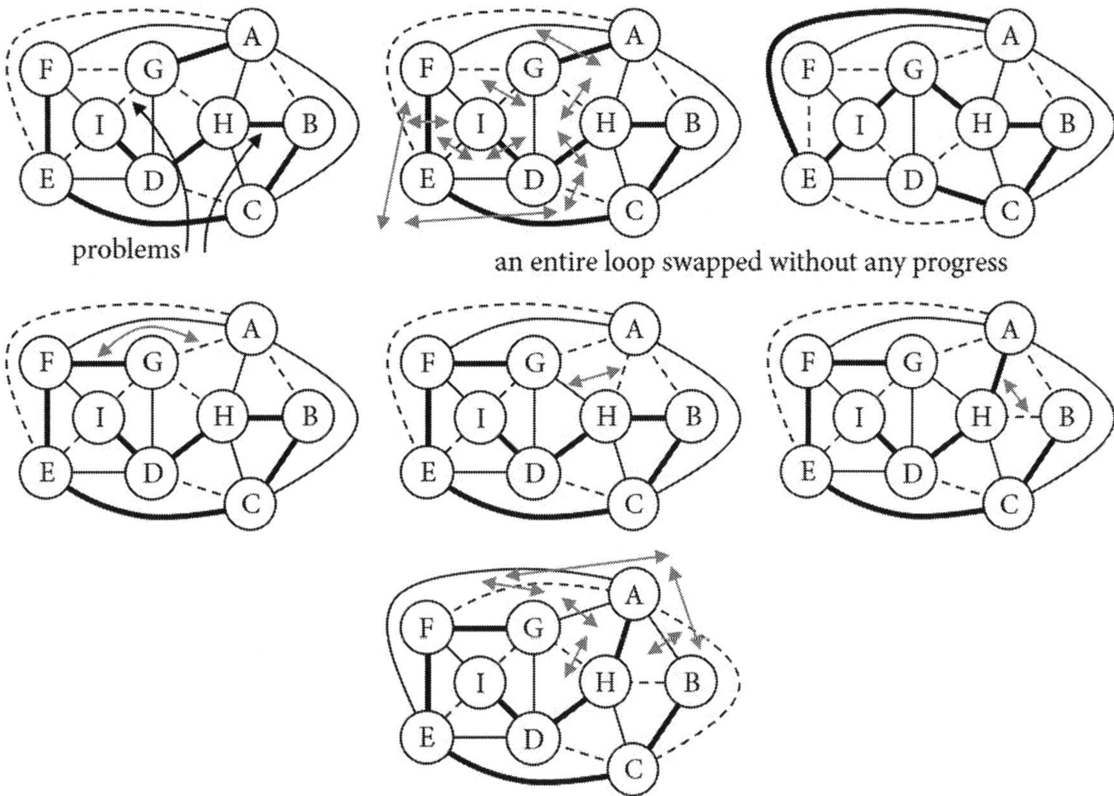

How does this help us recolor a MPG? Recall that each type of edge is associated with one of three possible divisions of Kempe chains: B-G and R-Y chains, B-R and G-Y chains, or B-Y and G-R chains. Fixing the edges so that every face has exactly one type of each edge will render the MPG four-colorable via the associated Kempe chains, as we'll see.

Suppose that a MPG is almost four-colored, except that two vertices of the same color are connected by an edge. We can apply what we've learned about edge swapping to recolor the MPG as follows:

- Draw one style of edges (like thick solid edges) for B-G and R-Y edges.
- Draw another style of edges (like solid edges) for B-R and G-Y edges.
- Draw a third style of edges (like dashed edges) for B-Y and G-R edges.
- Count the edges. If the counts are unequal, change edges as needed until there are equal numbers of each kind of edge. For example, if there are 10 thick solid edges, 7

solid edges, and 7 dashed edges, change one thick solid edge to a solid edge and one thick solid edge to a dashed edge so that there will be 8 of each kind.

- Apply edge swapping (as previously discussed) until every face has exactly one of each kind of edge.

- Color B-G and R-Y chains for the first style of edges (thick solid edges).

- Check that this coloring works. Another style of edges (such as solid edges) should have B-R and G-Y chains and the remaining style of edges (such as dashed edges) should have B-Y and G-R chains.

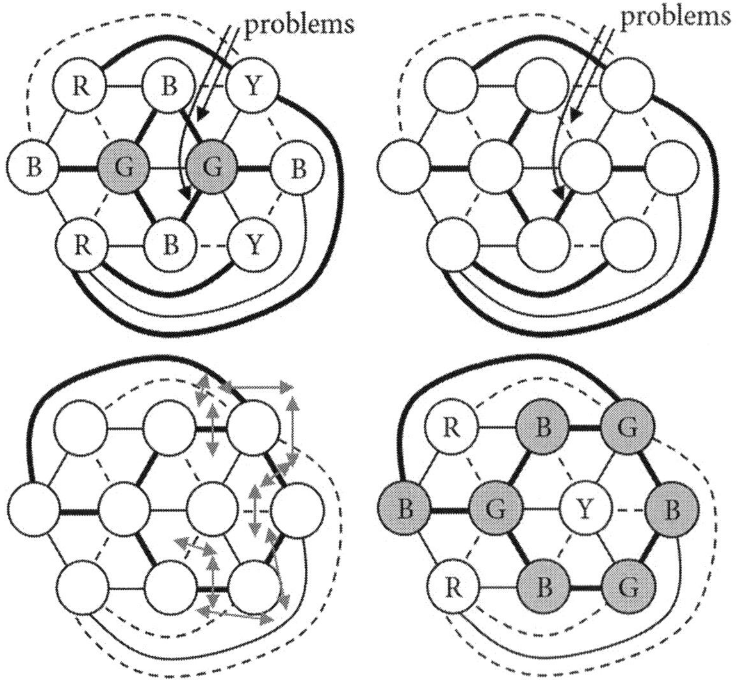

The example above illustrates our recoloring method:

- The original MPG was nearly four-colored; but two G's shared an edge (top left).

- First we drew thick solid edges for B-G and R-Y, solid edges for B-R and G-Y, and dashed edges for B-Y and G-R (top left). We left the G-G edge solid.

- This gave us 9 thick solid edges, 8 solid edges, and 7 dashed edges. To even this out, we changed one thick solid edge to a dashed edge. After this change, there were 8 thick solid edges, 8 solid edges, and 8 dashed edges (top right).

- The problem edges are now dashed and thick solid edges. They turned out to be part of the same dashed/thick loop; we swapped them as shown on the bottom left.
- Finally, we colored the vertices of thick solid edges with B-G and R-Y chains (bottom right).

A second example is shown below.

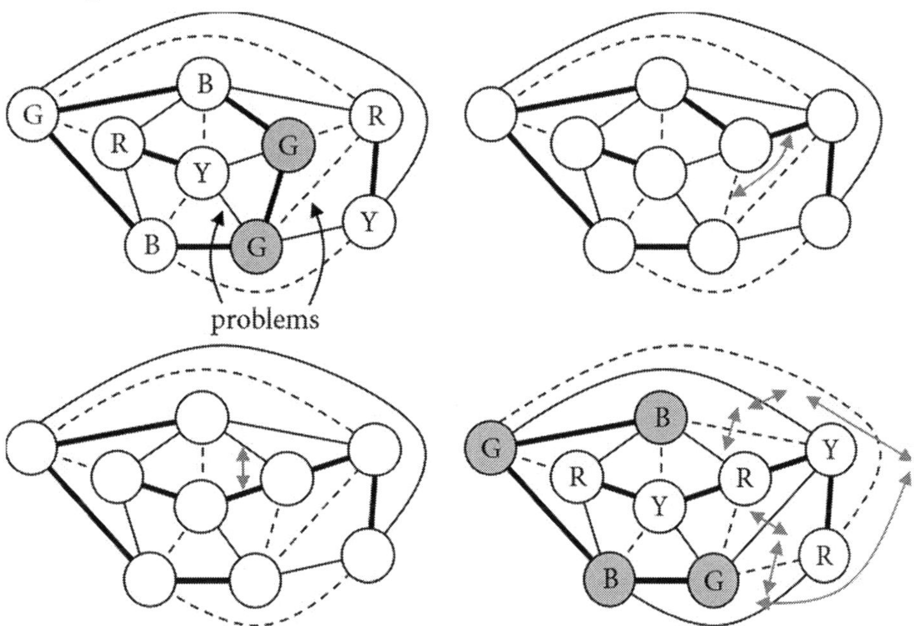

We recolored the previous example as follows:

- First we drew thick solid edges for B-G and R-Y, solid edges for B-R and G-Y, and dashed edges for B-Y and G-R (top left). We made the G-G edge a thick solid edge so that there would be 7 edges of each kind.
- The problem faces are separated by thick solid edges that form a closed loop.
- We first swapped a solid thick edge with a dashed edge (top right), and then swapped a solid thick edge with a solid edge (bottom left). This breaks up the closed loop of thick solid edges.
- This allowed us to swap solid and dashed edges according to the gray arrows (bottom right).
- Finally, we colored the vertices of thick solid edges with B-G and R-Y chains.

We'll break our method down visually to help you better see how it works. When we leap over alternate solid and dashed edges, for example, these faces form a chain. The internal edges of the chain are the two edges that we're leaping over (solid and dashed below); the external edges are the third kind of edge (thick solid below). If we swap two kinds of edges as we leap over them, each swap only changes internal edges; the external edges remain the same. In the example below, we swapped a series of solid and dashed edges along a chain of faces, which changed all of the internal edges without disturbing the external edges.

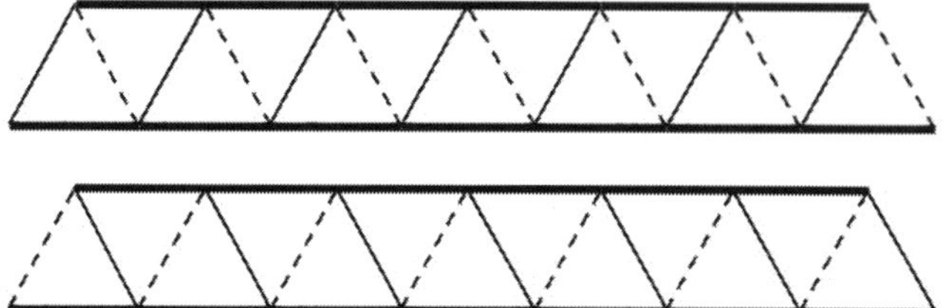

In the example below, the left face has two solid edges while the right face has two dashed edges. We swapped solid and dashed edges to correct this problem. Since the external edges are unaffected by this, we are able to solve the problem without causing a new problem.

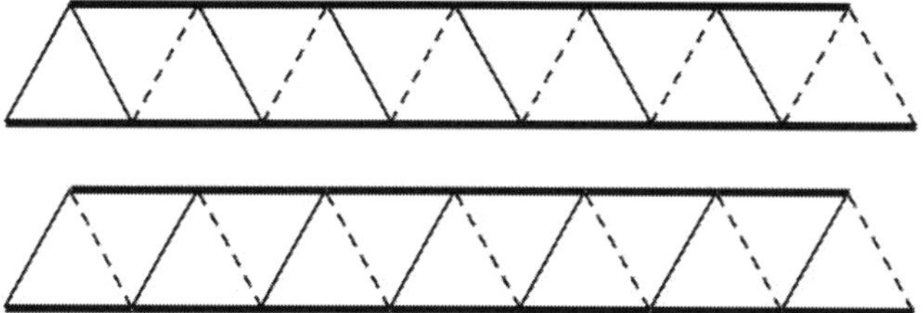

We are only showing part of the complete chain above. A complete chain of faces forms a loop like those shown below. The diagrams below show that the same principle still applies.

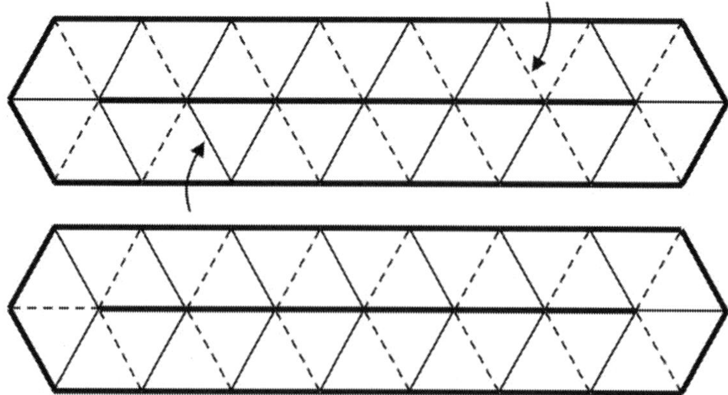

We're only looking at parts of complete MPG's, but these chains and loops illustrate the main underlying principles.

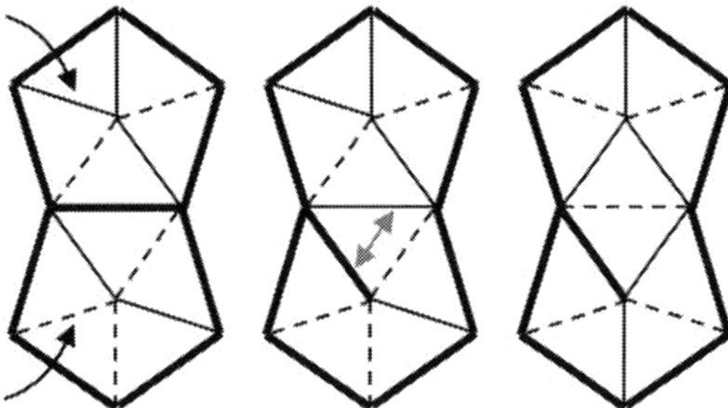

In the left diagram above, the problem faces lie in separate loops. We first swapped a thick solid edge with a solid edge to open a closed loop of thick solid edges, which allowed us to swap solid and dashed edges. In this example, we're not finished yet. We next need to swap thick solid edges with either solid or dashed edges outside of the remaining loop of thick solid edges; we need to see the complete MPG to do that. However, these diagrams illustrate the principle of breaking open a closed loop to swap edges.

Although our method isn't the most efficient way to recolor a MPG, it is a strategy that we may apply in general (when there are only two problem faces):

- Identify the repeated edges in the two problem faces.
- Are the problem edges separated by a closed loop of the third type of edge?
- If not, the problem faces are part of the same chain. Simply swap edges along that chain until the problem is solved.
- If yes, you will need to perform more than one type of swap. One of the swaps will need to involve the third type of edge in order to break open the closed loop.
- Beware of swaps that create new closed loops that separate the target faces. Not all closed loops are a problem. A closed loop needs to be opened if it separates two faces for which you wish to swap edges.
- (If you want to swap two edges separated by a closed loop and if the closed loop is made up of one of those kinds of edges, you may not need a separate swap to open the closed loop; you may be able to open the loop as part of the swap.)

Let's discuss an alternative to our edge-swapping recoloring method. Instead of working with three kinds of edges, we can apply our face-pairing concept from Chapter 22.

- With edge-swapping, once every face has one edge of each kind, the MPG will be four-colored. However, a complicated case may involve multiple steps, take some careful planning, and each step may involve a domino effect with numerous swaps. Edge-swapping isn't particularly efficient, but once every face has one edge of each kind, the coloring is finished.

- With face-pairing, it's generally easier and quicker to group faces into pairs (compared to dividing three kinds of edges among faces). However, face-pairing has two challenges. Recall from Chapter 22 that we need to avoid island faces and closed loops with an odd number of diagonal edges. We need to meet both of these challenges in order for the coloring to be finished.

Suppose that a MPG has all of its faces paired except for two island faces, like the example shown below on the left. There is a simple way to pair these island faces: Any island face can be moved around by sliding it along pairs of faces, as shown below. Simply move one island face (or both) in this manner until they are paired.

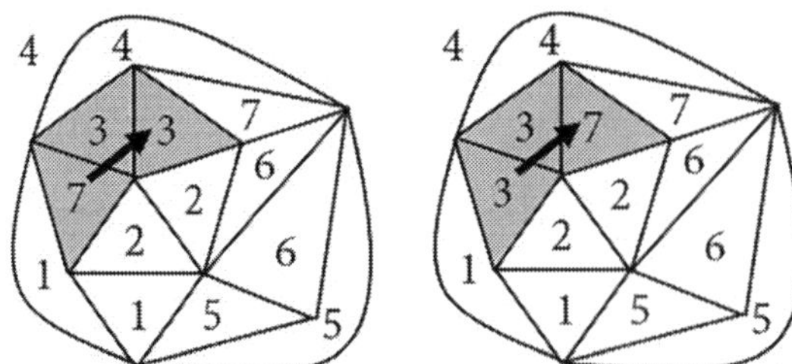

In the example above, we slid the 7 past a pair of 3's. In this way, the pair of 3's stays together. No other pairs are disturbed. This allows us to pair the island faces without unpairing any of the other faces.

We can similarly apply the concept of sliding faces to break open a closed loop with an odd number of diagonal edges, as illustrated on the following page.

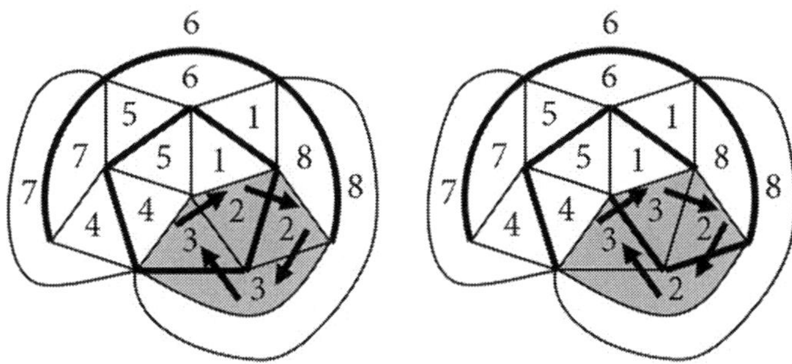

In the example above, a pair of 2's and a pair of 3's form a loop in a way that allows us to slide both the 2's and the 3's one face clockwise without disturbing any of the pairings. The thick solid closed loop opens up when we do this.

Let's look at an example that applies our face-pairing recoloring method.

- The MPG below on the left is nearly four-colored, but has two G's connected by an edge. (We used this same MPG earlier in an edge-swapping example.)

- The B-G and R-Y chains separate the MPG into 7 quadrilaterals, labeled 1-7. The last two faces (labeled 8) are island faces (shaded gray).

- We slid the top 8 past the pair of 5's as shown by arrows in the right graph (these faces are shaded gray). Now all of the faces are paired into quadrilaterals.

- The 8 thick solid edges in the right graph are diagonals of 8 quadrilaterals.

- We colored the vertices with B-G and R-Y chains along the thick solid edges. Note that the B on the left edge of the right graph is part of a B-G chain that only has one vertex (we could have colored it G instead of B).

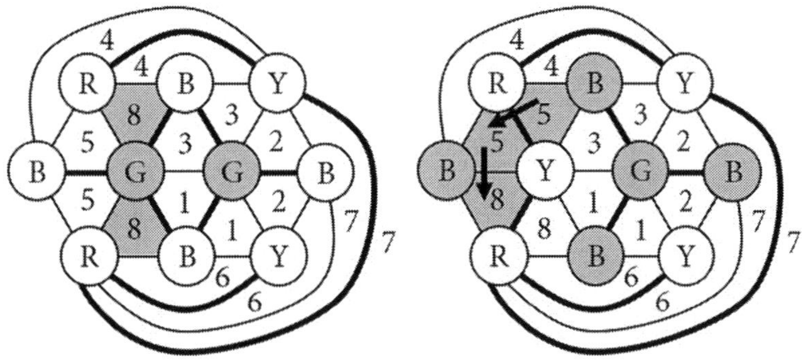

Following is another example that applies our face-pairing recoloring method.

- The MPG below on the left is nearly four-colored, but has two G's connected by an edge. (We also used this MPG earlier in an edge-swapping example.)

- The B-G and R-Y chains separate the MPG into 7 quadrilaterals, labeled 1-7. This example doesn't have any island faces.

- The problem with the original coloring is that five thick solid edges form a closed loop (the faces enclosed by this loop are shaded gray on the left).

- We slid the 1's and 2's clockwise (shown by the arrows on the right; these faces are shaded gray) in order to break open the closed path of thick solid edges.

- The 7 thick solid edges in the right graph are diagonals of 7 quadrilaterals.

- We colored the vertices with B-G and R-Y chains along the thick solid edges.

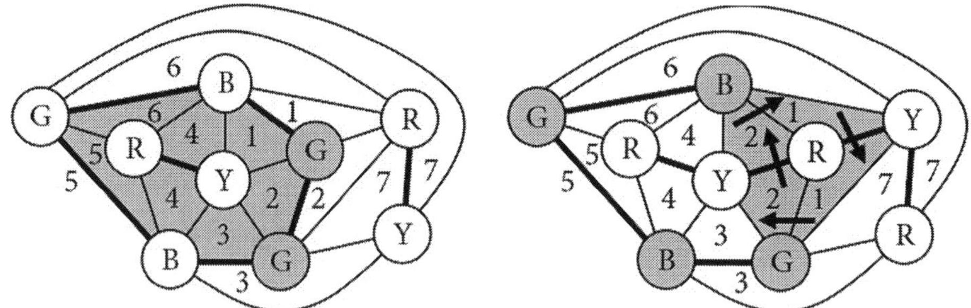

In Chapter 27, we'll consider another feature which may help to reduce (or perhaps even avoid) the need for recoloring a MPG.

Chapter 24 Exercises

1. For the MPG on the left below, at least one face has two edges of the same kind. Identify the faces that have two edges of the same kind. Use our edge-swapping method to redraw the MPG on the right below such that every face has exactly one of each kind of edge. (A simple way to make three kinds of edges by hand is to draw ++++'s and >>>>'s over some of the existing edges. You can do that instead of trying to draw thick and dashed edges.)

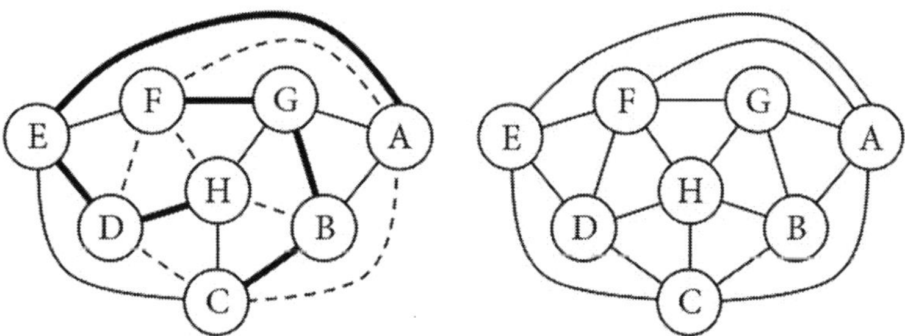

2. For the MPG on the left below, at least one face has two edges of the same kind. Identify the faces that have two edges of the same kind. Use our edge-swapping method to redraw the MPG such that every face has exactly one of each kind of edge.

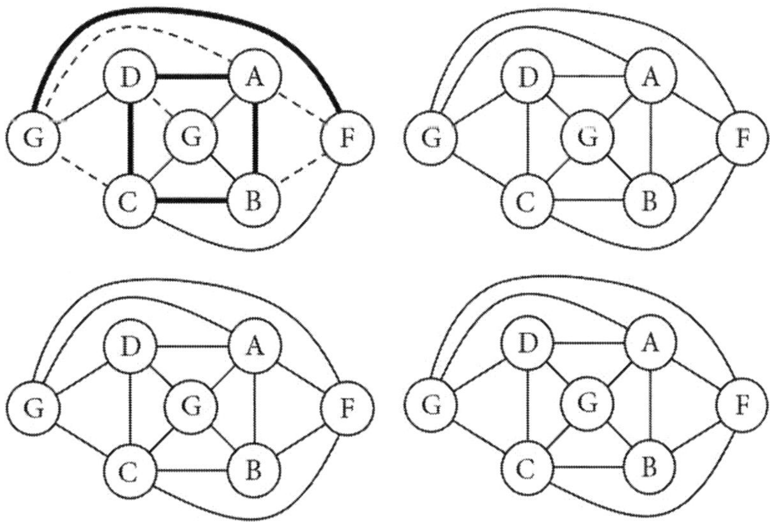

3. The MPG on the left below isn't yet four-colored; a pair of Y's are connected by an edge.

Use our edge-swapping method to recolor the MPG so that it is properly four-colored.

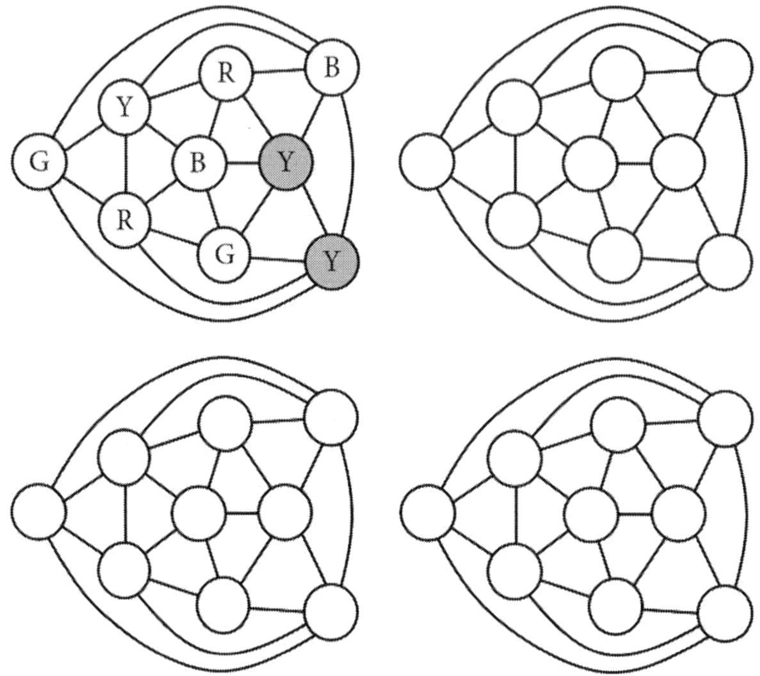

4. The MPG on the left below isn't yet four-colored; a pair of Y's are connected by an edge.

Use our edge-swapping method to recolor the MPG so that it is properly four-colored.

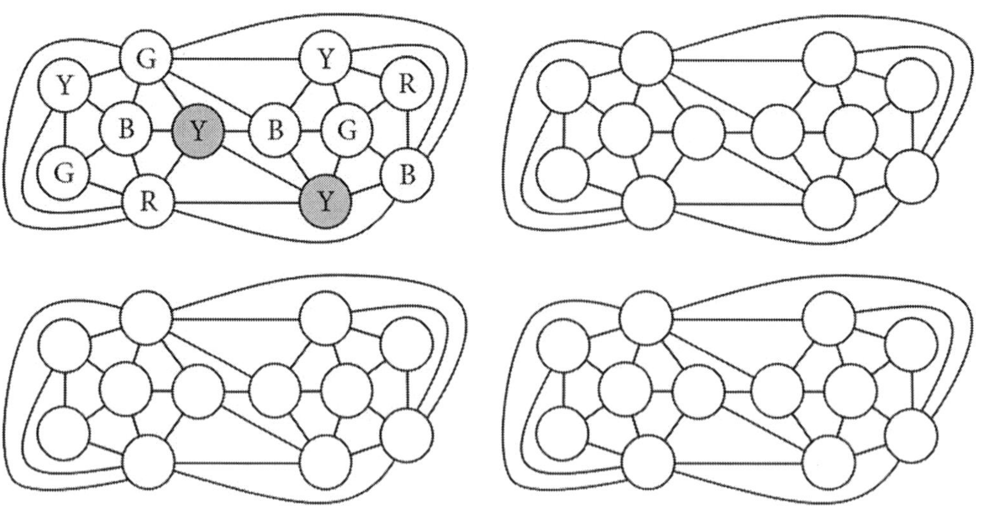

5. The MPG on the left below was formed by vertex splitting (Chapter 18). The graph needs to be recolored in order to accommodate this new vertex. Use our edge-swapping method to recolor the MPG so that it is properly four-colored.

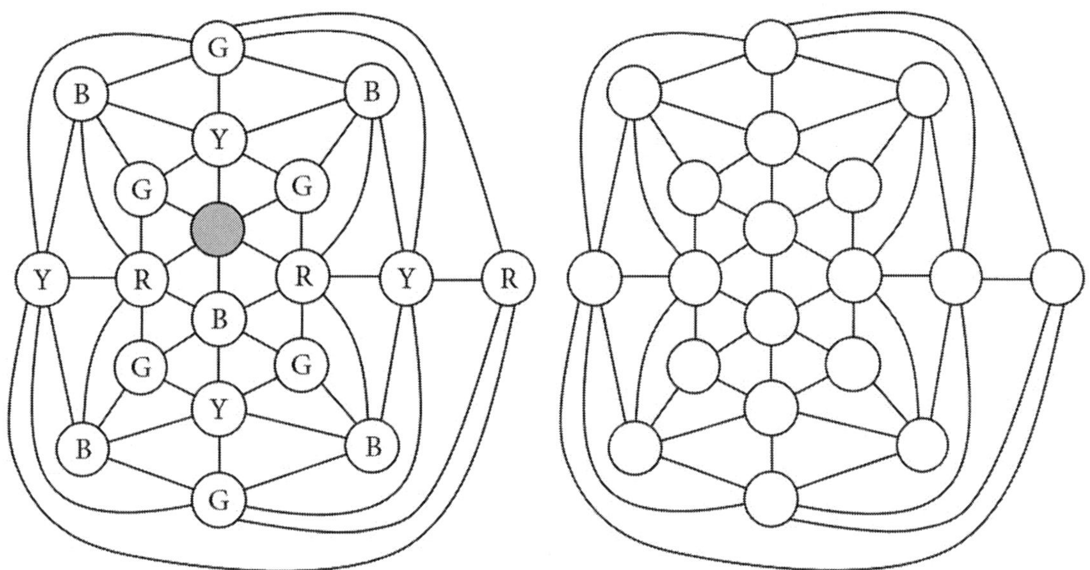

6. The MPG on the left below was properly four-colored. The middle MPG was formed by turning a color-changing quadrilateral switch (Chapter 19): edge G-Y was turned to B-B. The middle MPG needs to be recolored to accommodate this change. Use our edge-swapping method to recolor the MPG so that it is properly four-colored.

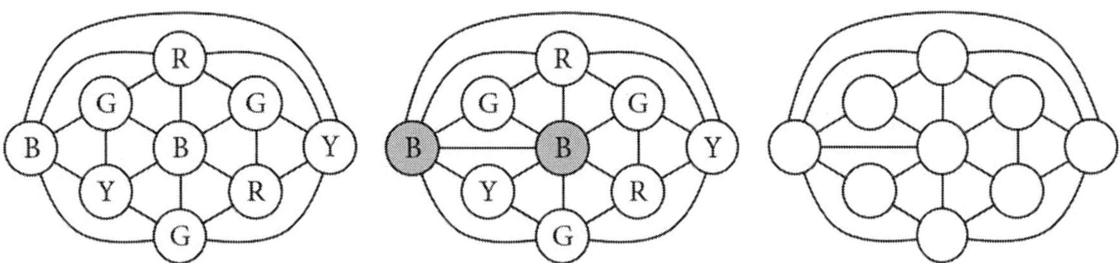

7. There are two island faces in the MPG on the left below. Use our face-sliding method to relabel the faces so that there aren't any island faces. (For this exercise, it will be okay if an odd number of diagonal edges form a closed loop. The focus of this problem is island faces.)

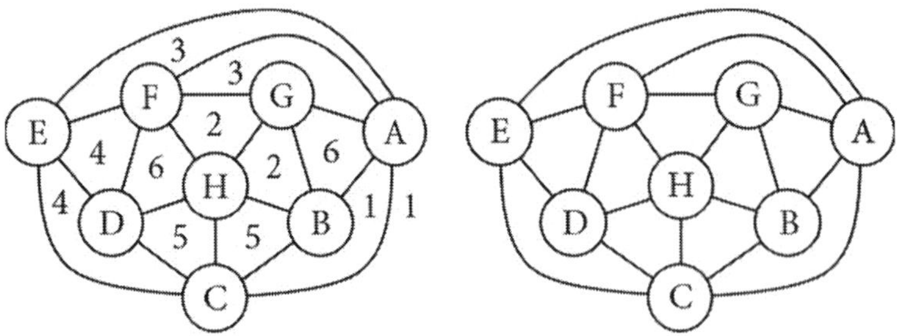

8. The MPG on the left below has an odd number of diagonal edges in a closed loop. Use our face-sliding method to break open this closed loop (without creating any island faces or new odd closed loops of diagonal edges).

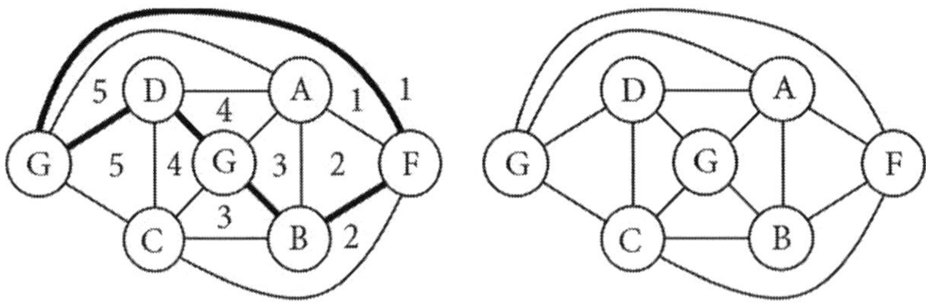

9. The MPG on the left below isn't yet four-colored; a pair of Y's are connected by an edge. Use our face-sliding method to recolor the MPG so that it is properly four-colored.

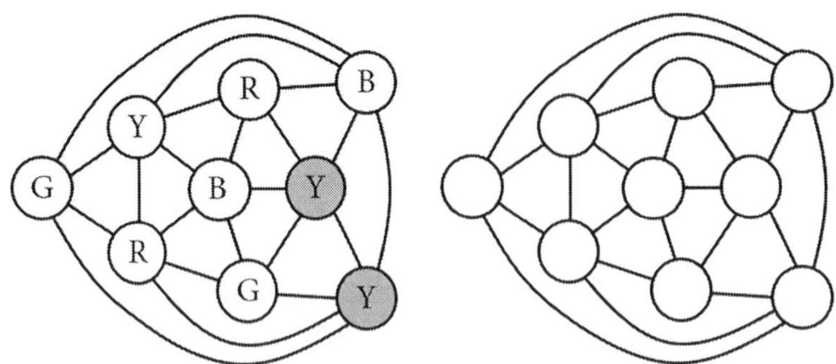

10. The MPG on the left below isn't yet four-colored; a pair of Y's are connected by an edge.

Use our face-sliding method to recolor the MPG so that it is properly four-colored.

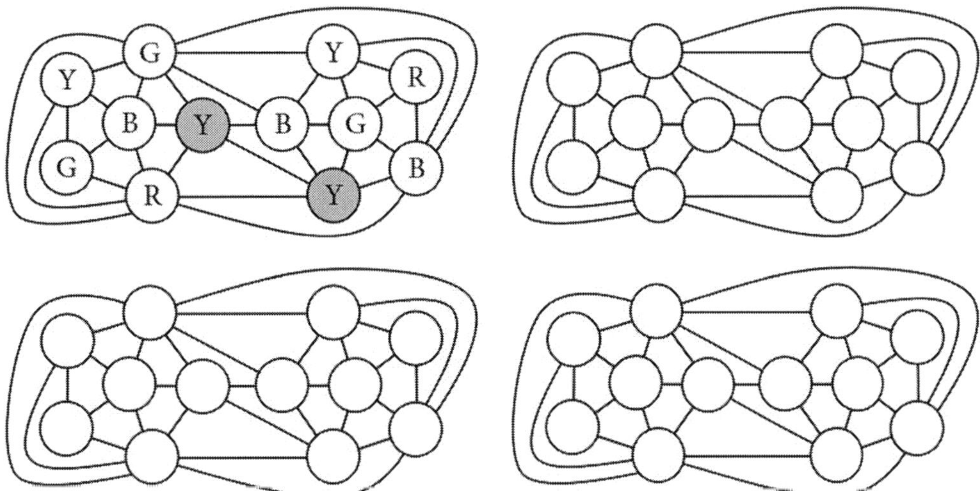

Challenge problem 1: Prove that our edge-swapping method will always work to recolor a MPG provided that:

- the MPG has the same number of each type of edge to begin with
- there are exactly two problem faces

Your proof should allow for the case where the problem faces are separated by a closed loop. If you're unable to prove this, show and describe what makes this proof challenging.

If the MPG doesn't have the same number of each type of edge to begin with, we can always reassign edges to distribute them equally (as discussed in this chapter). Can you generalize your proof to accommodate this? (If you couldn't formulate a proof for the previous part of the problem, discuss this in principle supposing that such a proof exists.) Explain.

Can you generalize your proof to accommodate more than two problem faces? Explain.

Can you apply your proof (or generalized proof) to prove the four-color theorem? Explain.

Note: The answer key doesn't include answers to the challenge problems. These problems are intended to encourage you to think about the ideas.

Challenge problem 2: Try to draw an example MPG with exactly two island faces where our face-sliding method won't work to pair all of the faces (without yet worrying about odd closed loops of diagonal edges). Now try to draw an example MPG with a closed loop of an odd number of diagonal edges where our face-sliding method won't work to break open the closed loop (without creating any island faces or new odd closed loops of diagonal edges).

If you're able to come up with an example of either case, can you find a way to expand our method to include other techniques besides face-sliding that would work for more (or perhaps all) cases? If you're unable to come up with an example for either case, can you prove that our face-sliding method will always work? If not, describe the challenge of doing this.

Note: The answer key doesn't include answers to the challenge problems. These problems are intended to encourage you to think about the ideas.

25 Kempe's Problem Revisited

Recall Kempe's argument for a vertex with degree five and also recall the problem pointed out by Heawood (Chapter 7). Kempe considered chains like the diagram shown below on the left; Heawood pointed out a subtle problem that can occur if the chains cross like the diagram below on the right. Kempe considered different types of chains like B-R and B-G, but Kempe didn't think to draw three different types of edges or that each face would have one of each type of edge in a four-colored MPG. As in Chapter 7, the diagrams below are just portions of a MPG; many vertices and edges aren't shown below.

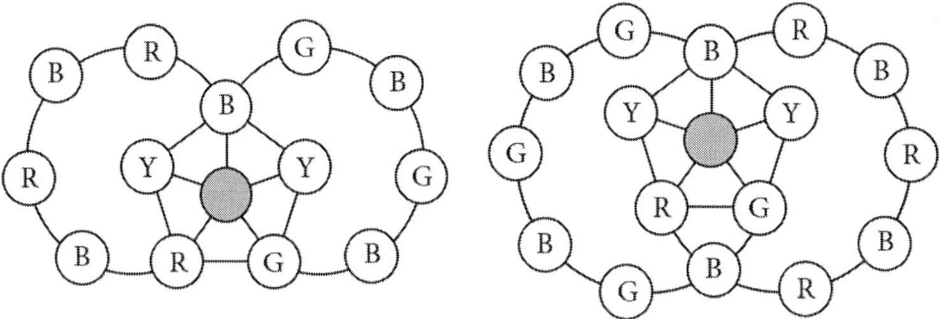

Let's redraw these diagrams using thick solid edges for B-G and R-Y chains, solid edges for B-R and G-Y chains, and dashed edges for B-Y and G-R chains.

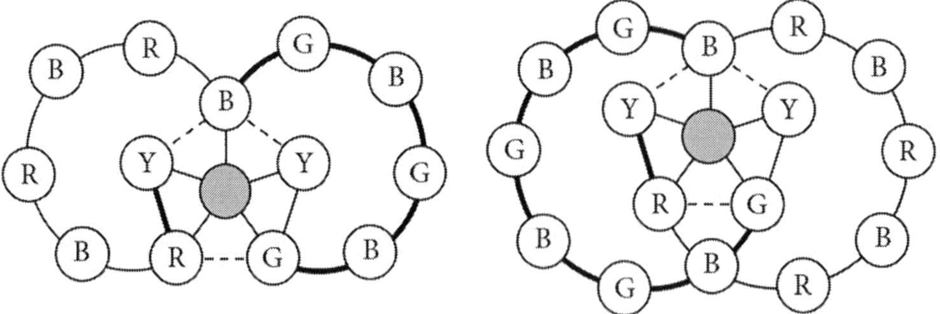

The above diagrams are incomplete; we still need to determine the styles of the radial edges (the edges connecting the vertex shaded gray, which has degree five). When we choose the styles of the radial edges, we run into a problem. We can't choose them so that every face has exactly one edge of each kind. One way to choose the edges is shown on the next page.

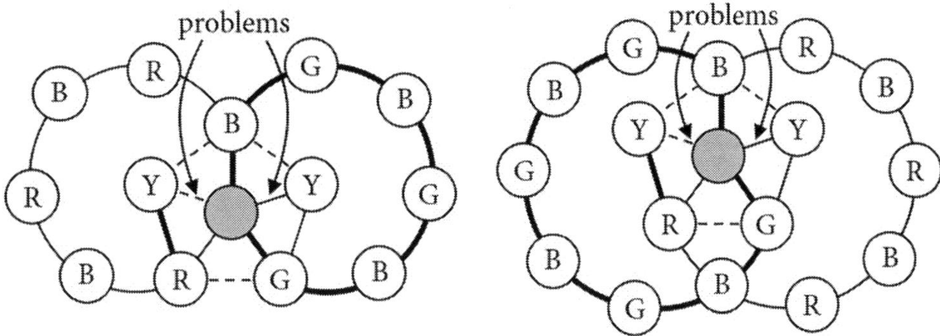

Note that it if we leave the edges along the pentagon alone and only modify the radial edges, it isn't possible to make every face have exactly one edge of each kind. Try it! One way to see this is to look at the radial edge connecting to G. The radial edge connecting to G can't be solid or dashed without making one of its faces have two solid or two dashed edges, yet if the radial edge connecting to G is a thick solid edge then the radial edge connecting to Y will make a face with two edges of the same kind. It doesn't matter how we assign the radial edges: no matter what, at least one face will have two edges of the same kind (until we start swapping edges, in which one of the edges along the pentagon will need to change).

Consider the diagrams above. The underlying problem is that the problem faces have two solid or two dashed edges, and these problem faces are separated by a closed loop of thick solid edges. As we discussed in the previous chapter, in order to recolor the MPG, we need to break open this closed loop of thick solid edges. We can do this by swapping one of the solid or dashed edges with a thick solid edge, as illustrated below.

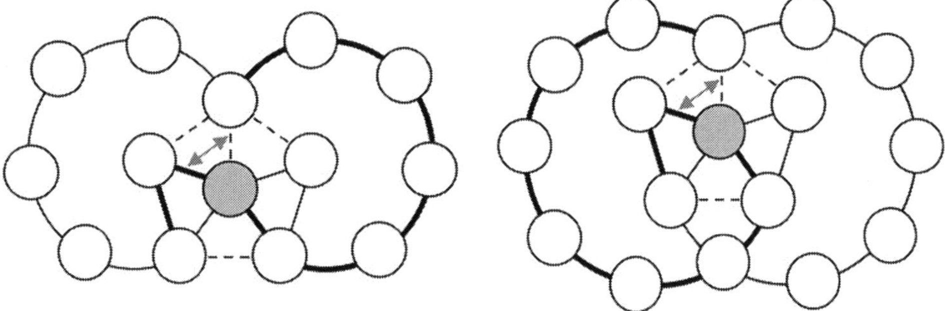

Now the problem faces aren't separated by a closed loop of the same kind of edge. We can recolor the graph using our edge swapping technique from the previous chapter.

In this way, we can recolor any vertex with degree five. We'll see this in the example that follows with the Fritsch MPG. You can also explore this with the Errera MPG in an exercise at the end of the chapter.

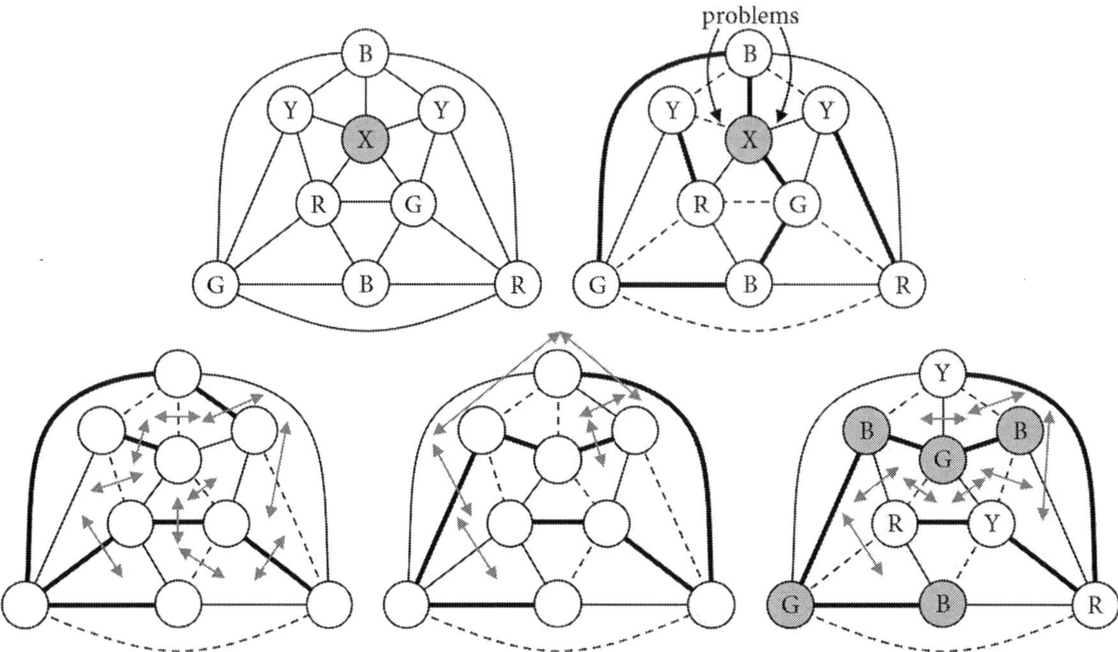

There is more than one way to recolor the graph above. Our solution above involves only three sets of swaps, though each set involves several edges. Other possible solutions involve fewer edges, but instead involve more sets of swaps. The solution can also get a little tricky due to multiple possible closed loops. Note that the original MPG on the top right already has two closed loops: one thick solid closed loop and another dashed closed loop. We broke both closed loops open in our solution above. A third closed loop could come about if you swap a solid edge into the top radial edge (B-X in the original graph) early in the solution. Any of these loops can cause a problem with recoloring attempts if they are not opened.

- We swapped thick solid edges and dashed edges to open the thick solid closed loop and the dashed loop, without introducing any new loops (bottom left).
- Next we swapped thick solid edges with solid edges (bottom center).
- Then we swapped solid edges with dashed edges (bottom right).
- Finally, we colored the thick solid edges with B-G and R-Y chains.

Let's look at the previous example in terms of our face-sliding recoloring method (Chapter 24) instead of our edge-swapping recoloring method. As before, we need to recolor the graph in order to accommodate vertex X. In the left graph below, we identified the B-G and R-Y edges as diagonals of quadrilaterals in order to pair the faces labeled 1-5. In order to pair the last two faces (labeled 6-7), we also treated B-X and G-X as diagonals of quadrilaterals. The underlying problem with the left graph is the closed loop of five thick solid edges. The arrows in the right graph show how we solved this problem with our face-sliding method. We recolored the graph with B-G and R-Y Kempe chains.

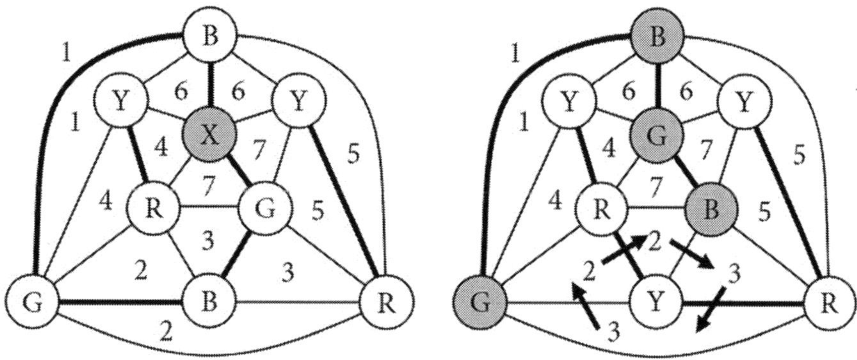

Chapter 25 Exercises

1. The Errera graph below is nearly four-colored, except for the vertex labeled X. The MPG needs to be recolored in order to accommodate vertex X. Use our edge-swapping method to recolor the MPG so that it is properly four-colored. (The top right graph below helps to visualize the crossed B-G and B-Y chains; multiple B's are used for both chains.) The blank graph is drawn six more times on the next page in case your solution involves more steps.

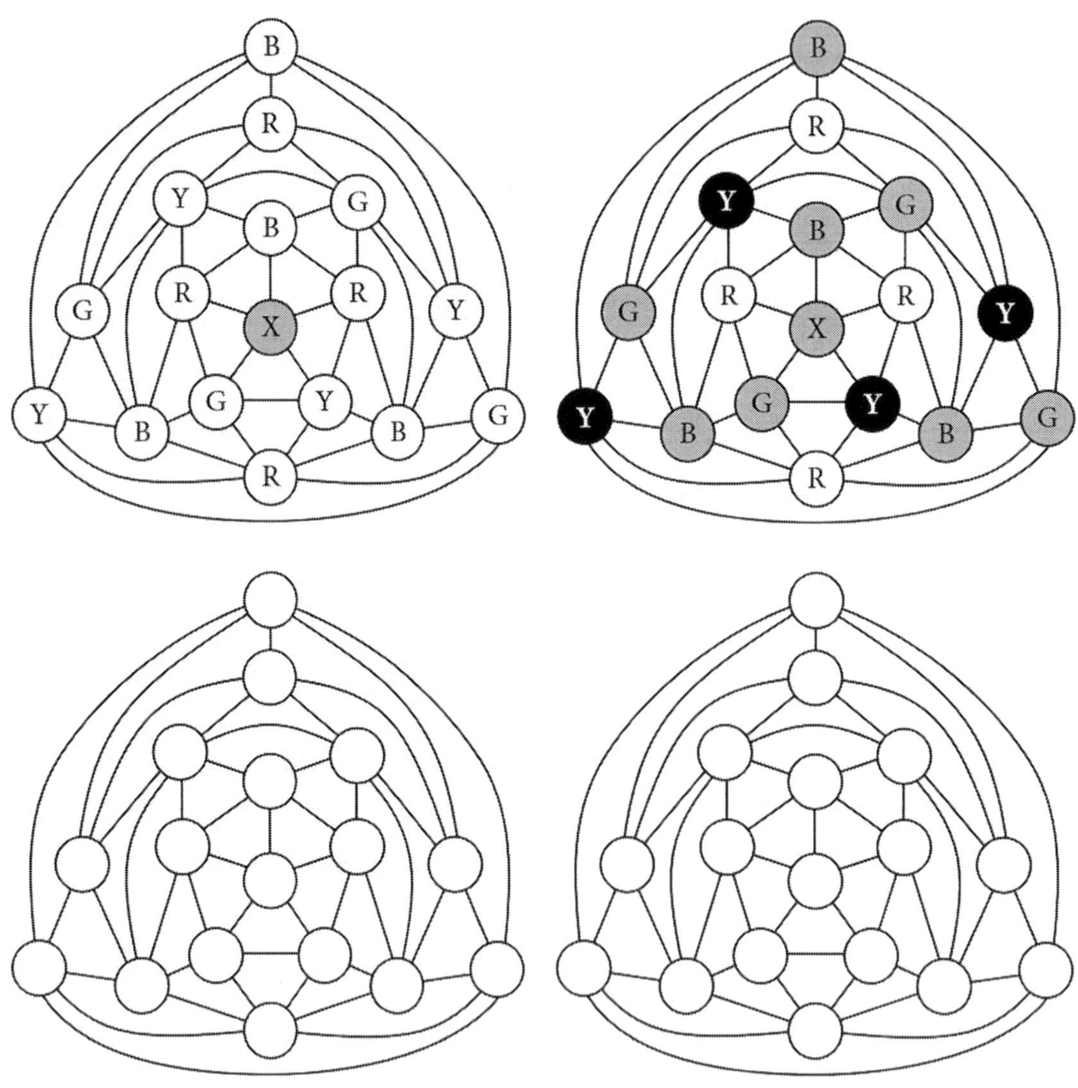

2. The Errera graph below is the same as from the previous problem. The difference is that this problem is asking you to use our face-sliding method to recolor the MPG.

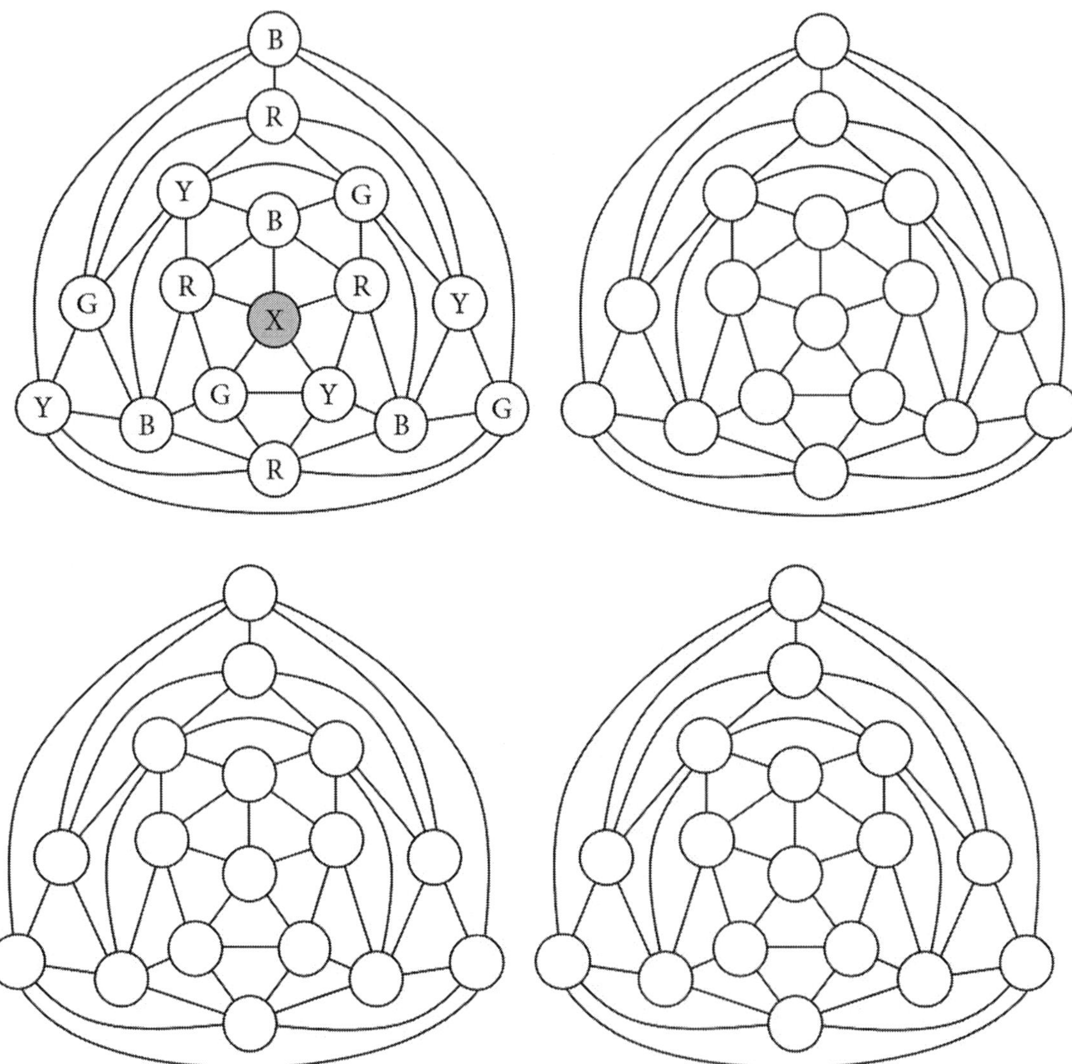

Challenge problem 1: Reformulate Kempe's argument for a vertex with degree four in terms of three types of edges instead of four colors of vertices.

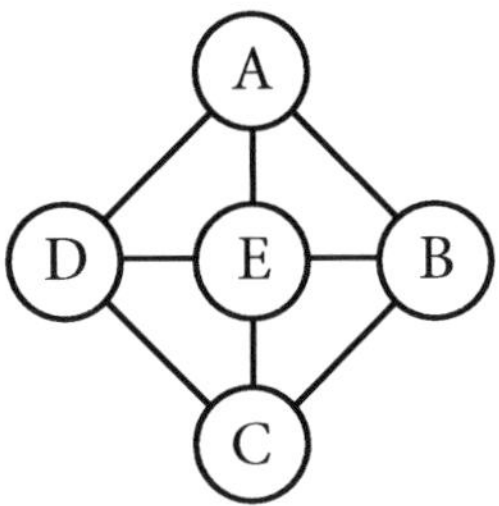

Challenge problem 2: Compare our edge-swapping method and our face-sliding method with the concept of reversing the colors of separated Kempe chains. Are these methods all equivalent? Explain. If not, which methods offers advantages over other methods, and what are these advantages?

Can you prove that the edge-swapping method (or the face-sliding method) will work for any possible recoloring of a planar graph that is four-colored except for one vertex labeled X? If so, prove this and use this proof to prove the four-color theorem. If not, describe the hurdles that make this approach challenging.

Note: The answer key doesn't include answers to the challenge problems. These problems are intended to encourage you to think about the ideas.

26 Degrees of Separation

Every face on a MPG shares an edge with three other faces. Every face thus has three nearest neighbors. This is the first degree of separation.

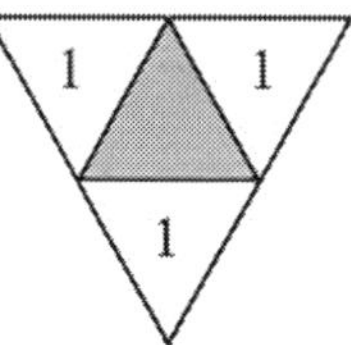

Each face's three nearest neighbors share a face with two other faces. In this way, there can be up to 6 faces in the second degree of separation, as shown below on the left. Why does it say "up to 6"? Because two nearest neighbors can share a mutual face, as shown below on the right. Note that we're numbering faces and not vertices; we're not yet drawing the circles for the vertices (so that we can focus on the faces for now). Also note that the numbers in the faces aren't colors; these numbers represent the degree of separation of each face relative to a particular face (the shaded face in the diagrams below).

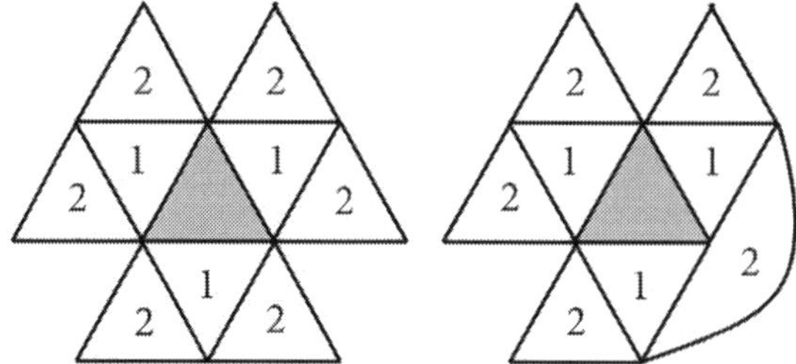

Let's look at the "outside" face which represents the infinite area outside and number every face according to its degree of separation from the outside face. Note that this is different from how we've been numbering faces in previous chapters. Right now we're numbering the degrees of separation (not pairing faces). It's easy to tell the difference: There are exactly two labels for each number when we're pairing faces; when you see more than two labels of the same number, we're numbering the degrees of separation. Also note that the word "degree" has multiple meanings. The degree of a vertex tells you how many edges connect to it, whereas a degree of separation tells you how many moves it takes to reach one face from another face. Henceforth, we will use the degree of separation specifically to indicate how many moves it takes to reach each interior face from the outside face.

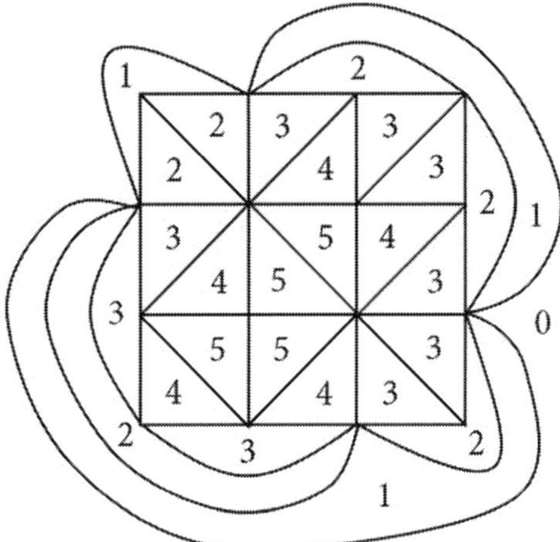

As shown above, the degrees of separation form layers. The first layer of every MPG has three 1's; since a MPG is triangulated, the outside face (like every other face) shares an edge with exactly three other faces. The highest numbers represent the deepest layer. Note that the numbers for a layer are not always contiguous; for example, none of the 4's in the graph above shares an edge with any of the other 4's.

The degrees of separation indicate how a MPG could be made. Imagine that the MPG had just 4 vertices originally (so it was K_4); this is shown as layer 1. The full MPG could be made one vertex at a time by splitting vertices (Chapter 18), working our way inward from layer 1 to the deepest layer. Each time we split vertices, the MPG gains 1 vertex, 3 edges, and 2 faces. As discussed in Chapter 18, the original vertex splits into two vertices, and the edge connecting the two vertices is the diagonal of a quadrilateral; the two new faces are the faces of this quadrilateral.

The diagram that follows shows how two faces in the deepest layer can form by vertex splitting. The thick solid edge is the diagonal of the quadrilateral containing the two new faces. The new MPG (on the right) has 1 more vertex, 3 more edges, and 2 more faces than the original MPG (on the left).

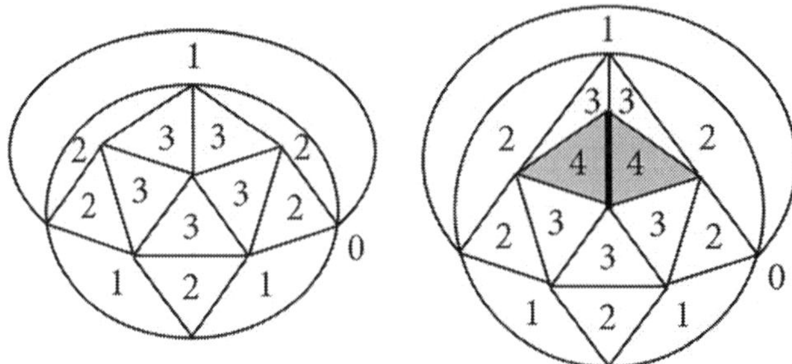

This illustrates how any MPG can be built "outside in." We can begin with K_4 (the three 1's of the outer layer plus face 0, which is the infinite area outside). The first time that we split vertices, we form a pair of 2's. The next time that we split vertices, we could form another pair of 2's, or we could form a 2 and a 3. Each 3 either forms with another 3, or it forms with a 2 or a 4. The last pair to form lies in the deepest layer.

This is one way that we can build a MPG (though it isn't the only way). This way is useful because if we visualize the MPG being built mostly according to the degrees of separation, it makes it easy to pair the faces. First we will use numbers to indicate the degrees of separation, and then we will use numbers to pair the faces. Again, it should be easy to tell the difference because only in the latter case will the numbers come in pairs.

We'll pair faces starting in the deepest layer and working our way outward. The left graph below shows degrees of separation, while the right graph shows paired faces. We first paired the faces in layer 4. When we paired the faces in layer 3, one of the 3's got paired with a 2. Similarly, for layer 2, two of the 2's paired with 1's. The final pair included a 1 and a 0. The thick solid edges below are the diagonals of quadrilaterals.

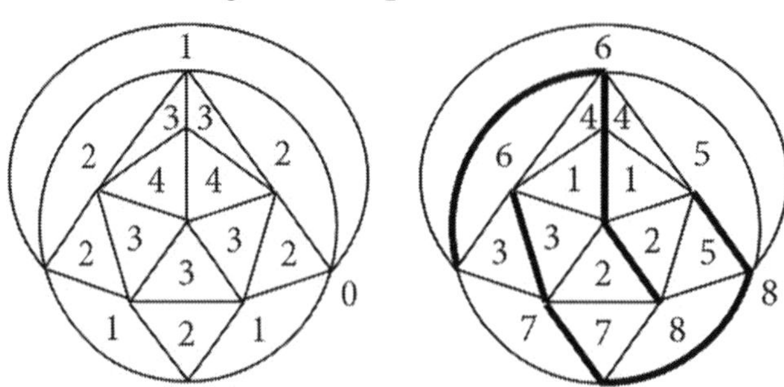

When we let the degrees of separation guide the pairing process in this manner, it helps to recognize how the process will end.

- The last pair will always be a 0 and a 1.
- Each of the other two 1's will always pair with a 2.

When you reach layers 3 and 2, it may be helpful to have this ending in mind. By planning for this ending, it might help to avoid making an island face.

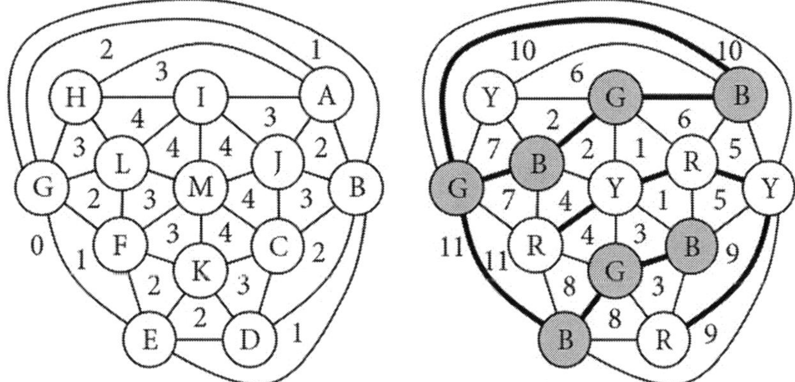

The example above shows how we can combine the concepts of degrees of separation and face pairing to color a MPG. We first numbered the degrees of separation from the outside to the inside (left figure). Next we paired faces starting with the deepest layers (the 4's) and working outward (right figure). These paired faces form quadrilaterals; the diagonals of the quadrilaterals appear as thick solid edges in the right graph. The thick solid edges form separated sections of Kempe chains. We colored these chains B-G and R-Y. (Note that the lone Y at the top left is a R-Y chain with a single vertex. We could have colored it R instead.)

Chapter 26 Exercises

1. The same graph appears twice below. Label the degrees of separation from the outside to the inside on the left graph. Use this to pair the faces from the inside to the outside on the right graph. Use the paired faces of the right graph to four-color the MPG.

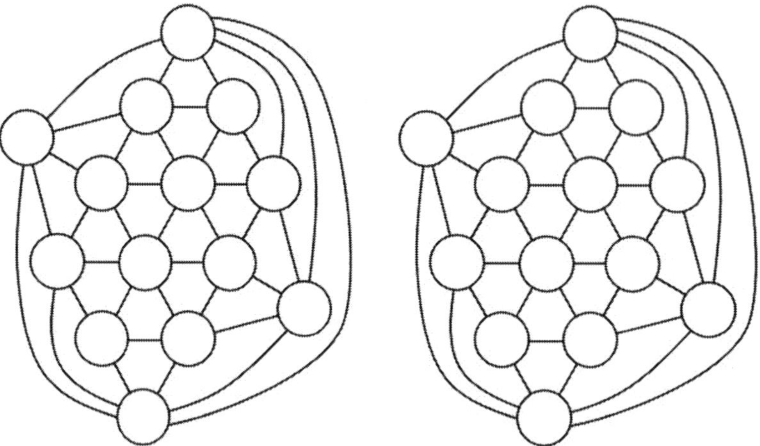

2. The same graph appears twice below. Label the degrees of separation from the outside to the inside on the left graph. Use this to pair the faces from the inside to the outside on the right graph. Use the paired faces of the right graph to four-color the MPG.

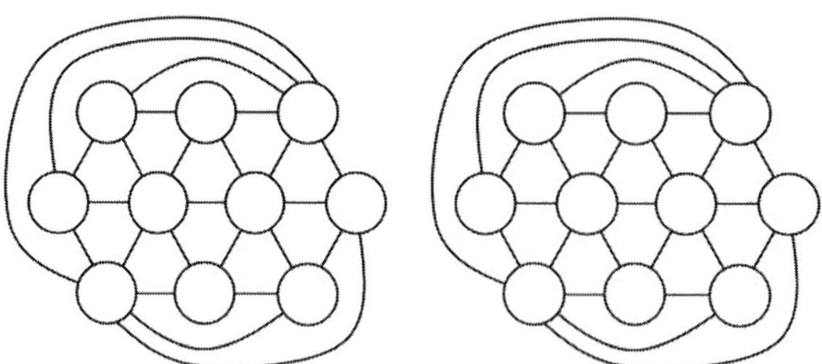

27 A Handwaving "Proof" of the 4CT

This chapter provides a novel handwaving argument for why the four-color theorem holds true. In its current form, this is not a formal or rigorous proof, although if Step 6 could be strengthened then the argument could be transformed into a proof. As it is, this argument will help to understand why the four-color theorem is true. This handwaving argument is much simpler and far more accessible than the only currently known proof [Ref. 14] which was computer-assisted.

Definitions:

- A **complete tetrapartite graph** is what we referred to as an U4CG in Chapter 16. It is a complete multipartite graph with four sets of vertices, where every vertex in one set connects to every vertex from the other sets, but where no vertex connects to the other vertices in its own set. The diagram that follows shows an example of a complete tetrapartite graph with $K = L = 3$ and $M = N = 2$. One set of vertices is colored blue, one set is colored green, one set is colored red, and the last set is colored yellow.

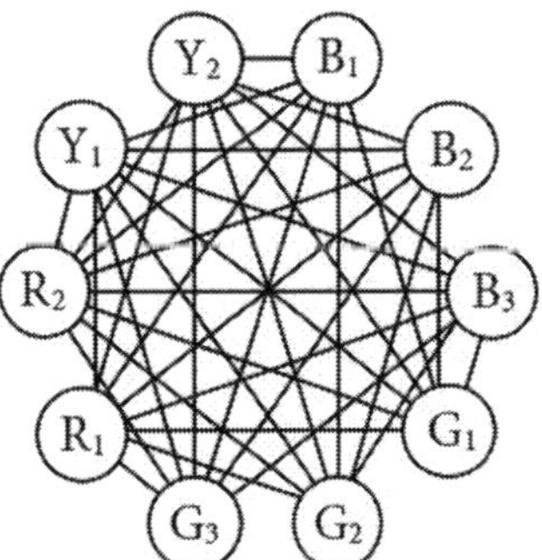

- A **maximal K_5-less graph** is a graph that does not have any K_5 subgraphs, for which adding a single edge between any two vertices (without making a double edge) would create a K_5 subgraph. (It is similar to how a "maximal planar graph" is a planar graph where adding a single edge between any two vertices, without making a double edge, would create a nonplanar graph.)

Step 1. Every complete tetrapartite graph is a maximal K_5-less graph, but not every maximal K_5-less graph is a complete tetrapartite graph.

- A complete tetrapartite graph is a maximal K_5-less graph because if you add one new edge (which isn't a double edge) it is guaranteed to form a K_5 subgraph. The new edge will join two vertices which previously had the same color. Each of these same-color vertices connects to all of the vertices of the other colors, which makes at least one K_5 subgraph (which is a set of five vertices where each vertex connects to the other four vertices). For example, suppose that a new edge connects Y_1 to Y_2. In this case, the $B_1G_1R_1Y_1$ and $B_1G_1R_1Y_2$ subgraphs (which are each K_4's) make a $B_1G_1R_1Y_1Y_2$ subgraph (which is a K_5). Of course, K_5 is not four-colorable; either Y_1 or Y_2 needs to change to a fifth color to avoid having an edge join two yellows.

- The diagram below shows an example of a complete tetrapartite graph where one new edge is added to form a K_5 subgraph. The added edge connects two red vertices (R_1 and R_2); one red was changed to color X since the new graph is not four-colorable. This forms multiple K_5 subgraphs, such as B_1-G_1-R_1-Y_1-X and B_1-G_1-R_1-Y_2-X.

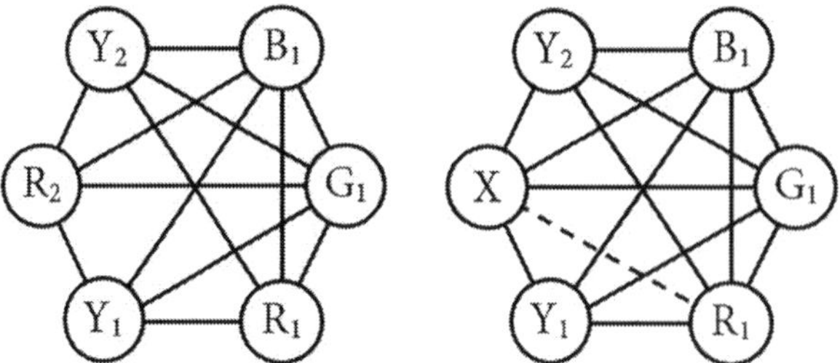

- Although every complete tetrapartite graph is a maximal K_5-less graph, the converse is not true; not every maximal K_5-less graph is a complete tetrapartite graph. The graph shown on the following page is a maximal K_5-less graph that isn't a complete tetrapartite graph. It is a maximal K_5-less graph because it doesn't have a K_5 subgraph and you can't add an edge (that isn't a double edge) without creating a K_5 subgraph, yet it isn't a complete tetrapartite graph because it isn't four-colorable.

- The graph on the following page doesn't have any K_5 subgraphs because there isn't a set of five vertices where all five vertices connect to all four of the other vertices.

For example, X_1, B_1, G_1, R_1, and Y_1 don't form a K_5 subgraph because X_1 doesn't connect to Y_1. Similarly, no set of five consecutive vertices form a K_5 subgraph since the first and last vertices of the set aren't connected. Any set of five nonconsecutive vertices will also not form a K_5 subgraph because at least one vertex will not connect to all four of the others. For example, B_1, Y_1, and R_2 aren't part of a K_5 subgraph because no other vertex connects to all three of these vertices. This shows that the graph below is K_5-less.

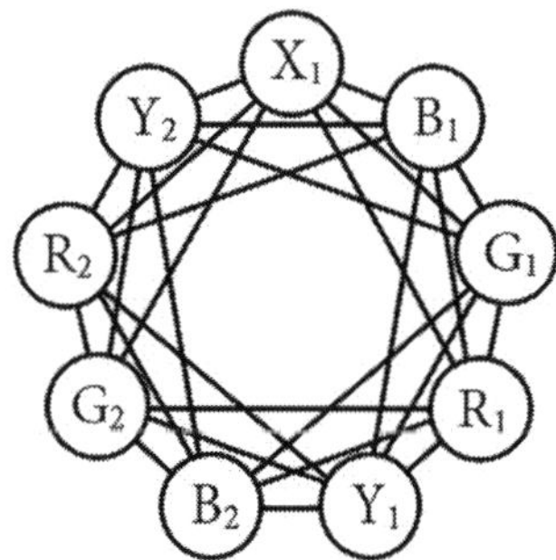

- The graph above is a maximal K_5-less graph because, in addition to being K_5-less, it isn't possible to add an edge (other than a double edge) without forming a K_5 subgraph. Adding edge B_1B_2 would make $B_1G_1R_1Y_1B_2$ a K_5 subgraph, adding B_1G_2 would make $B_1X_1Y_2R_2G_2$ a K_5 subgraph, adding G_1G_2 would make $G_1R_1Y_1B_2G_2$ a K_5 subgraph, adding G_1R_2 would make $G_1B_1X_1Y_2R_2$ a K_5 subgraph, adding R_1R_2 would make $R_1Y_1B_2G_2R_2$ a K_5 subgraph, adding R_1Y_2 would make $R_1G_1B_1X_1Y_2$ a K_5 subgraph, adding Y_1Y_2 would make $Y_1B_2G_2R_2Y_2$ a K_5 subgraph, adding Y_1X_1 would make $Y_1R_1G_1B_1X_1$ a K_5 subgraph, and adding B_2X_1 would make $B_2G_2R_2Y_2X_1$ a K_5 subgraph. Any other edge would make a double edge, which shows that the graph above is a maximal K_5-less graph.

- Every complete tetrapartite graph is four-colorable because it has four sets of vertices where no vertices within the same set are connected by an edge, which allows each set of vertices to be a different color. However, not every K_5-less graph is four-colorable (since not every K_5-less graph is a complete tetrapartite graph). For example, the maximal K_5-less graph above isn't four-colorable. To see this, note that B_1, G_1, and R_1 must all be different colors since they form a triangle. Then Y_1 is a fourth color because $B_1G_1R_1Y_1$ is a K_4 subgraph. A $G_1R_1Y_1B_2$ K_4 subgraph forces B_2 to be blue, a $R_1Y_1B_2G_2$ K_4 subgraph forces G_2 to be green, a $Y_1B_2G_2R_2$ K_4 subgraph forces R_2 to be red, and a $B_2G_2R_2Y_2$ K_4 subgraph forces Y_2 to be yellow. Since X_1 connects to B_1, G_1, R_1, and Y_2, a fifth color is needed for X_1. This shows that the graph on the previous page isn't a complete tetrapartite graph (since a complete tetrapartite graph must be four-colorable, but this graph isn't). This bullet point and the two previous bullet points combined together demonstrate that the graph on the previous page is a maximal K_5-less graph, but isn't a complete tetrapartite graph.

- Step 1 effectively states that the set of complete tetrapartite graphs is a subset of the set of maximal K_5-less graphs. That is, any complete tetrapartite graph is a maximal K_5-less graph, but a given maximal K_5-less graph may or may not be a complete tetrapartite graph.

Step 2. Every K_4-less graph can be made by removing edges from *some* maximal K_5-less graph.

- If we add edges to a K_4-less graph, at some point it will no longer be K_4-less and if we continue to add edges it will eventually have enough K_4 subgraphs that it is no longer possible to add another edge without making a K_5 subgraph. In this way, any K_4-less graph can be transformed into a maximal K_5-less graph by adding enough edges to it (in such a way as to avoid creating any K_5 subgraphs).
- The process described in the previous bullet point can then be carried out in reverse to show how the K_4-less graph can be formed by removing edges from a maximal K_5-less graph.
- The example on the next page shows a K_4-less graph on the left which was transformed into the maximal K_5-less graph on the right by adding edges. To see that the right graph is a maximal K_5-less graph, note that adding edge AE would make K_5 subgraphs (such as ABCDE), and similarly with edges BF and CG (which make BCDEF and CDEFG, for example) and edges CH and EH (which make HABCD and DEFGH).

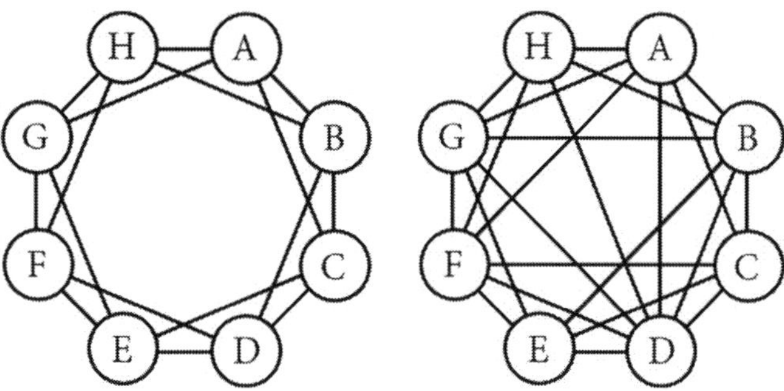

Step 3. A complete tetrapartite graph with K blue vertices, L green vertices, M red vertices, and N yellow vertices contains KLMN K_4 subgraphs. Every vertex is involved in a K_4 subgraph.

- KLMN means K times L times M times N. This is the number of K_4 subgraphs. For example, a complete tetrapartite graph with 5 blue vertices, 4 green vertices, 3 red vertices, and 2 yellow vertices contains $5(4)(3)(2) = 20(6) = 120$ K_4 subgraphs. These include $B_1G_1R_1Y_1$, $B_1G_1R_1Y_2$, $B_1G_1R_2Y_1$, $B_1G_1R_2Y_2$, $B_1G_1R_3Y_1$, $B_1G_1R_3Y_2$, $B_1G_2R_1Y_1$, and so on, until reaching $B_5G_4R_3Y_2$. The graph for this example appears below.

- The simplest complete tetrapartite graph, for which $K = L = M = N = 1$, is K_4. Every other complete tetrapartite graph has multiple K_4 subgraphs.

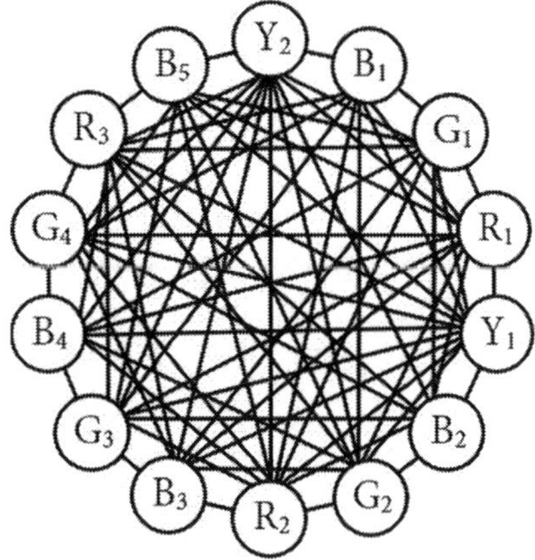

Step 4. For a complete tetrapartite graph with K blue vertices, L green vertices, M red vertices, and N yellow vertices, each blue vertex participates in LMN K_4 subgraphs, each green vertex participates in KMN K_4 subgraphs, each red vertex participates in KLN K_4 subgraphs, and each yellow vertex participates in KLM K_4 subgraphs.

- For example, if $K = 5$, $L = 4$, $M = 3$, and $N = 2$ (like the graph shown on the previous page), each blue vertex participates in $LMN = 4(3)(2) = 24$ K_4 subgraphs, each green vertex participates in $KMN = 5(3)(2) = 30$ K_4 subgraphs, each red vertex participates in $KLN = 5(4)(2) = 40$ K_4 subgraphs, and each yellow vertex participates in $KLM = 5(4)(3) = 60$ K_4 subgraphs. (There are a total of $KLMN = 5(4)(3)(2) = 120$ K_4 subgraphs in this example. Note that 24, 30, 40, and 60 do not add up to 120. Rather, the numbers work like this: Since there are 5 blue vertices and each blue vertex participates in 24 K_4 subgraphs, the 5 blue vertices are involved in all 120 K_4 subgraphs. Similarly, each green vertex participates in 30 K_4 subgraphs so that green vertices are involved in all 120 K_4 subgraphs. The red and yellow vertices are similarly involved in all 120 K_4 subgraphs.)

- Except for extreme cases (with a small number of vertices like $V = 6$ or where most of the values of K thru N are very small like $K = 5$, $L = 1$, $M = 1$, and $N = 1$), every vertex of a complete tetrapartite graphs participates in several K_4 subgraphs.

Step 5. If a PG can be made by removing edges from a complete tetrapartite graph, the values of K thru N will each be less than one-half of V.

- Recall that PG stands for "planar graph" even if the graph isn't drawn in planar form; it really stands for a "graph that can be drawn as a planar graph."

- We demonstrated this in the solution to Exercise 4 in Chapter 16. The main idea is that every edge in the MPG is the diagonal of a quadrilateral since every edge is shared by two triangles, and no more than two of the four vertices of any such quadrilateral can be the same color. As a result, no single color can have as many as $V/2$ vertices.

- Since any PG can be formed by removing edges from a suitable MPG, and since a four-coloring that works for a MPG will also work for a PG (as explained in Chapter 3), it follows that the previous bullet point must apply to all PG's, not just to MPG's.

- For example, consider a complete tetrapartite graph with $V = 14$. Although we can make a complete tetrapartite graph with $K = 11$ and $L = M = N = 1$ or a complete tetrapartite graph with K = 7, L = 4, M = 2, and N = 1, neither of these can make a MPG by removing edges from them. We can make K = 6, L = 6, M = 1, N = 1, we can make K = 6, L = 4, M = 2, and N = 2, and many other combinations where all of the values of K thru N are less than 7.

- The value of Step 5 is that we don't need to consider extreme cases like K = 17, L = 1, M = 1, and N = 1 if our goal is to attempt to make all possible MPG's by removing edges from complete tetrapartite graphs. In that case, we won't need to consider any case where a single value of K thru N is as large as V/2. The most extreme (in terms of unevenness) we will need to consider is when K and L are just slightly below one-half and M = N = 1, such as K = 9, L = 9, M = 1, and N = 1.

Step 6. Every K_4-less PG can be made by removing edges from *some* complete tetrapartite graph.

- This is the only step that will involve some handwaving. We will not prove this step formally. We will make an argument to try to explain why it is so. (A more rigorous demonstration of this step, if it could be achieved, could transform this argument into an actual proof.)
- We will use ideas from Steps 1-5 in our argument.
- First, note how Step 6 is modifying the statement made in Step 2. According to Step 2, which is easily seen to be true, every K_4-less graph can be made by removing edges from some maximal K_5-less graph. Step 6 modifies this statement by replacing "every K_4-less graph" with "every K_4-less PG" and replacing "some maximal K_5-less graph" with "some complete tetrapartite graph." Recall from Step 1 that complete tetrapartite graphs are maximal K_5-less graphs (but not all maximal K_5-less graphs are complete tetrapartite graphs).
- The main idea is that except for a very small number of vertices (like 4 thru 9), which can easily be checked by hand, since a complete tetrapartite graph (being a maximal K_5-less graph) has several K_4 subgraphs (equal to KLMN according to Step 3) with every vertex participating in several K_4 subgraphs (equal to LMN, KMN, KLN, or

KLM according to Step 4), it is necessary to remove several edges involving many of the vertices in order to make a K_4-less graph from a complete tetrapartite graph with the same number of vertices. This means that the difference between a K_4-less graph and a complete tetrapartite graph is very significant. In contrast, for any maximal K_5-less graph that isn't a complete tetrapartite graph, there will be multiple complete tetrapartite graphs rather similar to it. Since every complete tetrapartite graph is a maximal K_5-less graph, for any K_5-less graph that isn't a complete tetrapartite graph, you can always find complete tetrapartite graphs that resemble the maximal K_5-less graph far more than they resemble a K_4-less graph that could be made by removing edges from the maximal K_5-less graph.

- Since Step 2 states that every K_4-less graph can be made by removing edges from some maximal K_5-less graph, it follows that every K_4-less PG can be made by removing edges from some maximal K_5-less graph (since every K_4-less PG is a K_4-less graph). For Step 6, we just need to narrow the last part of this statement down from "some maximal K_5-less graph" to "some complete tetrapartite graph." Since every complete tetrapartite graph is a maximal K_5-less graph according to Step 1, we just need to show that there isn't a K_4-less PG that could only be made by removing edges from a maximal K_5-less graph that isn't a complete tetrapartite graph. The main idea stated in the previous bullet point addresses this issue. To reiterate, for any maximal K_5-less graph which isn't a complete tetrapartite graph, there will complete tetrapartite graphs that are fairly similar to it in comparison to how these graphs contrast with the K_4-less PG. To make any K_4-less PG from a maximal K_5-less graph (whether or not it is a complete tetrapartite graph), it will generally be necessary to remove many edges involves many of the vertices. These many removed edges involving many of the vertices make it possible to compensate for the small difference between maximal K_5-less graphs that are fairly similar (for the specific case where it is known that a K_4-less PG can be made by removing edges from a maximal K_5-less graph that isn't a complete tetrapartite graph and it needs to be determined whether or not the K_4-

less PG can also be made by removing edges from a complete tetrapartite graph), as discussed in the next bullet point.

- It may help to consider the process in reverse like we did for Step 2. Let's begin with a K_4-less PG and proceed to add edges to it in order to form a maximal K_5-less graph. As mentioned previously, in general we will need to add many edges involving many of the vertices. This gives us freedom in which edges to add. There are several different maximal K_5-less graphs that we can form by adding edges to a K_4-less PG. We just need *one* of these several different maximal K_5-less graphs to be a complete tetrapartite graph.

- Another point that aids this argument is that for a given number of vertices, there are multiple complete tetrapartite graphs. For example, for $V = 24$ vertices, you can draw a complete tetrapartite graph with $K = L = M = N = 6$, with $K - 10$, $L - 8$, $M = 4$, and $N = 2$, with $K = L = 9$ and $M = N = 3$, and many other combinations. For a set value of V, we're not relying on a single complete tetrapartite graph to work. We have several to choose from.

- Let's look at some examples with numbers. It's pretty easy to check that any K_4-less PG with a small number of vertices (like up to 12 or so) can be made by removing edges from a complete tetrapartite graph. When $V = 4$, we get $K = L = M = N = 1$ and the corresponding complete tetrapartite graph is itself K_4, the same as a MPG with $V = 4$, so any PG with $V = 4$ can clearly be made by removing edges from this complete tetrapartite graph. For values of V from 5 to around 12, it is easy to check that a MPG with the same number of vertices can be made by removing edges from a complete tetrapartite graph, such that any K_4-less PG can also be made this way.

- For $V = 12$, a complete tetrapartite graph could have $K = L = M = N = 3$, it could have $K = L = 5$ and $M = N = 1$, or any combination in between these extremes. (From Step 5, we don't need to worry about any values of K thru N as large as 6.) For $K = L = M = N = 3$, there are $KLMN = 3^4 = 81$ K_4 subgraphs and every vertex participates in 27 K_4 subgraphs. For the other extreme, $K = L = 5$ and $M = N = 1$, there are KLMN = 25 K_4 subgraphs, every blue and green vertex participates in 5 K_4 subgraphs, and every red and yellow vertex participates in all 25 K_4 subgraphs. For K = L = M = N = 3, the complete tetrapartite graph has 54 edges, while for K = L = 5 and M = N = 1 the complete tetrapartite graph has 46 edges (see Chapters 16-17). In comparison, a MPG with V = 12 has E = 3V – 6 = 30 edges, so that 16 to 24 edges need to be removed from a complete tetrapartite graph with 12 vertices in order to form a MPG (and even more edges may need to be removed in order to form a K_4-less PG). Thinking about the process in reverse, imagine starting with a K_4-less PG with 12 vertices and adding 16-24 (or more) edges to it in order to make a maximal K_5-less graph. There are so many ways to do this that many different maximal K_5-less graphs can be made this way. We just need one of these to be a complete tetrapartite graph. Given that any maximal K_5-less graph that isn't a complete tetrapartite graph will have some complete tetrapartite graphs that are fairly similar to it, with a choice of where to put 16 or more edges, it is unthinkable that it won't be possible to make a complete tetrapartite graph this way.

- As the number of vertices increases, the numbers become increasingly compelling. For example, for V = 40 and K = L = M = N = 10, there are KLMN = 10^4 = 10,000 K_4 subgraphs and every vertex participates in 1000 K_4 subgraphs. This complete tetrapartite graph has 600 edges. A MPG with V = 40 has E = 3V – 6 = 114 edges. This means that 486 edges would be removed from this complete tetrapartite graph to form a MPG (and at least as many edges would be removed to form a K_4-less PG).

- In the following example, the graph on the left is a complete tetrapartite graph with V = 8 vertices, where K = 3, L = 2, M = 2, and N = 1. There are KLMN = 3(2)(2)(1) = 12 K_4 subgraphs and E = KL + KM + KN + LM + LN + MN = 3(2) + 3(2) + 3(1) + 2(2) + 2(1) + 2(1) = 6 + 6 + 3 + 4 + 2 + 2 = 23 edges. The graph on the right is a K_4-less PG (even though it is not shown in planar form) that can be formed by removing 7 edges from the complete tetrapartite graph on the left. For the left graph, every one of the 8 vertices participates in at least 4 (and the yellow vertex participates in all 12) K_4 subgraphs. For the graph on the right, no vertex participates in a K_4 subgraph.

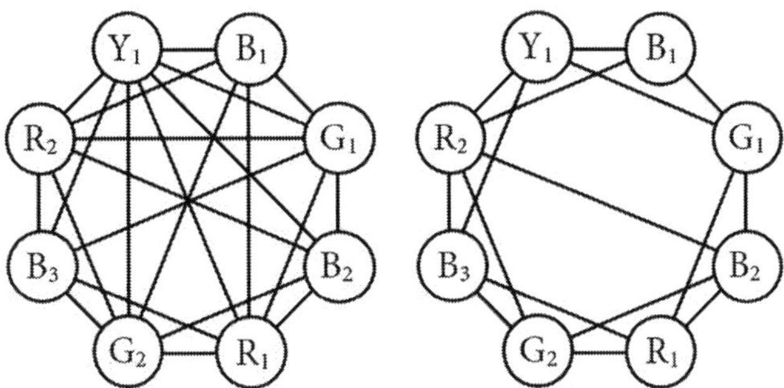

- In order to disprove Step 6 and also disprove the four-color theorem, you would need to find a K_4-less PG where it isn't possible to add edges to it in such a way as to form a complete tetrapartite graph. For such a hypothetical K_4-less PG, adding edges to it could only form maximal K_5-less graphs that aren't complete tetrapartite graphs. Let's look at a visual example that may help to convince you that no such K_4-less PG exists. The left graph below is a K_4-less PG with $V = 9$ vertices and $E = 21$ edges. We added 7 edges (BE, BG, CF, CG, DF, DH, and EH) to make the maximal K_5-less graph shown on the right. It is K_5-less since neither G_1 nor R_1 connect to X_1 and G_2 doesn't connect to R_2 (so no GR pair is part of any K_5) and it is maximal since adding edge AD, AG, BF, BH, CH, CI, DG, or FI would create a K_5 subgraph. The right graph is not a complete tetrapartite graph since it is not four-colorable. Once we set the colors B_1, G_1, and R_1 (which must be three different colors), the colors of Y_1, R_2, B_2, B_3, and G_2 follow, at which point the last vertex must be a fifth color (X_1).

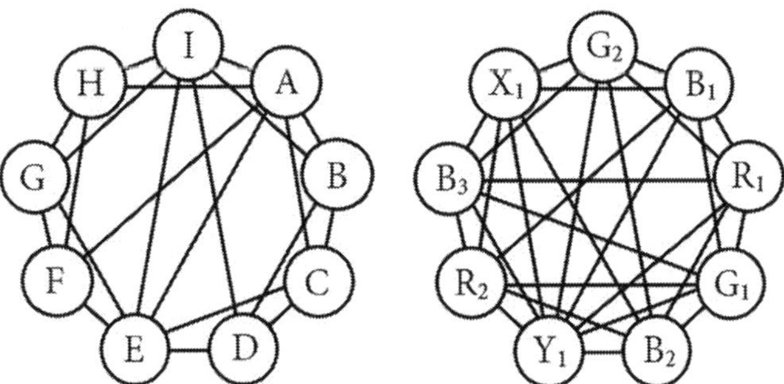

- However, that is just one of multiple ways to add edges to the left graph to form a maximal K_5-less graph. The middle and right graphs on the following page show two more ways to add edges to the left graph to form maximal K_5-less graphs (there are yet other ways, too). The middle and right graphs are complete tetrapartite graphs.

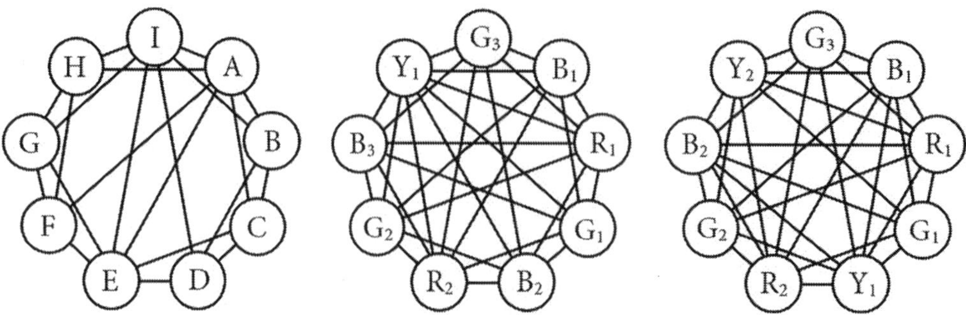

The diagrams above show that for this example (which has a mere 9 vertices), there are multiple ways to form a complete tetrapartite graph from a K_4-less graph. Even though the first attempt (on the previous page) led to a maximal K_5-less graph that wasn't a complete tetrapartite graph, there were at least two ways to make a complete tetrapartite graph. (If you visualize this in reverse, it means that we could start with either complete tetrapartite graph and remove edges to form the given K_4-less PG.) This example (with just 9 vertices) shows two complete tetrapartite graphs which are both fairly similar to each other and are also fairly similar to a maximal K_5-less graph which isn't a complete tetrapartite graph; they are far more similar to one another than they are to the K_4-less PG. The middle complete tetrapartite graph has 29 edges and the right complete tetrapartite graph has 30 edges, so in this example we added 8 to 9 edges to the K_4-less PG to form a complete tetrapartite graph. For K_4-less PG's with more vertices, there will be many more edges to work with; this example with 9 vertices has a relatively small number of edges, yet these 8-9 edges provided plenty of combinations for how to add edges to the K_4-less PG; even though a maximal K_5-less graph that wasn't a complete tetrapartite graph was encountered, at least two of the multiple variations available offered enough flexibility to make complete tetra-partite graphs as well. As the number of vertices increases, the number of edges added grows rapidly, which offers more flexibility and more combinations.

- Finally, let's note the importance of Step 5, which states that for Step 6 we only need to consider cases where values of K thru N are each less than one-half of V. One way that this helps is that we don't need to worry about cases where K thru N have one very large value and three very small ones like K = 17 and L = M = N = 1. Such an extreme lopsided distribution would put a strain on where to add the edges to a K_4-less graph to form a complete tetrapartite graph. Step 5 states that we don't need to worry about cases like K = 17 and L = M = N = 1 because for V = 20, the highest K (or L, M, or N) can be is 9 (since it must be less than V/2 = 20/2 = 10). The most lopsided distribution we can have is K = L = 9 and M = N = 1.

Step 7. Every MPG that does not contain a ST does not contain a K_4 subgraph.

- Recall that MPG stands for "maximal planar graph." A MPG is a special kind of PG in that a MPG is triangulated, meaning that every face has three edges (Chapter 6).
- Recall that ST stands for "separating triangle." We discussed ST's in Chapter 12.
- We discussed Step 7 in Chapter 12. Any K_4 subgraph in a MPG contains at least one ST. In the simplest case, like the graph shown below, the ST contains a single vertex. However, in general the ST may contain numerous vertices, like the examples given in Chapter 12.

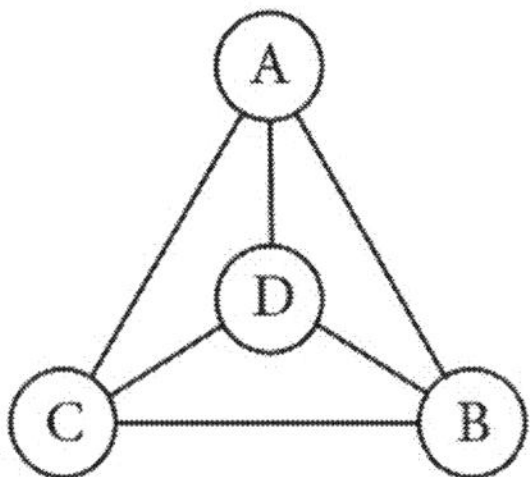

Step 8. Every MPG that does not contain a ST can be made by removing edges from *some* complete tetrapartite graph.

- Step 8 simply combines Steps 6-7 together.
- According to Step 6, every K_4-less PG can be made by removing edges from some complete tetrapartite graph. According to Step 7, every MPG that does not contain a ST does not contain a K_4 subgraph (which is what we have referred to as K_4-less). Since a MPG is a kind of PG, it follows that every MPG that does not contain a ST can be made by removing edges from some complete tetrapartite graph.

Step 9. Every complete tetrapartite graph is four-colorable.

- A complete tetrapartite graph is what we referred to as a U4CG (which stands for "ultimate four-colorable graph") in Chapter 16.

- A complete tetrapartite graph has four sets of vertices, where every vertex in one set connects to every vertex from the other sets, but where no vertex connects to the other vertices in its own set. Each of the four sets of vertices may thus have a different color, since no vertices with the same color will be connected by an edge.

Step 10. Every MPG that does not contain a ST is four-colorable.

- Step 10 combines Steps 8-9 together.

- Step 10 applies the same logic that motivates the idea of triangulation (Chapter 3). If a graph is four-colored and edges are removed from the graph, the same coloring works for the new graph with edges removed. We discussed this idea in Chapter 3 and included an example (in the context of triangulation).

- According to Step 9, every complete tetrapartite graph is four-colorable. According to Step 8, every MPG that does not contain a ST can be made by removing edges from some complete tetrapartite graph. It follows that every MPG that does not contain a ST is four-colorable.

Step 11. Every MPG is four-colorable.

- Step 11 follows from Step 10. We discussed Step 11 in Chapter 12 and included an example.

- As discussed in Chapter 12, any MPG that has a ST can be split into two separate MPG's. If each of the smaller MPG's is individually four-colorable, it follows that the original MPG is also four-colorable. If every MPG which does not contain a ST is four-colorable, it follows that every MPG is four-colorable.

Step 12. Every PG is four-colorable.

- Step 12 follows from Step 11. Step 12 applies the same logic that motivates the idea of triangulation (Chapter 3) and the same logic that we applied in Step 10.

- Any PG which is not a MPG can be formed by removing edges from some MPG.

- If a MPG is four-colored and edges are removed from the MPG, the same coloring works for the new PG with edges removed. We discussed this idea in Chapter 3 and included an example.
- Step 12 states the four-color theorem.

Steps 1-12 do not form a formal or rigorous proof of the four-color theorem since Step 6 involves a handwaving argument, but they should convince you that the theorem is true. Following are a couple of additional notes:

- **<u>Exercise 2, Exercise 4, and Challenge Problem 1 will help you appreciate Step 6.</u>**
- Complete tetrapartite graphs with more than five vertices are nonplanar. Although they do not contain K_5 subgraphs, they may have subgraphs that are subdivisions of K_5 or $K_{3,3}$ (in accordance with Kuratowski's theorem; see Chapter 6). When Steps 1-12 mention K_5-less graphs, they mean graphs that do not contain K_5 subgraphs; they are not concerned with subdivisions.
- When Steps 1-12 mention adding edges to a graph, they mean adding edges in such a way as not to create any double edges. Suppose, for example, that vertex D already connects to vertex J. If a new edge is added from D to J, that would then create a double edge. Steps 1-12 do not allow adding edges to form double edges.

Chapter 27 Exercises

1. For each graph below, state whether or not it is:

- a K_5-less graph

- a maximal K_5-less graph

- a complete tetrapartite graph.

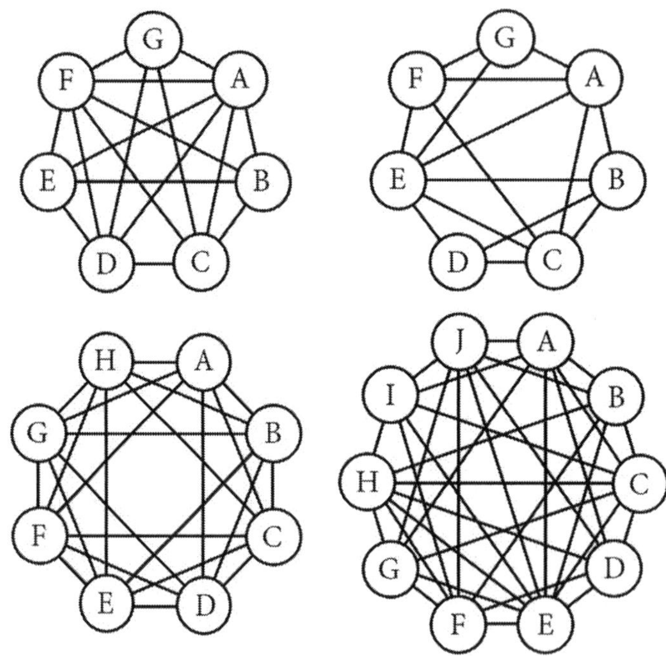

2. Draw a graph with 8 vertices which is a maximal K_5-less graph, but which isn't a complete tetrapartite graph. **Exercises 2, 4, and Challenge Problem 1 will help you appreciate Step 6.**

3. Draw a graph which does not contain a K_5 subgraph, but which does contain a subgraph that is a subdivision of K_5, which is four-colorable. (Of course, this graph will be nonplanar.)

Draw a graph which does not contain a K_5 subgraph, but which does contain a subgraph that is a subdivision of K_5, which is not four-colorable.

4. Add edges to the K_4-less PG below to make each of the following nonplanar graphs:

- a maximal K_5-less graph that isn't a complete tetrapartite graph
- a complete tetrapartite graph

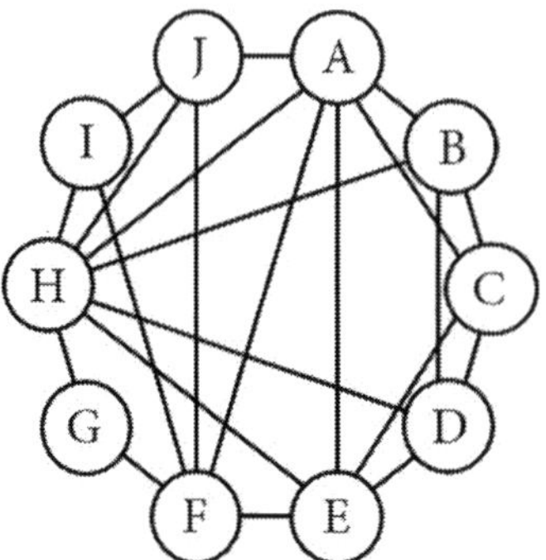

5. A complete tetrapartite graph has 8 blue vertices, 6 green vertices, 5 red vertices, and 3 yellow vertices. How many K_4 subgraphs does it contain? For a vertex of each color, indicate how many K_4 subgraphs such a vertex participates in.

6. For each MPG below, show that it can be four-colored by removing edges from a suitable maximal tetrapartite graph.

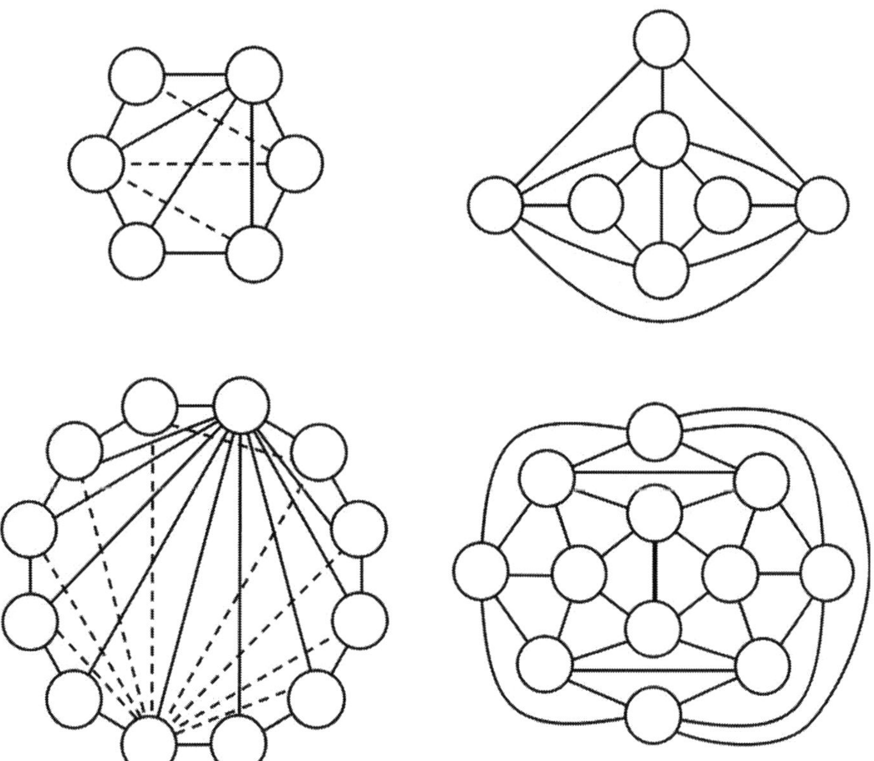

Challenge problem 1: Is it possible to draw a graph with 7 vertices that is a maximal K_5-less graph, yet which is not a complete tetrapartite graph? If yes, draw an example. If no, show that it isn't possible.

If you believe the answer is yes, check your solution thoroughly and carefully. When trying to draw a maximal K_5-less graph (with any number of vertices) that isn't a complete tetrapartite graph, it's really easy to inadvertently draw a graph that (A) contains a K_5 subgraph, (B) is actually four-colorable, or (C) makes it possible to add an edge without making a K_5 subgraph (or a double edge). Rest your eyes, refresh your brain, and check again. Inspect your graph for a possible K_5 subgraph (a K_5-less graph can't have one), recolor the graph to ensure that it isn't four-colorable (if it is four-colorable, it's a tetrapartite graph, but we're trying to make a graph that isn't a tetrapartite graph), and try adding every possible edge to check that every single added edge (which isn't a double edge) always makes a K_5 subgraph (otherwise it wouldn't be "maximal").

Note: The answer key doesn't include answers to the challenge problems. These problems are intended to encourage you to think about the ideas.

Challenge problem 2: Devise a three-coloring argument that parallels the 12-step hand-waving argument for the four-color theorem given in this chapter. Your argument should mention tripartite graphs and discuss distinctions between classes of similar graphs (such as how the 12-step argument of this chapter discussed the distinction between maximal K_5-less graphs and complete tetrapartite graphs). A "tripartite graph" is similar to a tetrapartite graph except that it has three sets of vertices instead of four.

Explain how your argument relates to the graph below which does not contain a K_4 subgraph, yet isn't three-colorable. (Note that we saw this same graph in the last problem of Chapter 10.)

Challenge problem 3: Either come up with a way to strengthen the argument given for Step 6 in the 12-step argument of this chapter or describe what makes this very challenging.

Note: The answer key doesn't include answers to the challenge problems. These problems are intended to encourage you to think about the ideas.

Challenge problem 4: For each statement (A) thru (Z) below, determine if the statement is:

- definitely true
- definitely false
- probably true
- probably false
- possibly true or false, but difficult to determine

If your answer is "definitely true," either prove the statement or offer a supportive argument.

If your answer is "definitely false," provide a counterexample to demonstrate this. For any

other answer, describe what makes the answer difficult to determine and explain the reason

for your answer.

(A) Every complete tetrapartite graph is K_5-less.

(B) Every complete tetrapartite graph contains a K_4 subgraph.

(C) Every graph that contains a K_5 subgraph is not a complete tetrapartite graph.

(D) Every maximal planar graph is K_4-less.

(E) Every planar graph is K_5-less.

(F) Every K_4-less graph can be made by removing edges from a K_5-less graph.

(G) Every K_5-less graph can be formed by removing edges from a complete tetrapartite graph.

(H) Every planar graph can be made by removing edges from a K_5-less graph.

(I) Every K_4-less planar graph can be made by removing edges from a complete tetrapartite

graph.

(J) Every planar graph can be made by removing edges from a maximal planar graph.

(K) Every complete tetrapartite graph is nonplanar.

(L) Every K_5-less graph is planar.

(M) Every graph that contains a K_4 subgraph is planar.

(N) Every graph that contains a K_5 subgraph is nonplanar.

(O) Every maximal planar graph that contains a separating triangle contains a K_4 subgraph.

(P) Every maximal planar graph that contains a K_4 subgraph contains a separating triangle.

(Q) Every maximal planar graph contains a separating triangle.

(R) Every maximal planar graph has a Hamiltonian cycle.

(S) Every maximal planar graph that does not contain a separating triangle has a Hamiltonian cycle.

(T) Every K_4-less graph has a Hamiltonian cycle.

(U) Every K_4-less graph is four-colorable.

(V) Every K_5-less graph is four-colorable.

(W) Every K_5-less graph is five-colorable.

(X) Every complete tetrapartite graph is four-colorable.

(Y) Every graph that does not contain a subgraph that is a subdivision of K_5 is four-colorable.

(Z) Every planar graph is four-colorable.

Note: The answer key doesn't include answers to the challenge problems. These problems are intended to encourage you to think about the ideas.

28 Random Notes

Topics

28.1 No Vertices with Degree Two

A MPG with at least 4 vertices can't have any vertices with degree two. This has to do with triangulation. Take any PG that has a vertex with degree two and fully triangulate the graph, and the resulting MPG won't have a vertex with degree two (if it has at least 4 vertices).

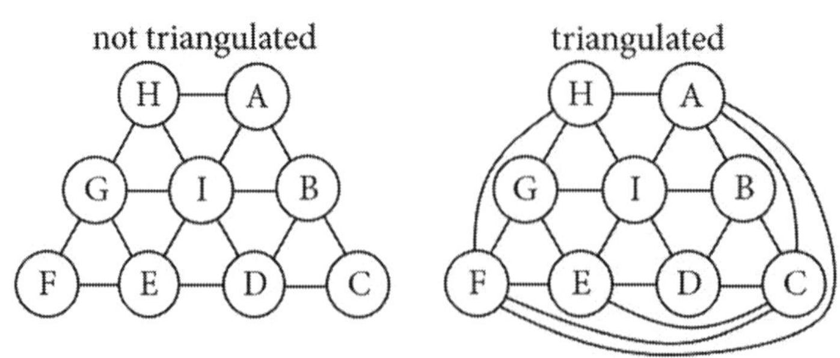

In the previous figure, the PG on the left had two vertices (C and F) which had been degree two. When the PG is triangulated, these vertices are no longer degree two, as shown in the example MPG on the right. You can't avoid this. For example, you can't connect an edge from B to D around C because that would result in a double edge (BD is already drawn).

Let's look at it another way. Imagine in the diagram below that we have zoomed in on a larger graph and are looking at face ABC, and that your goal is to leave vertex A degree 2. You can only have three vertices on the "outside" for the infinite area outside to be triangulated, so eventually vertices A thru E need to get "closed" in by adding more edges. You can't loop over A, closing it off, by adding a curve from B to C because B and C are already connected (you can't double connect them). If you connect C to D by looping over A, you will still need to connect A to D in order for every face to be a triangle. Similarly, if you connect C to F and F to B, you will still need to connect A to F in order for every face to be a triangle. You should be able to convince yourself that vertex A can't remain degree two.

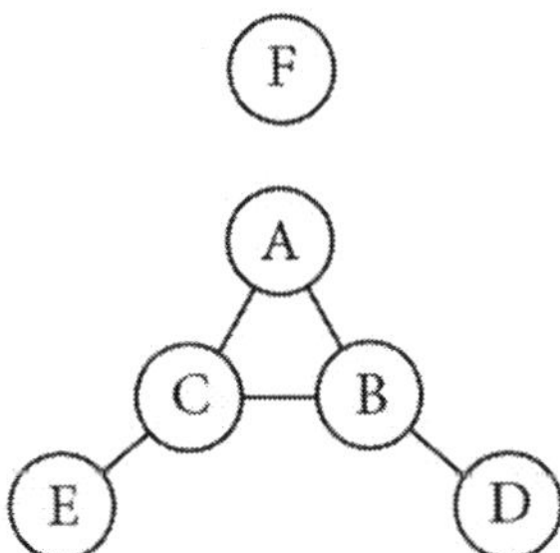

The only exception occurs when V < 4. That's because there aren't enough vertices available to create more than one triangle. For V = 3, the MPG is a triangle and each vertex has degree two. (For V = 2, which is technically not a MPG because it doesn't have a triangular face, the PG consists of two vertices joined by an edge, for which each vertex has degree one).

Another way to demonstrate this convincingly will be explored in the problems at the end of the chapter. Until then, can you think of a way to prove this convincingly?

28.2 Degree Three Vertices Don't Connect

If a MPG with at least 5 vertices has two or more vertices with degree three, these vertices with degree three don't connect to one another (they don't share any common edges). The only exception occurs when $V = 4$, shown below. You should recognize this graph as K_4. In this case, all four vertices are degree three.

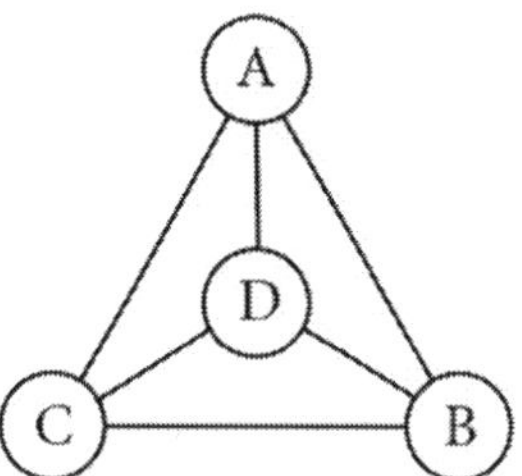

If you try to add any new vertices to this graph, no two vertices with degree 3 will be connected. For example, imagine adding a single vertex in the face ABD. To triangulate the graph, that new vertex would have to connect to A, B, and D, and then only vertex C and the new vertex would be degree 3, but they don't share an edge. A similar result occurs if you add a new vertex "outside" the graph since the infinite area outside must also be triangulated.

To help visualize why no two vertices with degree 3 can share an edge in a MPG, imagine trying to draw a graph with more than 4 vertices which doesn't satisfy this rule. Consider the diagrams below.

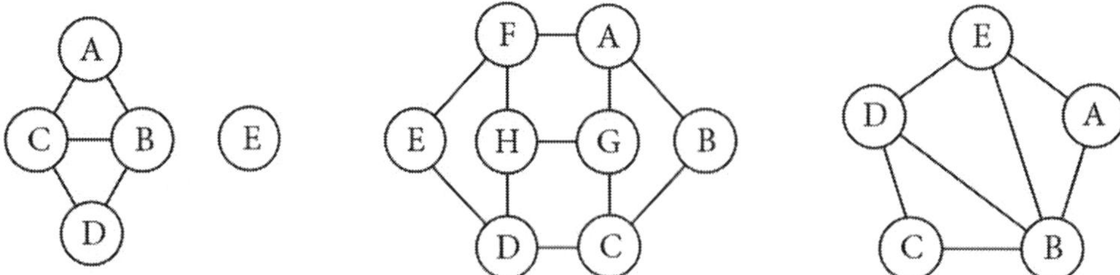

First consider the left diagram above (including the as-of-yet not connected E). Presently, B and C are degree 3, while A and D are degree 2. However, we must add edges to triangulate this graph. (Also try to imagine adding additional vertices.) Let's see if we can keep B and C both degree 3. If we connect A to D (even if we first add some intermediate vertex) to close off C, we won't be able to close off B if we are to use more than 4 vertices total (we can obviously join A to D and call it quits, but that just gives us the K_4, whereas our goal is to have V > 4). After joining A to D on the left, if we then join A to E to D, we must then connect B to E to triangulate the graph (and then B and C would no longer both be degree 3). You get the same problem if you try to join A to D on the right. There must be two vertices like A and D for a MPG because edge BC must be an edge for two triangular faces, so we can't simply do without A and D (since our goal is to have at least five vertices).

Now consider the middle diagram. G and H are both degree 3, connected in a different way than B and C had been in the left diagram. Right now, we would need to connect A to C or B to G and D to F or E to H, and we also need to connect either A to H or F to G and C to H or D to G. Even if we choose A to G (instead of B to G) and D to F (instead of E to H), we will need to add edges to G or H when we choose between A to H and F to G and between C to H and D to G. To make a MPG, G and H can't both remain degree 3.

In the right diagram, if you connect A to C, you will either need to connect A to D or C to E to triangulate the infinite area outside. If you erase edge BE, you will need to replace it with AD, otherwise you could connect AC and CE outside and close off both B and D; having to choose between BE and AD spoils this approach.

Let's look at it one more way. Imagine in the diagram on the following page that we have zoomed in on a larger graph and are looking at face ABC, and that your goal is to leave two of the three vertices A, B, and C with degree 3. Since you can only have three vertices on the "outside" for the infinite area outside to be triangulated, you will either need to connect A to D or B to F (or if there are other vertices, A or B still need to connect to one of them). Similarly, you either need to connect A to E or C to F, and B to E or C to D. You can't avoid adding an edge to two of the three vertices A, B, and C. For example, you can choose A to D on the right and A to E on the left, but then you still need to choose between B and C on the bottom. Vertices A, B, and C are all connected, but only one of these can be degree three.

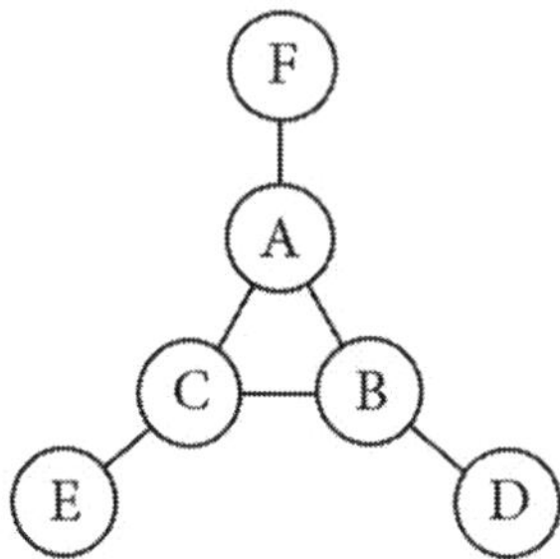

Can you think of a more convincing way to demonstrate this? We'll consider an alternative in the problems at the end of the chapter.

28.3 Degree Number Degeneracy

There are two structurally different MPG's with 6 vertices, as shown below. The left MPG has 6 vertices with degree 4, whereas the right MPG has two vertices with degree 5, two vertices with degree 4, and two vertices with degree 3. One way to label these graphs is 4-4-4-4-4-4 and 5-5-4-4-3-3 (or 5-4-3-5-3-4 if you want to match the order of the labels).

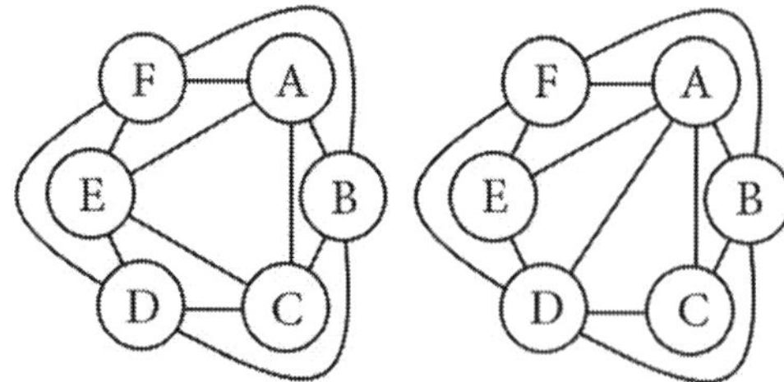

A problem with labeling a graph with its degree values is that there may be degeneracy. This means that there could potentially be two structurally different graphs with the same degree values. The example on the following page illustrates this with a PG (not a MPG). Each PG has degree values 3-2-2-2-1, but the left graph makes quadrilateral ABCD whereas the right graph makes triangle ABC. These graphs are clearly structurally different.

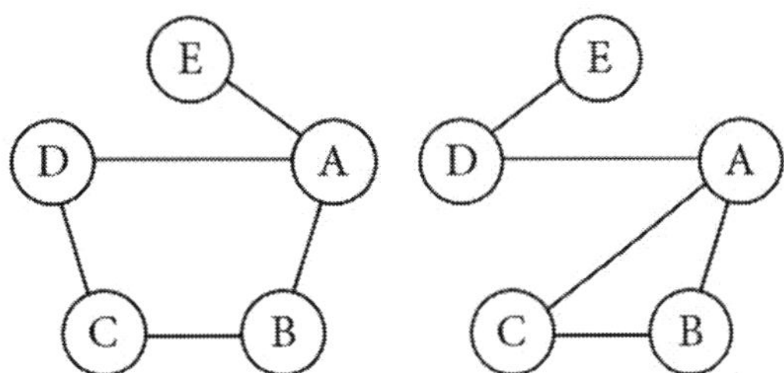

28.4 Maximum Degrees

Obviously, no single vertex can have a degree higher than $V - 1$ because there are only $V - 1$ other vertices for it to connect with.

No MPG can have more than two vertices with degrees equal to $V - 1$. Why? Try drawing any MPG that has two vertices with degree $V - 1$ and you will see that the edges from the $V - 1$ vertices close all of the remaining vertices so that there is no choice for how to draw the remaining edges.

The example below has $V = 9$ vertices and $E = 3V - 6 = 3(9) - 6 = 21$ edges. We first added $V - 1 = 8$ edges to vertex A. Next we added 7 edges to vertex E. (Why not 8 again? Edge AE had already been drawn from A.) After these $8 + 7 = 15$ edges are drawn, that only leaves $21 - 15 = 6$ edges left, and all 6 naturally fit in to triangulate the graph. Any possibility of choosing how to draw that remaining 6 edges was closed off by the first 15.

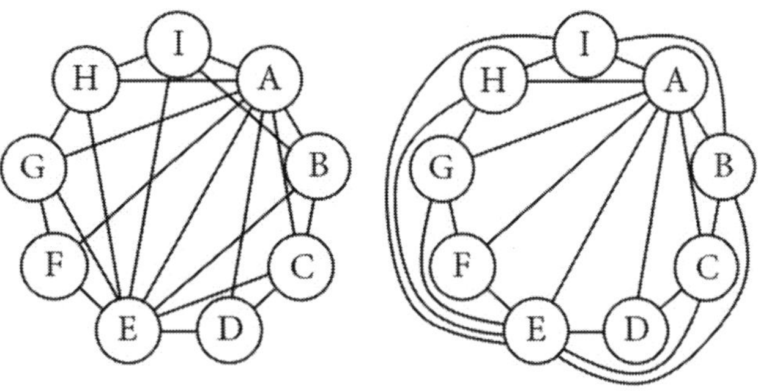

For any MPG with at least 4 vertices, there can't be more than two vertices with degree $V - 1$, and if there are two vertices with degree $V - 1$, there will also be two vertices with degree 3, and the remaining vertices will be degree 4 (like the previous example). (For a MPG with 4 vertices, every vertex is degree 3, which is also $4 - 1$. For a MPG with 3 vertices, no vertex can have a degree as high as 3.)

28.5 Attempts at Disproof

Attempting to disprove the four-color theorem can offer some insight into the nature of the theorem and may reveal some concepts that relate to the theorem. You are encouraged to get out some scratch paper and try creating your own maps in an effort to disprove the four-color theorem. You may learn something in the process.

A simple way to search for a possible disproof is to draw a vertical line to create a simple *map* (this isn't a *graph*) with color 1 to its left and color 2 to its right. Now add new regions to the vertical line. With colors 1 and 2 already in play, there are only two colors remaining. All you need to do is add a new region in such a way that it shares an edge with colors 1, 2, 3, and 4, and you would disprove the four-color theorem. However, any such attempt will be thwarted. Recoloring of previous regions may be needed, but it will always be possible to color the entire map in a way that satisfies the four-color theorem. Note how some colors need to change in the map below as new regions are added. This situation is *dynamic*.

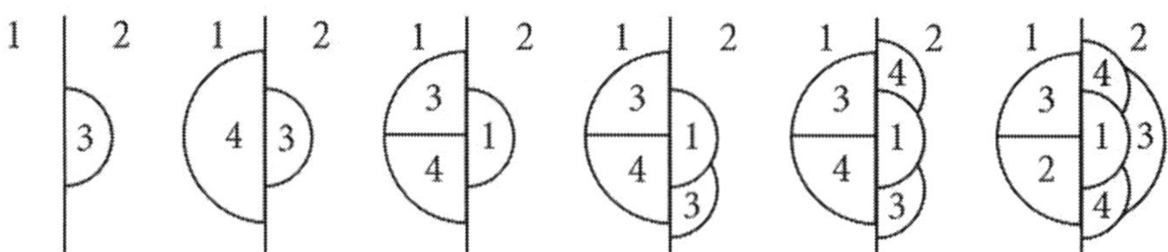

The example on the following page explores the effect of adding regions near the vertex of a map. Three regions (currently colored 1, 2, and 3) meet at a vertex. One by one we will add new regions near this vertex.

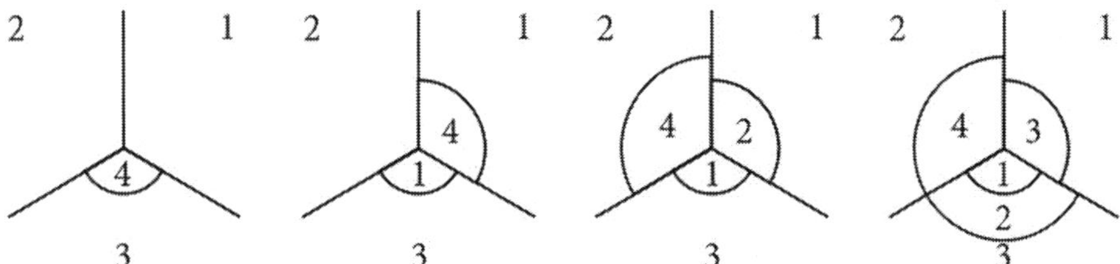

A common, yet futile, attempt is to continually add a new region so that it shares an edge with regions of three different colors. There are numerous ways to vary this approach, such as to create separate regions and later join them with a large region. We will first illustrate the basic idea with a simple *graph* (this diagram isn't a *map*).

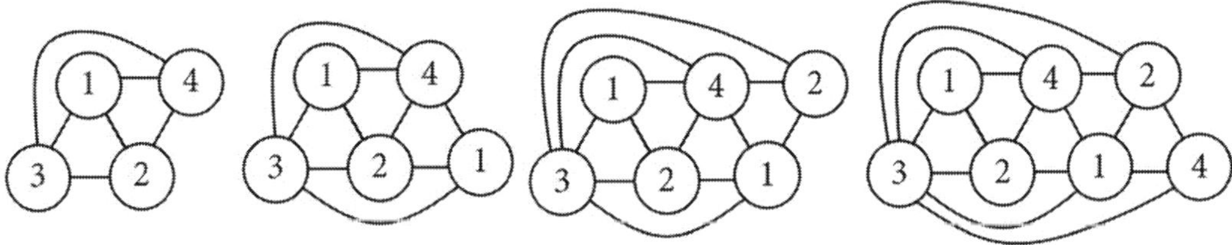

The simple idea illustrated above will obviously fail. When a new region connects to three regions of different colors, the new region simply uses the remaining color. We're simply nesting K_4's, which makes a trivially four-colorable graph (or map); see Chapter 11.

The next step is to add a new region that shares edges with more than three regions. The following *map* shows a simple example of this; the outer pentagon shares an edge with five regions. The inner pentagon also shares an edge with five regions. Yet this combination isn't enough to disprove the four-color theorem.

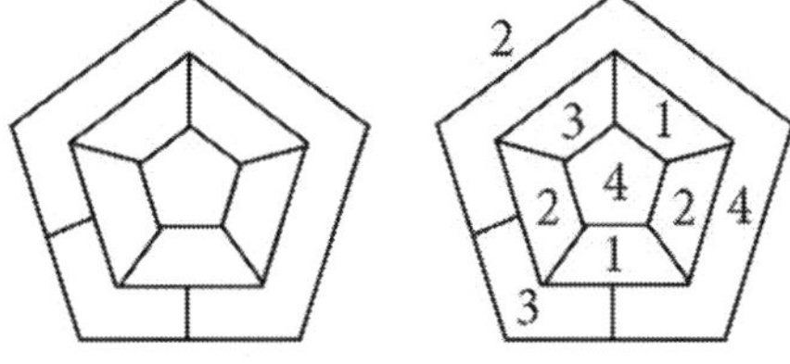

There are a few common ways that a "false disproof" may arise. A false disproof shows a map or graph that is either:

- partially colored using four colors in such a way that the remaining regions can't possibly be colored in a way that satisfies the four-color theorem (without recoloring)
- fully colored using five colors

One thing that is common to false disproofs is a flaw in the logic and reasoning. For example, one subtle yet incorrect assumption along the way can lead to a false disproof. Think through the logic, reasoning, assumptions, analysis, and conclusions carefully and thoroughly.

When coloring a map or graph, there are always choices to be made. You need to choose at least three colors to start with before you can reason where other colors can go. You must be careful where you put your initial colors. For example, if you begin by coloring regions with colors 1, 2, 3, and 4, but two of those regions don't share an edge, it could turn out that the only viable solution was when those two regions had the same color, in which case this incorrect starting assignment would lead to a false disproof.

Some maps or graphs involve making several choices. The logic and reasoning can be similar in some respects to a challenging sudoku puzzle. It sometimes takes a little trial and error to properly color a map or graph using only four colors. One wrong assumption along the way can lead to a false disproof.

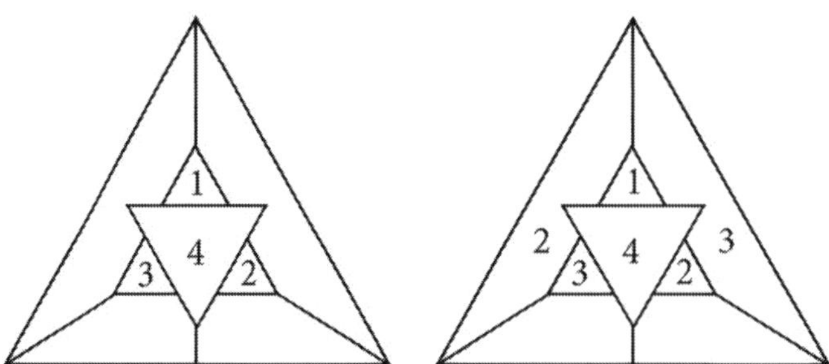

One example is shown above. On the *map* on the left, we began by coloring the central region with color 4. The three small triangles sharing an edge with it can't be color 4, so we made these colors 1, 2, and 3. This forces the right region to be color 3 (since it shares an edge with colors 1, 2, and 4) and the left region to be color 2 (since it shares an edge with colors 1, 3, and 4). The last two regions each share an edge with colors 2, 3, and 4, but the last two regions can't both be color 1 because they share an edge. It looks like a fifth color is needed.

Can you find the mistake in this false disproof?

In the previous example, there is a mistake in the initial assumption. The logic and reasoning are sound after that. The challenge with false disproofs is that the arguments often appear to be logical and well-reasoned, but there is always some flaw (sometimes a subtle one).

Here is the mistake in the previous false disproof. After labeling the central triangle 4, we correctly reasoned that the three small triangles couldn't be color 4, but we were incorrect when we assumed that they had to be colors 1, 2, and 3. All three small triangles can be the *same* color and still be different from color 4. The picture below shows how the previous map may be colored in a way that satisfies the four-color theorem.

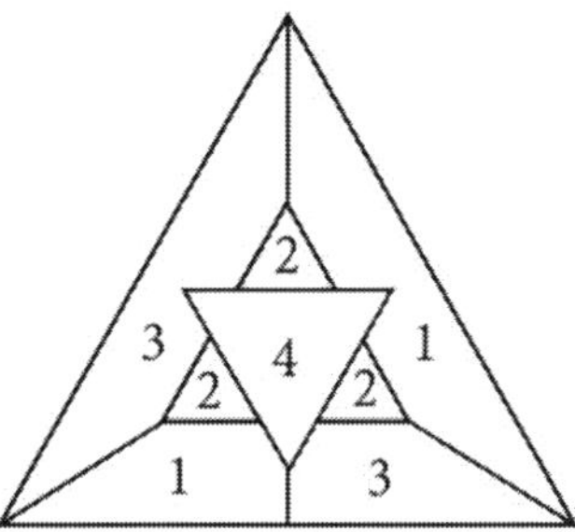

Following is another example of a false disproof. This time we will use a graph instead of a map. We will begin with a small graph and continue to add new regions to it.

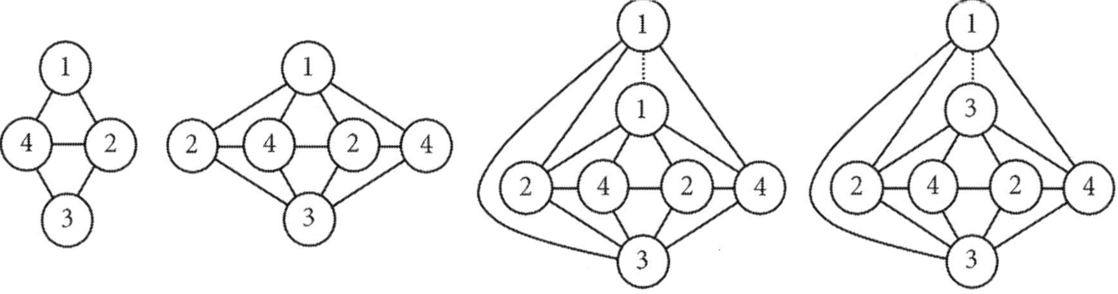

The dotted line in the diagram above is a problem; two regions with color 1 can't share an edge. This false disproof suggests that one of the color 1's needs to change to color 5. Can you find the flaw in this false disproof?

This example takes advantage of a dynamic situation. As discussed in Chapter 24, when new regions are added to a map or graph, it may be necessary to recolor some of the previous regions. In this case, the first region colored 1 could instead be color 3, as shown below.

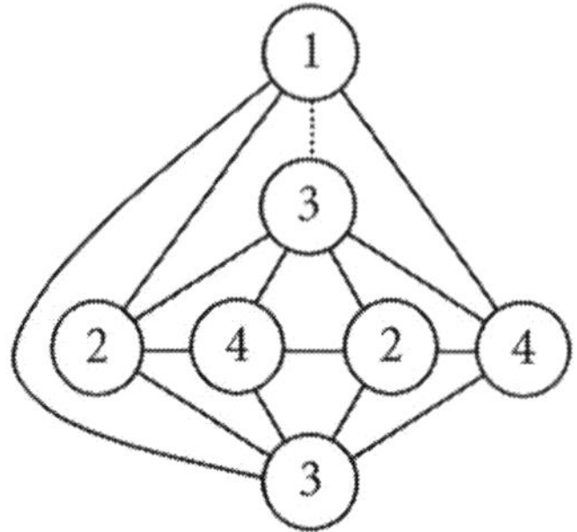

28.6 The Currently Accepted Proof of the 4CT

Despite several attempts to prove the four-color theorem by hand, as of the publication of this book, the only proof of the four-color theorem with significant acceptance uses computer calculations [Ref. 14]. The original proof by Kenneth Appel and Wolfgang Haken (1976) involved nearly 2000 of what were termed "reducible configurations." Checking each of these 2000 configurations for "reducibility" proved to be a very tedious task, requiring the use of a computer program. Neil Robertson, Daniel Sanders, Paul Seymour, and Robin Thomas (1996) showed that this can be diminished to about 600 configurations, but this is still more tedious than can be plausibly checked by hand (as each configuration itself is rather complex).

The basic concepts of the currently known computer-assisted proof include reducibility, unavoidable configurations, and discharging. If you wish to learn more about the computer-assisted proof of the four-color theorem, a good place to start is Birkhoff's paper on the reducibility of maps [17]. This involves a ring of connected regions, which separates a map (or PG) into three parts: the part inside the ring, the ring itself, and the part outside the ring. A ring with four or five regions is easy to reduce to a much simpler map (or PG), but as the number of regions in the ring grows large, this becomes much more complex.

28.7 Other Surfaces

The four-color theorem applies to planar graphs. As we showed in Chapter 14, it also applies to graphs drawn on the surface of a sphere. It also applies to graphs drawn on the surface of a cylinder. However, it doesn't apply to graphs drawn on *any* surface. For example, for graphs drawn on the surface of a single-holed ring torus (shaped like a donut), 7 colors may be needed instead of 4. (The next time you're at the front of a long line at a donut shop, be sure to say, "I'd like one chocolate covered single-holed ring torus with sprinkles, please." Hopefully, you can also get your coffee in a Klein bottle, and if you decide to draw graphs on the surface of the Klein bottle, you may need 6 colors). See [Ref. 15].

28.8 3D Space

If you attempt to generalize the four-color theorem to 3D space instead of the 2D plane, you will run into an interesting problem. Go outside and imagine throwing a bunch of pillows around. Let the pillows represent 3D regions in 3D space. Now imagine connecting edges to the pillows. You can use string to make the edges. You can wrap these strings around like intertwined spaghetti, if needed. If you have 1,000,000 pillows, and if you can find enough string, you can tie every pillow to every other pillow so that you would need 1,000,000 different color pillows in order to avoid having one pillow connect to another pillow of the same color. No matter how many regions you have, you can add edges in such a way that every region needs to be a different color. This shows that the number of colors needed in 3D space is unbounded; there is no limit. If there are V regions, for the most general case you would need V different colors.

28.9 Three-Color Theorems

There are a number of three-color theorems, such as:

- Any map or graph that only consists of exterior regions is three-colorable, like the *map* (not graph) shown below.

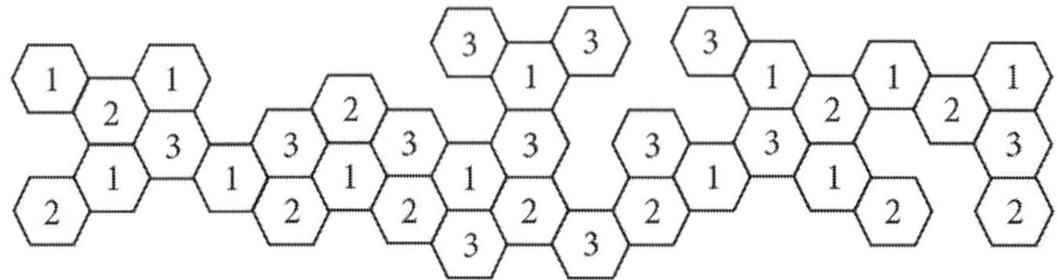

- Any map or PG can be four-colored in such a way that its exterior regions are three-colored. Any new region may be added that completely surrounds the continent of a map or which connects to every outer region of a graph. The four-color theorem then requires that these exterior regions be three-colorable.

- If one vertex of a graph with V vertices has degree V − 1, the other V − 1 vertices must be three-colorable.

- If a MPG has a HC and if the MPG is divided into two separate PG's at its HC (so that each PG includes the HC), each PG is three-colorable. See the Challenge Problem from Chapter 14.

Although each of these three-color theorems can be proven from the four-color theorem, it is also possible to prove them independently of the four-color theorem. We'll let you think about this in one of the challenge problems at the end of the chapter.

28.10 Two-Coloring

Two-coloring is possible with certain types of maps and their corresponding PG's, such as a checkerboard or a simple chain. The examples below are *maps* (not *graphs*).

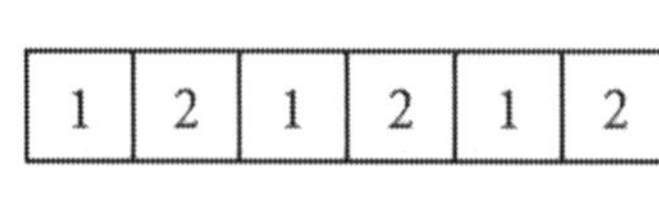

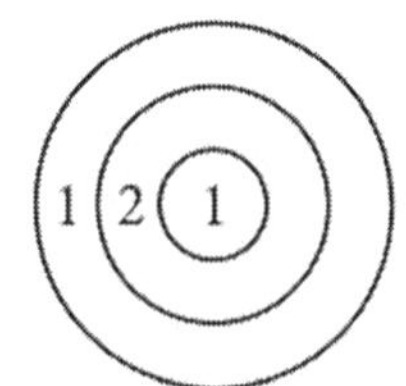

For a simple chain that forms a closed loop, two-coloring of its exterior regions is possible if the chain consists of an even number of exterior regions, but a third color is required for the exterior regions if the chain consists of an odd number of exterior regions. (Either way, at least three colors are needed to also color its interior regions.)

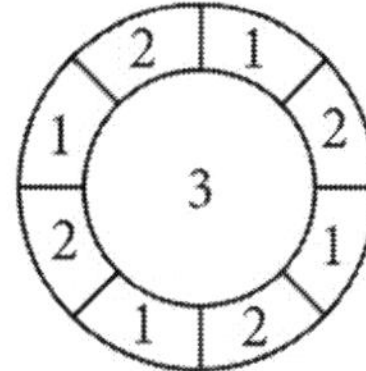 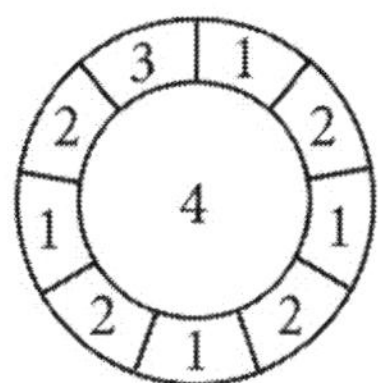

For pie slices, two-coloring is similarly possible if there are an even number of slices, but a third color is required if there are an odd number of slices. If the unbounded surrounding area is also to be colored, the map on the left below would actually require three-coloring. For two-coloring that includes the unbounded surrounding area, see the one-coloring map that follows and add a second color for the unbounded surrounding area.

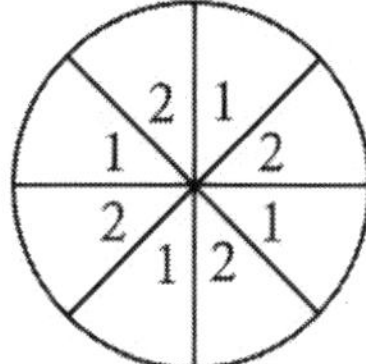 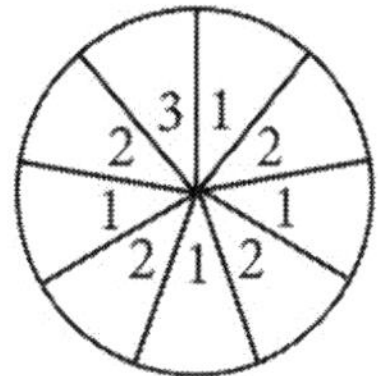

The concept of one-coloring is interesting. For a map with multiple regions, one-coloring requires that no two regions share an edge. If two regions of a map share a vertex, but don't share an edge, they may be the same color. An example of one-coloring for a map is shown below, but this requires that we don't color the unbounded surrounding area. Think about what the corresponding graph would look like.

28.11 A Couple of Cool Graphs… Just Because

Feel free to four-color the graphs that follow.

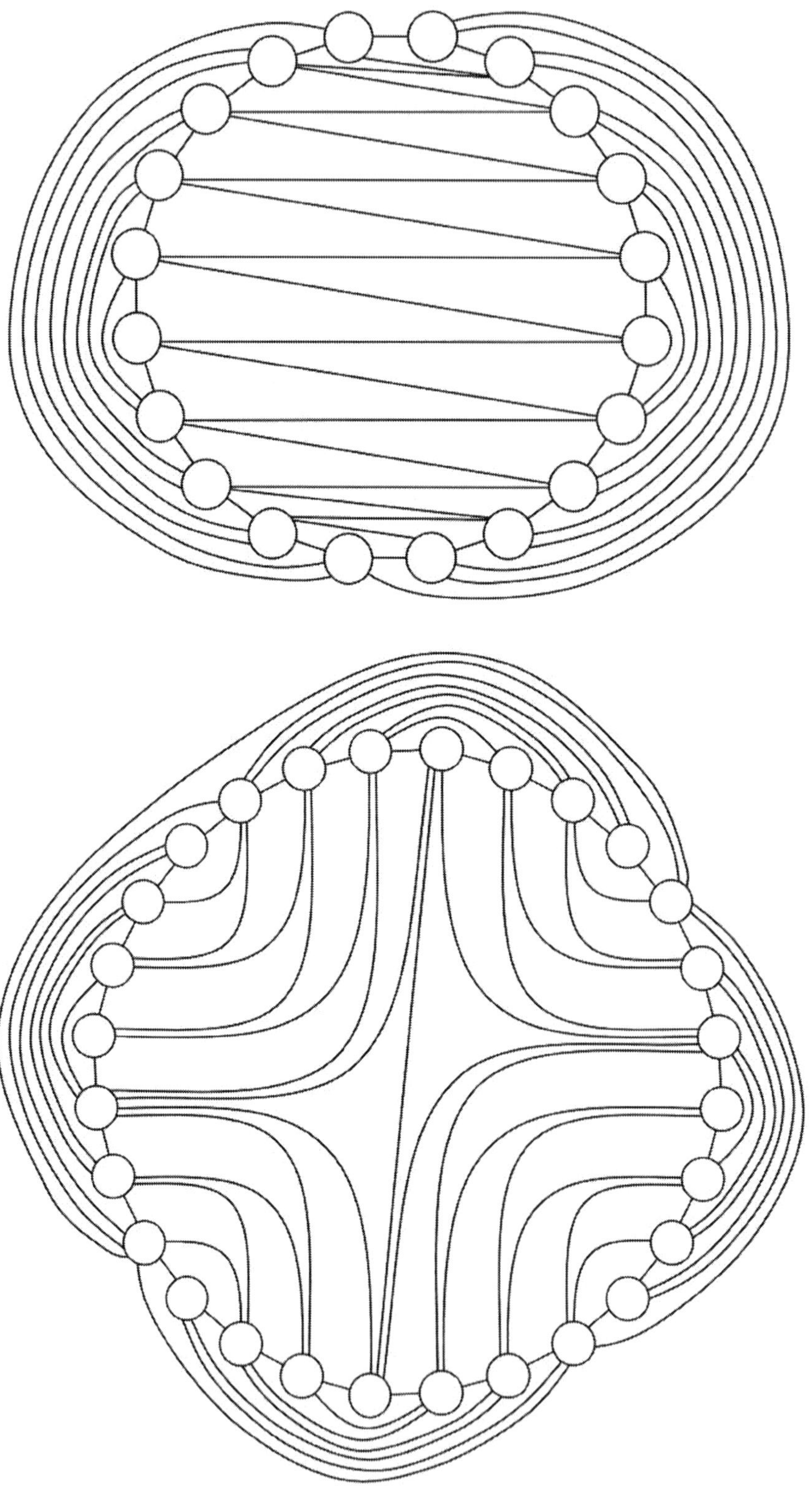

Chapter 28 Exercises

1. The graph below has a serious "problem" compared to all of the other graphs in this book. Can you figure out what it is?

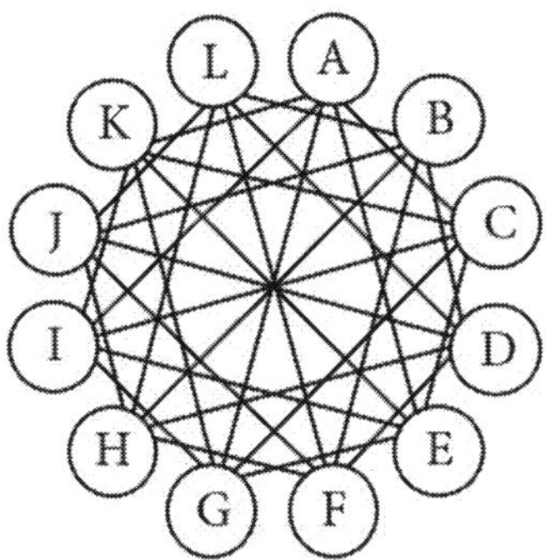

As a separate issue (from the "problem" that you need to figure out), does this graph have the right number of edges to be a MPG? Is it a MPG? Is it planar? (Whether or not the graph may be planar or a MPG, this has nothing to do with the more serious "problem.")

2. As we discussed in Chapter 19, every edge in every MPG is the diagonal of a quadrilateral like the one shown below (provided that the graph has at least 4 vertices). Use this fact to prove that every MPG with at least 4 vertices doesn't have any vertices with degree 2. (This has nothing to do with quadrilateral switching; just to do with the quadrilaterals. The only reason we referred to Chapter 19 is that happens to be where we introduced this fact, not because the solution has anything to do with quadrilateral switching.)

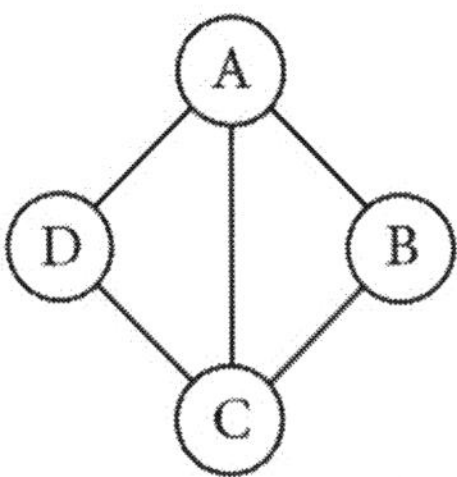

3. Use the quadrilaterals mentioned in Problem 2 (but again, just the quadrilateral shapes, not quadrilateral switching) to show that no two vertices with degree 3 in any MPG can be joined by an edge (if the MPG has at least 5 vertices).

4. Prove that if one vertex in a MPG has degree $V - 1$, the other $V - 1$ vertices must be three-colorable if the four-color theorem is true.

Prove that if two vertices in a MPG each have degree $V - 1$, the other $V - 2$ vertices must be two-colorable if the four-color theorem is true.

5. Imagine that we have a complete graph (Chapter 5) and that we wish to remove edges from the complete graph to form a MPG (which is one of the types of edge-removing that we considered in Chapter 17). The following questions relate to this. Express your answer in terms of V, which represents the number of vertices of the complete graph.

- What is the maximum number of edges that can be removed from any single vertex of a complete graph so that no vertices of the new graph have a degree less than 3? The reason behind this question is that no vertex of a MPG can have degree 2.

- What is the maximum number of edges that can be removed from any single vertex of a complete graph so that no vertices of the new graph have a degree less than 5? The reason behind this question is that in Chapter 11 we showed that we just need to prove the four-color theorem for MPG's that don't have any vertices with degree 3 or 4.

Challenge problem 1: Prove that any map or planar graph that doesn't have any interior regions is three-colorable without using the four-color theorem to prove it. Similarly, prove the three-color exterior theorem (that the exterior regions of any map or planar graph are always three-colorable) without using the four-color theorem to prove it.

Challenge problem 2: Is it possible for a MPG to have degree number degeneracy (like that discussed in Sec. 28.3)? If so, draw an example. If not, show or explain why.

Note: The answer key doesn't include answers to the challenge problems. These problems are intended to encourage you to think about the ideas.

Further Reading and References

(1) Francis Guthrie is credited as the first to ask how many colors are needed to color a map in 1852. Arthur Cayley and Augustus De Morgan helped draw attention to the problem.

(2) Alfred Kempe attempted to prove the four-color theorem in 1879 in his paper, "On the Geographical Problem of Four-Colors" (*American Journal of Mathematics*, Volume 2, No. 3, pp. 193-200), using the concept that has since become known as Kempe chains (Chapter 7). Kempe also showed that every PG has at least one vertex with degree five or less.

(3) Percy Heawood showed the subtle fallacy in Kempe's attempted proof in 1890 in his article, "Map-Colour Theorem" (*Quarterly Journal of Pure and Applied Mathematics*, Volume 24, pp. 332-338), which included a counterexample to Kempe's argument (see Chapter 7). A clear, accessible presentation of this idea is given by Timothy Sipka's 2002 article, "Alfred Bray Kempe's 'Proof' of the Four-Color Theorem" (*Math Horizons*, Volume 10). Heawood is also credited with proving the five-color theorem.

(4) Leonhard Euler presented his formula relating the number of vertices, edges, and faces of a polyhedron in 1752 in his article, "Elementa Doctrine Solidorum" (*Novi Commentarii Academiae Scientiarum Petropolitanae*, Volume 4, pp. 109-160) (now available in the Euler Archive, E230).

(5) The main idea behind the puzzle known as the "utility problem" is very old. It was posed as a problem in terms of basic utility services (such as gas, electricity, and water) by Henry Dudeney. His puzzle first appeared in 1913 in his article, "Perplexities, with Some Easy Puzzles for Beginners" (*The Strand Magazine*, Volume 46, p. 110). In 1917, the puzzle appears in Problem 251 – Water, Gas, and Electricity – in the book *Amusements in Mathematics*.

(6) Casimir Kuratowski presented the theorem which now bears his name on the distinction between planar and nonplanar graphs in 1930 in his paper, "Sur le Problème des Courbes Gauches en Topologie" (*Fundamenta Mathematicae*, Volume 15, pp. 217-283). Pontryagin proved the theorem earlier (in 1927-1928), but hadn't published his proof. This is why the theorem is also known as the Pontryagin-Kuratowski theorem. See also J.W. Kennedy, L.V. Quintas, and M.M. Syslo's article in 1985, "The Theorem on Planar Graphs" (*Historia Mathematica*, Volume 12, pp. 356-368).

(7) Klaus Wagner presented the theorem which now bears his name and which is similar to Kuratowki's theorem, except that the nature of his theorem works with minors rather than subdivisions (which can actually make a significant distinction in application), in 1937 in his article, "Über eine Eigenschaft der ebenen Komplexe" (*Mathematische Annalen*, Volume 114, pp. 570-590).

(8) The Fritsch graph provides a counterexample to Kempe's argument for four-coloring a vertex with degree five using a MPG with just 9 vertices. See the book by R. Fritsch and G. Fritsch, *The Four-Color Theorem* (Springer-Verlag, 1998). The Soifer graph also does this with 9 vertices, but the infinite area outside of the Soifer graph is a quadrilateral; you must add an edge to make it a MPG. See Alexander Soifer's article in 1997, "Map Coloring in the Victorian Age: Problems and History" (*Mathematics Competitions*, Volume 10, pp. 20-31). Although it isn't a MPG, the Soifer graph is technically a smaller graph than the Fritsch graph.

(9) The Errera graph provides a counterexample to Kempe's argument for four-coloring a vertex with degree five using a MPG with 17 vertices. Although this graph has more vertices than the Fritsch and Soifer graphs, it doesn't have any trivial vertices since no vertex has a degree less than five. See Alfred Errera's 1921 Ph.D. thesis, "Du Coloriage des Cartes et de Quelques Questions d'Analysis Situs" (Gauthier-Villars & Cie, Paris).

(10) The concept of the Hamiltonian cycle is named after Sir William Rowan Hamilton's 1857 pegboard puzzle called the Icosian game (which involves finding a Hamiltonian cycle on the edges of a dodecahedral graph, or equivalently, a dodecahedron). In 1856, Hamilton mentioned passing from corner to corner along an icosahedron or dodecahedron, in which he coined the term "Icosian calculus" (involving quaternions), in "Memorandum Respecting a New System of Roots of Unity" (*Philosophical Magazine*, Volume 12, p. 446).

(11) The Goldner-Harary graph is a MPG with just 11 vertices which doesn't have a HC. See A. Goldner and F. Harary's 1975 paper, "Note on a Smallest Nonhamiltonian Maximal Planar Graph" (*Bulletin of the Malaysian Mathematical Sciences Society*, Volume 6, pp. 41-42).

(12) Hassler Whitney showed in 1931 that any MPG without a ST has a HC in his article, "A Theorem on Graphs" (*Annals of Mathematics*, Volume 32, pp. 378-390). Various papers by subsequent authors have explored extensions of Whitney's theorem where MPG's with one or more ST's have HC's.

(13) Euler found criteria for the existence of an Eulerian cycle in his solution to the Seven Bridges of Königsberg, published in 1741 in the article, "Solutio Problematis ad Geometriam Situs Pertinentis (*Commentarii Academiae Scientarum Petropolitanae*, Volume 8, pp. 128-140) (now available in the Euler Archive, E53). The proof was completed by Carl Hierholzer, published posthumously in 1873 as "Über die Möglichkeit, einen Linienzug ohne Wiederholung und ohne Unterbrechung zu umfahren" (*Mathematische Annalen*, Volume 6, pp. 30-32).

(14) The only known proof of the four-color theorem (despite numerous attempts) that is currently accepted (though not at a rate of 100%) by the mathematics community involved nearly 2000 "reducible configurations" that required the assistance of a computer program. The following works relate to Kenneth Appel and Wolfgang Haken's 1977 proof:

- The 1977 two-part article, "Every Planar Map is Four-Colorable, I: Discharging and II: Reducibility" (*Illinois Journal of Mathematics*, Volume 21, pp. 429-567).

- An accessible 1977 article, "The Solution of the Four-Color Map Problem" (*Scientific American*, Volume 237, pp. 108-121).

- A 1986 article, "The Four Color Proof Suffices" (*The Mathematical Intelligencer*, Volume 8, pp, 10-20 and 58).

- The 1989 book, *Every Planar Map is Four-Colorable* (American Math Society).

Neil Robertson, Daniel Sanders, Paul Seymour, and Robin Thomas (1996) showed that this can be diminished to about 600 configurations, but this is still more tedious than can be plausibly checked by hand: "A New Proof of the Four Colour Theorem" (*Electronic Research Announcements of the American Mathematical Society*, Volume 2, pp. 17-25).

Before you read these papers, you may wish to start by reading Birkhoff's article on the reducibility of maps. (See Ref. 17.)

(15) The Heawood conjecture generalizes the N-color theorem to surfaces other than the plane, sphere, or cylinder [Ref. 3]. See G. Ringel and J.W.T. Youngs' 1968 article, "Solution of the Heawood Map-Coloring Problem," (*Proceedings of the National Academy of Sciences of the United States of America*, Volume 60, pp. 438-445) and Gerhard Ringel's 1974 book, *Map Color Theorem* (Springer-Verlag).

(16) If you would like to explore online, check out Wolfram's MathWorld resource (from Eric W. Weisstein):

http://mathworld.wolfram.com/Four-ColorTheorem.html

There is a good list of references within, which includes a page listing books about the four-color theorem.

(17) The computer-assisted proofs mentioned in Ref. 14 apply the concept of reducibility. A good place to begin learning about reducibility is George D. Birkhoff's 1913 article, "The Reducibility of Maps" (*American Journal of Mathematics*, Volume 35, pp. 115-128).

Answer Key

Chapter 1 Maps vs. Graphs

1. There is more than one way to draw the graph. The important details are which vertices are connected by edges:

- A connects to D, E, and H.
- B connects to D, E, and F.
- C connects to D, G, and H.
- D connects to A, B, C, E, F, G, and H.
- E connects to A, B, D, F, H, and I.
- F connects to B, D, E, G, and I.
- G connects to C, D, F, H, and I.
- H connects to A, C, D, E, G, and I.
- I connects to E, F, G, and H.

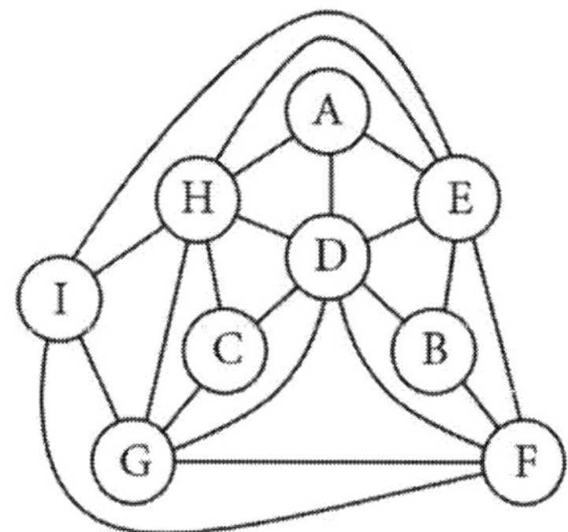

2. There is more than one way to draw the graph. The important details are which vertices are connected by edges:

- A connects to B, C, D, E, G, and K.
- B connects to A, C, E, F, and K.
- C connects to A, B, F, G, and K.
- D connects to A, E, G, H, and I.
- E connects to A, B, D, F, and H.
- F connects to B, C, E, G, H, and I.
- G connects to A, C, D, F, and I.
- H connects to D, E, F, I, and J.
- I connects to D, F, G, H, and J.
- J connects to H and I.
- K connects to A, B, and C.

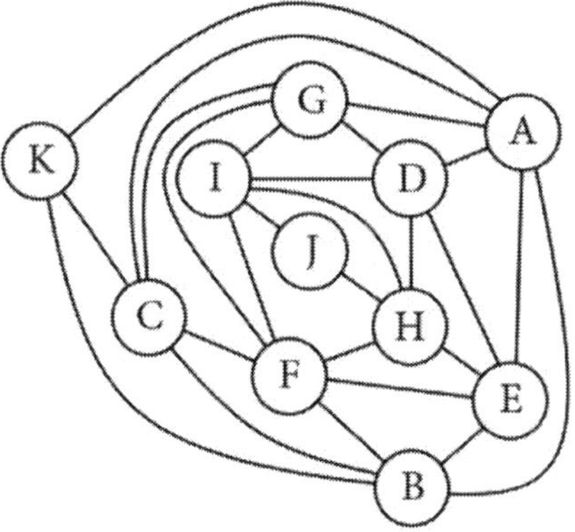

3. There are multiple ways to draw the maps.

- A connects to B, D, and E (but not C).

- B connects to A, C, and E (but not D).

- C connects to B, D, and E (but not A).

- D connects to A, C, and E (but not B).

- E connects to A, B, C, and D.

Note that A and C don't share an edge in the map on the left because they only touch at a vertex, and similarly for B and D.

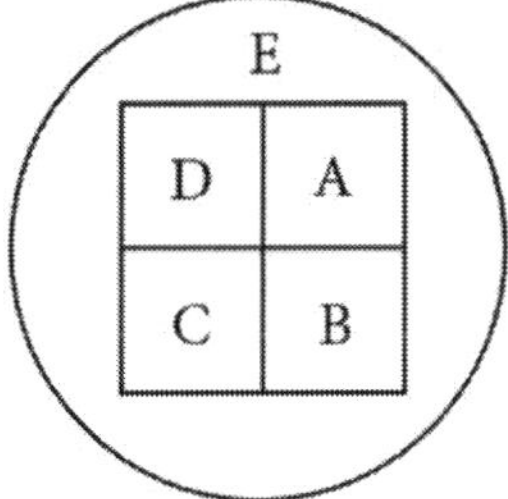

Chapter 2 The Four-Color Theorem

1. There is more than one possible solution. A sample solution is shown below. Whichever color you use for the unbounded surrounding area (the region outside), make sure that none of the exterior regions has the same color as that. Study one color at a time and check that no two regions of that color share an edge.

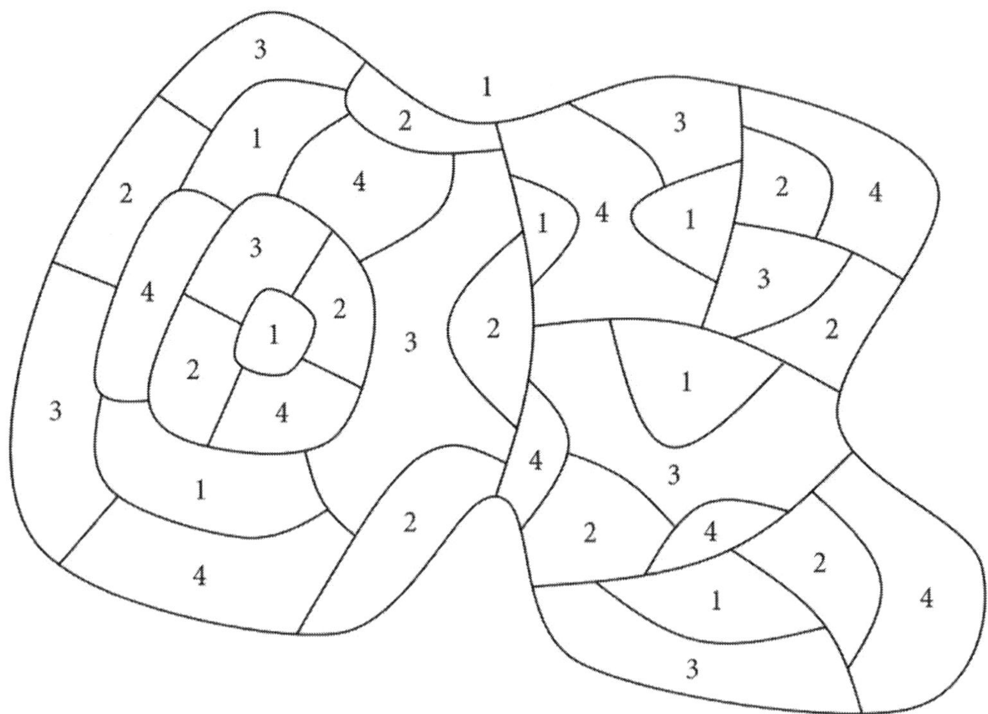

2. There is more than one possible solution. A sample solution is shown below. Study one color at a time and check that no two regions of that color share an edge.

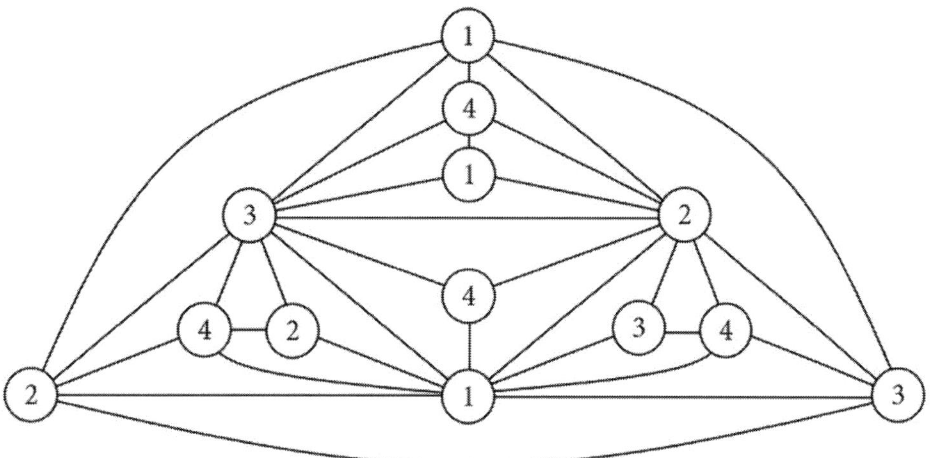

3. There is more than one possible solution. A sample solution is shown below. Study one color at a time and check that no two regions of that color share an edge.

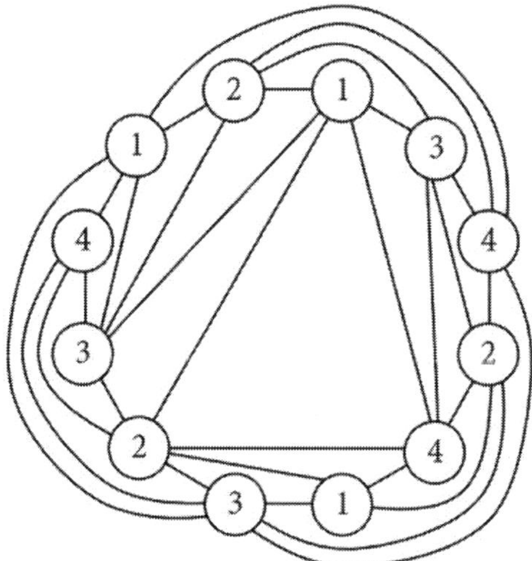

4. There is more than one possible solution. A sample solution is shown below. Study one color at a time and check that no two regions of that color share an edge.

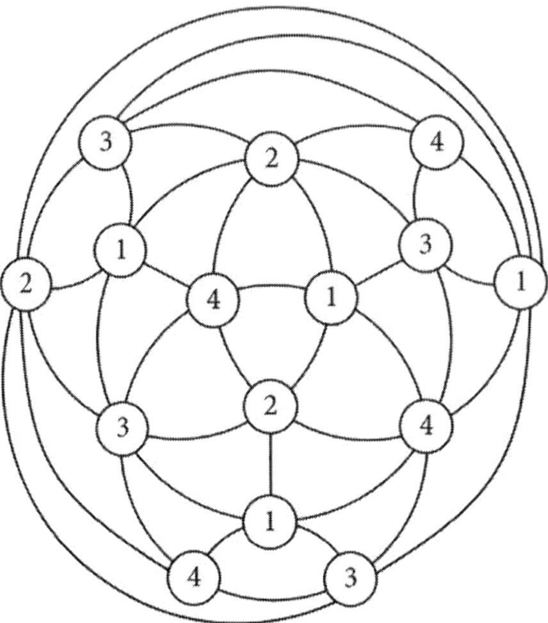

Chapter 3 Triangulation

1. The edges added to triangulate each PG are shown as thick dark lines or curves on the diagrams on the following page. (Note that the second graph is a MPG; this graph is already triangulated.) There is more than one answer, as described below.

- The left graph is originally a PG. To turn it into a MPG, you could add edge AD or you could add edge FC (but not both). As drawn on the next page, face ACD is the infinite area outside, but note that AD could be drawn differently such that face ADF would be the infinite area outside. (If instead you added edge FC, the infinite area outside is face ACF or CDF, depending on how you draw it.)

- The second graph is a MPG. (This graph is K_4, as we'll learn in Chapter 5.) The infinite area outside is face ABC.

- The third graph is originally a PG. To turn it into a MPG, you must add three edges. You could add AC, CF, and DF. There are many alternatives, such as AC, AD, and DF or as AE, BD, and BE. As drawn on the following page, the infinite area outside is face ACF, but if you added edges differently, the face corresponding to the infinite area outside will be different.

- The right graph is originally a PG. To turn it into a MPG, you must add four edges. You can't draw CI on the outside instead of AD because CI is already drawn on the inside. Similarly, you can't draw FI on the outside instead of DG. You can't draw GI inside instead of FH because GI is already drawn outside. Similarly, you can't draw DI inside instead of CE because DI is already drawn outside. This solution only has one alternative to what is shown on the following page: You could instead draw DG so that it loops around the long way. As drawn below, the infinite area outside is face DGI (but for the alternative way of drawing DG, it would be DFG).

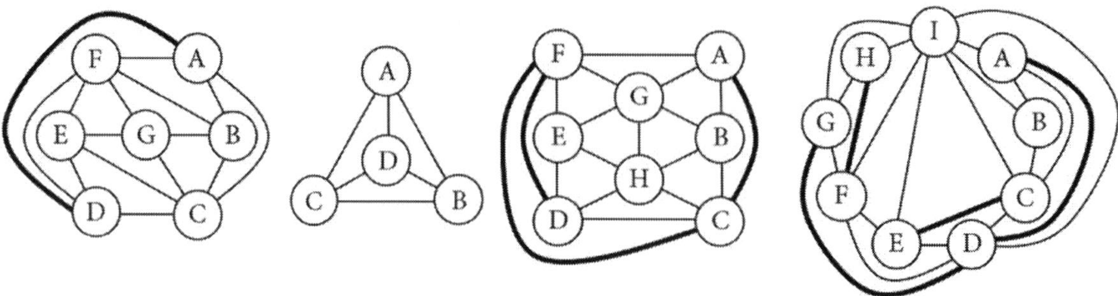

2. You must add four edges to triangulate the graph. Note that there are alternatives to what is shown below. For example, you could instead connect the left middle vertex to the right middle vertex outside (instead of the bottom left and top right). Note that with any alternative drawing the coloring of the graph may be much different from what is shown. Even for the graph shown below, there are multiple solutions for how to color the graph. Study one color at a time and check that no two regions of that color share an edge. The same coloring works for both graphs. There are more ways to color the PG on the right than the MPG on the left. Removing edges reduces the restrictiveness of the coloring (not always, as it is possible to remove an edge from a MPG and obtain a PG with the same number of colorings).

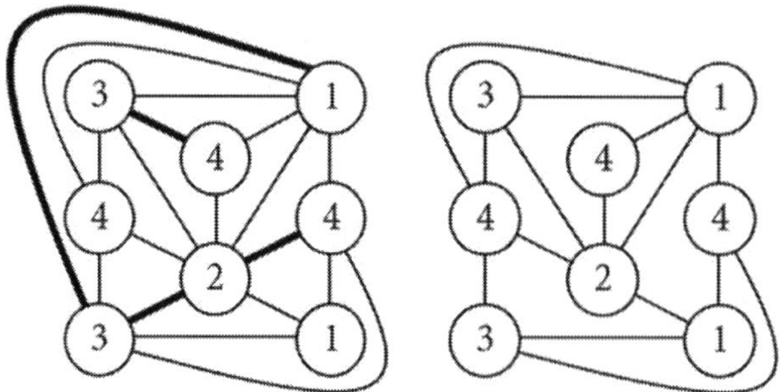

3. There is more than one way to draw the graph. The important details are which vertices are connected by edges. In the original PG:

- A connects to B, C, D, and E.
- B connects to A, C, and E (but not D).
- C connects to A, B, D, and E.
- D connects to A and C only (not B or E).
- E connects to A, B, and C (but not D).

You must add edge BD to triangulate the graph. With the added edge, B connects to D. There is more than one way to draw the map for the MPG. It will differ from the original map (for the PG) in that B will connect to D.

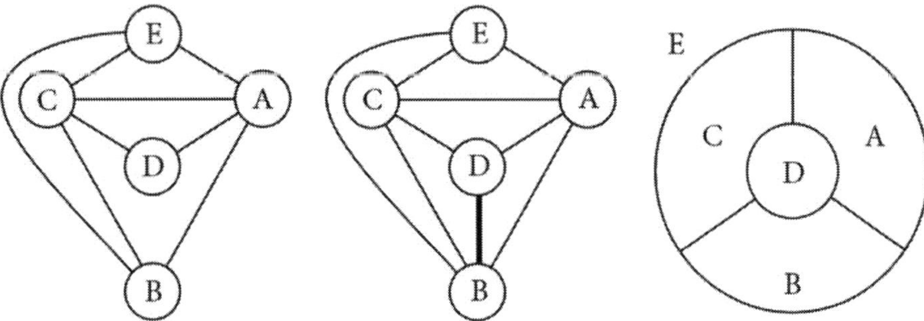

4. The second and third graphs are isomorphic. (One difference with the left graph is that A connects to E.) In the second and third graphs:

- A connects to B, C, and F.
- B connects to A, C, E, and F.
- C connects to A, B, D, E, and F.
- D connects to C, E, and F.
- E connects to B, C, D, and F.
- F connects to A, B, C, D, and E.

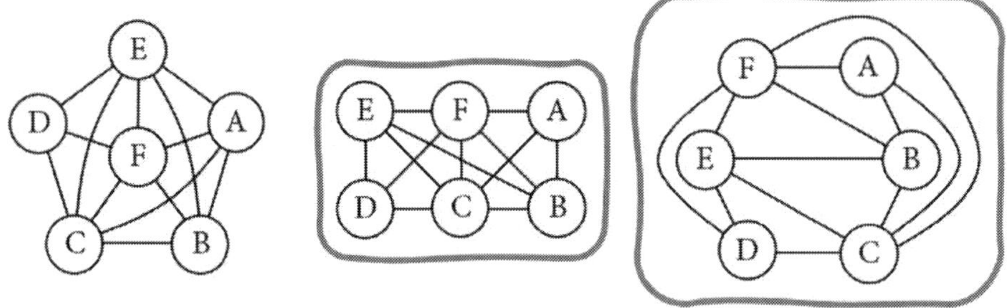

5. The first and fourth graphs are isomorphic. In the first and fourth graphs, A connects to C and D, B connects to D and E, C connects to A and E, D connects to A and B, and E connects to B and C.

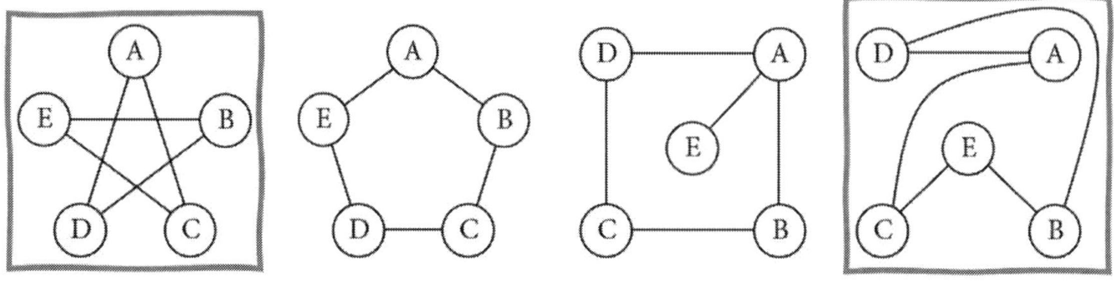

6. One condition is that all of the map's vertices must have degree three. Why? The faces of a graph correspond to the vertices of a map. In order for every face of a graph to have three edges (to have triangular faces), every vertex of the map must have three edges. Another condition is that every region must share an edge with at least three other regions (unlike the map given in Problem 3, where region D only shares edges with regions A and C). Are these two conditions sufficient? If not, what other condition(s) must be met?

Chapter 4 Euler's Formula

1. The map on the left has 6 vertices (the corners of the square, the point where the two diagonals cross inside, and the apex of the roof), 12 edges (the four sides of the square, four edges inside the square, two lines forming the roof, and the two curved edges), and 8 faces (regions A, B, C, D, E, F, G, and H). The graph on the right has 8 vertices (A, B, C, D, E, F, G, and H), 12 edges (AB, AD, AH, BC, BE, CD, CF, DG, EF, EH, FG, and GH), and 6 faces (EFGH, ABEH, BCFE, CDGF, AHGD, and the face ABCD corresponding to the infinite area outside). This map and graph are duals of each other since V, E, F = 6, 12, 8 for the map is swapped with V, E, F = 8, 12, 6 for the graph. The map corresponds to an octahedron and the graph corresponds to a cube. (Another feature of the dual relationship between polyhedra is that each face of a cube has 4 edges and 3 faces meet at each vertex, whereas each face of an octahedron has 3 edges and 4 faces meet at each vertex.) For the map, V + F = 6 + 8 = 14 and E + 2 = 12 + 2 = 14. For the graph, V + F = 8 + 6 = 14 and E + 2 = 12 + 2 = 14. The ratio of F to E for the graph is F/E = 6/12 = 1/2 = 0.5. Since F/E is less than 2/3, this PG isn't a MPG. (It should be obvious that this PG isn't triangulated.)

In dual polyhedra, in addition to swapping the roles of vertices and edges, symmetries are swapped. The Schläfli symbol {p,q} represents a polyhedron where each vertex has q regular p-sided polygonal faces. The symbol {q,p} then represents its dual polyhedron. For example, compare {4,3} for a cube and {3,4} for an octahedron. As another example, compare {5,3} for a dodecahedron and {3,5} for an icosahedron.

This definition of dual polyhedra incorporates the number of edges in each face and the number of faces meeting at each vertex. In comparison, when we say that a graph is dual to a map, it is in a more crude sense than when we say that two polyhedra are duals.

2. The graph has 7 vertices (A, B, C, D, E, F, and G), 15 edges (AB, AE, AF, AG, BC, BD, BE, BG, CD, CE, DE, DF, DG, EF, and FG), and 10 faces (ABG, BCD, BDG, DEF, DFG, AEF, AFG, BCE, CDE, and the face ABE corresponding to the infinite area outside). V + F = 7 + 10 = 17 and E + 2 = 15 + 2 = 17. Since F/E = 10/15 = 2/3, this PG is a MPG.

3. The graph has 6 vertices (A, B, C, D, E, and F), 9 edges (AB, AC, AE, BC, BF, CD, CE, DE, and EF), and 5 faces (ABC, ABFE, BCDEF, CDE, and the face ACE corresponding to the infinite area outside). V + F = 6 + 5 = 11 and E + 2 = 9 + 2 = 11. Since F/E = 5/9 which rounds to 0.56 (which is less than 2/3 which rounds to 0.67), this graph is not a MPG. It is a PG, but not a MPG.

We must add 3 edges to make the graph a MPG. One way to do this is shown below. (One alternative is to draw BE instead of AF. Another is to draw BD and BE instead of CF and DF. However, you can't draw CE on the inside because that edge is already drawn on the outside.) The MPG will have 6 vertices (A, B, C, D, E, and F), 12 edges (as drawn below: AB, AC, AE, AF, BC, BF, CD, CE, CF, DE, DF, and EF), and 8 faces (as drawn below: ABC, ABF, BCF, CDF, DEF, AEF, CDE, and the face ACE corresponding to the infinite area outside). For the MPG, V + F = 6 + 8 = 14, E + 2 = 12 + 2 = 14, and F/E = 8/12 = 2/3.

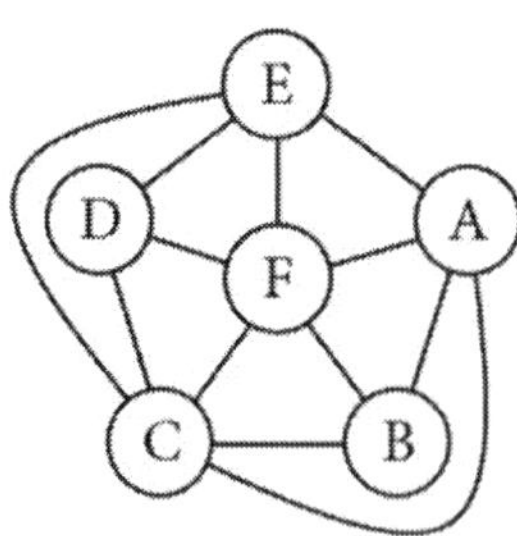

4. To get the number of faces F from the number of edges for this graph, divide E by 4 because each face has 4 edges and then multiply this by 2 because each edge is shared by 2 faces: F = 2E/4 which reduces to F = E/2 (since 2/4 = 1/2), from which it follows that F/E = 1/2 = 0.5 (divide both sides of F = E/2 by E to get F/E = 1/2). The ratio of F/E for this graph is smaller than the ratio of F/E for a MPG because 1/2 = 0.5 is smaller than 2/3, which rounds to 0.67.

5. For a MPG, E = 3V − 6 = 3(15) − 6 = 45 − 6 = 39. Plug V = 15 and E = 39 into Euler's formula, V + F = E + 2, to get 15 + F = 39 + 2 = 41. Subtract 15 from both sides to get F = 26. Our final answers are: E = 39 and F = 26. Check the answers: V + F = 15 + 26 = 41 and E + 2 = 39 + 2 = 41. Also, F/E = 26/39 = 2/3.

6. For a MPG, 2E = 3F = 3(10) = 30. Divide both sides by 2 to get E = 15. Plug F = 10 and E = 15 into Euler's formula, V + F = E + 2, to get V + 10 = 15 + 2 = 17. Subtract 10 from both sides to get V = 7. Our final answers are: V = 7 and E = 15. Check the answers: V + F = 7 + 10 = 17 and E + 2 = 15 + 2 = 17. Also, F/E = 10/15 = 2/3.

7. For a MPG, E = 3V − 6 so that 21 = 3V − 6. Add 6 to both sides to get 27 = 3V. Divide by 3 to get 9 = V. For a MPG, 2E = 3F so that 2(21) = 3F, which becomes 42 = 3F. Divide by 3 to get 14 = F. Our final answers are: V = 9 and F = 14. Check the answers: V + F = 9 + 14 = 23 and E + 2 = 21 + 2 = 23. Also, F/E = 14/21 = 2/3.

8. Plug V = 18 and E = 48 into V + F = E + 2 to get 18 + F = 48 + 2 = 50. Subtract 18 from both sides to get F = 32. The ratio of F/E = 32/48 = 2/3 so this graph could be a MPG. (As we'll learn in Chapter 6, it isn't necessarily a MPG, but it could be.) Since all MPG's are PG's, it could also be a PG.

9. Plug V = 32 and F = 54 into V + F = E + 2 to get 32 + 54 = 86 = E + 2. Subtract 2 from both sides to get 84 = E. The ratio F/E = 54/84 = 9/14 (divide the numerator and denominator each by 6 to see that 54/84 reduces to 9/14). Since 9/14 rounds to 0.64, this ratio is less than 2/3 which rounds to 0.67. Therefore, this graph can't be a MPG. Since F/E is less than 2/3, this graph could be a PG.

10. (There is no need to use V + F = E + 2.) Since E = 96 and F = 72, the ratio F/E is 72/96 = 3/4 = 0.75. Since 3/4 = 0.75, this ratio is greater than 2/3 which rounds to 0.67. Therefore, this graph can't be a MPG. Since F/E is greater than 2/3, this graph can't be a PG. (A MPG has the maximum possible ratio of F/E for any PG.)

11. Since every vertex has D = 5, we may use the formula V = 12/(6 – D) to get V = 12/(6 – 5) = 12/1 = 12. For a MPG, E = 3V – 6 = 3(12) – 6 = 36 – 6 = 30. Plug V = 12 and E = 30 into Euler's formula, V + F = E + 2, to get 12 + F = 30 + 2 = 32. Subtract 12 from both sides to get F = 20. Our final answers are: V = 12, E = 30, and F = 20. Check the answers: V + F = 12 + 20 = 32 and E + 2 = 30 + 2 = 32. Also, F/E = 20/30 = 2/3.

12. Note that the average degree is D = 5.5. Use the formula V = 12/(6 – D) to get 12/(6 – 5.5) = 12/0.5 = 24. For a MPG, E = 3V – 6 = 3(24) – 6 = 72 – 6 = 66. Plug V = 24 and E = 66 into Euler's formula, V + F = E + 2, to get 24 + F = 66 + 2 = 68. Subtract 24 from both sides to get F = 44. Our final answers are: V = 24, E = 66, and F = 44. Check the answers: V + F = 24 + 44 = 68 and E + 2 = 66 + 2 = 68. Also, F/E = 44/66 = 2/3.

13. The total number of degrees of the vertices is $2(8) + 2(3) + (V - 4)(4) = 16 + 6 + 4V - 16$, where V is the total number of vertices. This simplifies to $6 + 4V$ (since 16 cancels out). Divide the total number of degrees, $6 + 4V$, by the total number of vertices, V, to get the average degree, which is $D = (6 + 4V)/V$. Now use the formula $V = 12/(6 - D)$, where D is the average degree. Multiply both sides by $6 - D$ to get $(6 - D)V = 12$, which becomes $6V - DV = 12$. Subtract 12 from both sides and add DV to both sides to get $6V - 12 = DV$. Recall that the average degree is $D = (6 + 4V)/V$. Plug this into $6V - 12 = DV$ to get $6V - 12 = 6 + 4V$ (since $V/V = 1$). Add 12 to both sides and subtract 4V from both sides to get $2V = 18$. Divide both sides by 2 to get $V = 9$. Plug $V = 9$ into $E = 3V - 6$ to get $E = 3(9) - 6 = 27 - 6 = 21$. For a MPG, $3F = 2E$. Divide both sides by 3 to get $F = 2E/3 = 2(21)/3 = 42/3 = 14$. Our final answers are: $V = 9$, $E = 21$, and $F = 14$. Check the answers: $V + F = 9 + 14 = 23$ and $E + 2 = 21 + 2 = 23$. Also, $F/E = 14/21 = 2/3$.

Chapter 5 Complete Graphs and Bigraphs

1. Let $V = 12$ in the formula $V(V - 1)/2 = 12(11)/2 = 132/2 = 66$ for a complete graph with 12 vertices. There will be 66 handshakes.

2. Let $X = 9$ and $Y = 7$ in the formula $XY = 9(7) = 63$ for a complete bipartite graph with one set of 9 vertices and one set of 7 vertices. There will be 63 handshakes.

3. One way to redraw K_4 as a PG is shown below. The two graphs are isomorphic since in each case every vertex connects to all three of the other vertices. K_4 is a MPG because it is a PG and because it has the right number of edges according to Euler's formula for a MPG (Chapter 4). Set $V = 4$ in $E = 3V - 6$ to get $E = 3(4) - 6 = 12 - 6 = 6$, which agrees with $E = V(V - 1)/2 = 4(3)/2 = 12/2 = 6$.

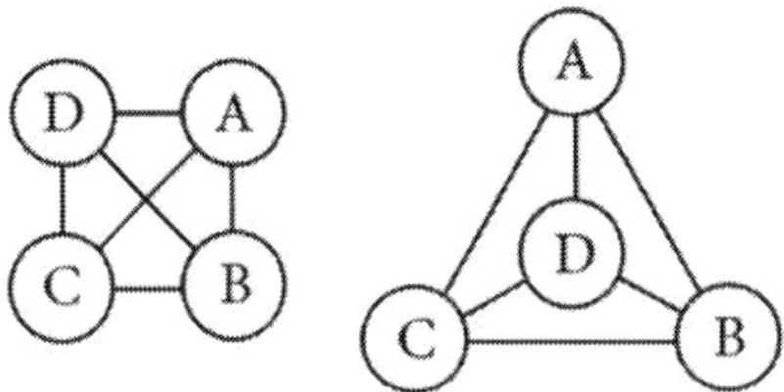

4. If one edge is removed from K_5, the new graph will be a PG. The graph on the right below shows how the new graph (with one edge removed) is a PG. The middle graph and right graph are isomorphic since they have the same edge-sharing: A connects to B, C, D, and E; B connects to A, C, D, and E; C connects to A, B, and D (but not E); D connects to A, B, C, and E; and E connects to A, B, and D (but not C). The new graph (with one edge removed) is a MPG because it is a PG and because it has the right number of edges according to Euler's formula for a MPG (Chapter 4). Set $V = 5$ in $E = 3V - 6$ to get $E = 3(5) - 6 = 15 - 6 = 9$. A complete graph with $V = 5$ vertices has $E = V(V - 1)/2 = 5(4)/2 = 20/2 = 10$ edges. Remove one edge from K_5 and it will have 9 edges, the same as a MPG with 9 vertices.

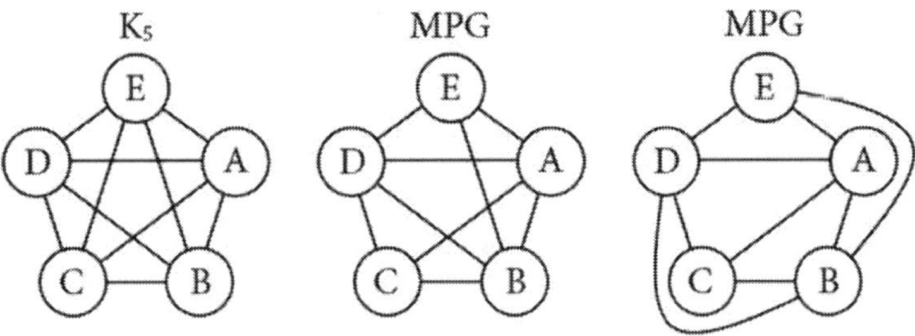

5. Set $V = 6$ in $E = V(V - 1) = 6(5)/2 = 30/2 = 15$ to see that K_6 has 15 edges. To see how many edges a MPG with $V = 6$ needs to have, use the formula $E = 3V - 6 = 3(6) - 6 = 18 - 6 = 12$. Subtract $15 - 12 = 3$ to see that we must remove 3 edges from K_6 to form a MPG. The second graph from the left below shows one way to remove 3 edges from K_6 to form a MPG: this MPG has two vertices with degree 5, two with degree 4, and two with degree 3. The third graph from the left below shows another way to remove 3 edges from K_6 to form a MPG: this MPG has six vertices with degree 4, which is structurally different from the other MPG. The two MPG's are redrawn below to show that they are indeed PG's. The graph on the right below shows how it is possible to remove 3 edges from K_6, yet not form a MPG. In Chapter 6, we'll learn a few methods to see that the graph on the right isn't a MPG.

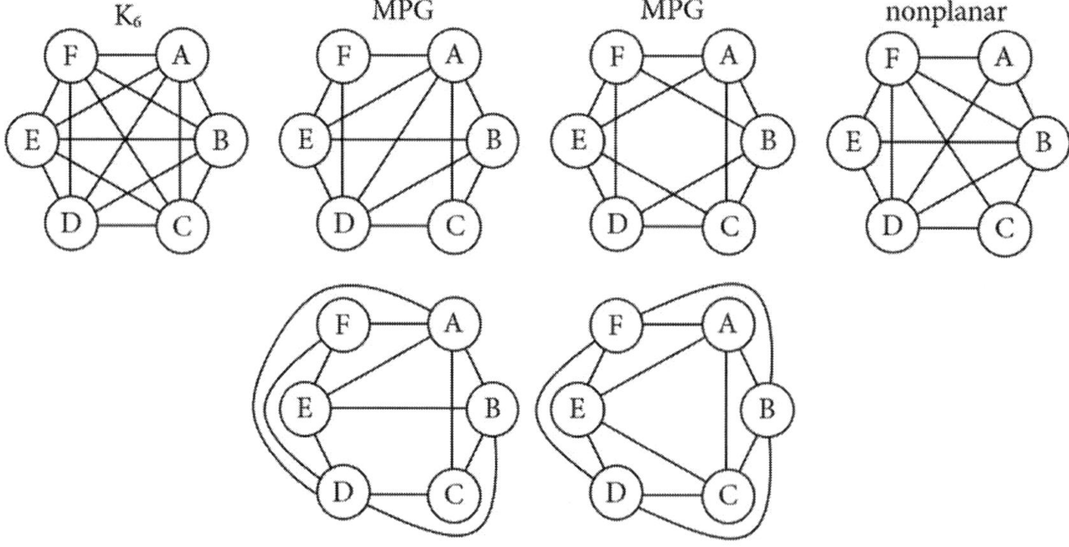

6. A complete graph with $V = 9$ vertices has $E = V(V - 1)/2 = 9(8)/2 = 72/2 = 36$ edges. According to the formulas from Chapter 4, a MPG with $V = 9$ vertices has $E = 3V - 6 = 3(9) - 6 = 27 - 6 = 21$ edges. If we subtract these, we get $36 - 21 = 15$. We need to remove 15 edges from a complete graph with 9 vertices in order to form a MPG with 9 vertices. The resulting graph won't necessarily be a MPG; it depends on which 15 edges are removed. The previous problem shows that removing the right number of edges may or may not result in a MPG; it depends on which edges are removed.

7. A complete graph with V vertices has $E = V(V - 1)/2$ edges. A MPG with V vertices has $E = 3V - 6$ edges according to Chapter 4. (These two values of E are different. You might prefer to call them E_C and E_M instead of using the same symbol E for both.) Subtract these expressions to determine how many edges need to be removed from a complete graph to form a MPG. Some algebra is involved, especially the distributive property. Note, for example, that $V(V - 1) = V^2 - V$ and that $-(3V - 6) = -3V - (-6) = -3V + 6$. Factoring is also used. For example, $-V/2 - 3V = -V(1/2 + 3) = -7V/2$ (note that $3 + 1/2 = 3.5 = 7/2$). The last step below involves the foil method backwards. Note that $(V - 3)(V - 4) = V^2 - 3V - 4V + 12 = V^2 - 7V + 12$.

$$\frac{V(V - 1)}{2} - (3V - 6) = \frac{V^2 - V}{2} - 3V + 6 = \frac{V^2}{2} - \frac{V}{2} - 3V + 6$$

$$= \frac{V^2}{2} - V\left(\frac{1}{2} + 3\right) + 6 = \frac{V^2}{2} - \frac{7V}{2} + 6 = \frac{V^2}{2} - \frac{7V}{2} + \frac{12}{2}$$

$$= \frac{V^2 - 7V + 12}{2} = \frac{(V - 3)(V - 4)}{2}$$

Let's check this answer using numbers for $V = 10$. A complete graph has $E = 10(9)/2 = 90/2 = 45$ edges. A MPG has $E = 3(10) - 6 = 30 - 6 = 24$ edges. The difference is $45 - 24 = 21$ edges. This agrees with the formula $(V - 3)(V - 4)/2 = (10 - 3)(10 - 4)/2 = 7(6)/2 = 42/2 = 21$ edges. As we observed in the previous problem, the resulting graph with the edges removed may or may not be a MPG; it depends on which edges are removed.

8. K_5 has $E = V(V - 1)/2 = 5(4)/2 = 20/2 = 10$ edges. $K_{2,3}$ has $E = XY = 2(3) = 6$ edges. Subtract these to see that $10 - 6 = 4$ edges need to be removed from K_5 to form $K_{2,3}$. The diagram below shows one way to remove 4 edges from K_5 to form $K_{2,3}$. In the graph on the right, set P includes A and B, while set Q includes C, D, and E. Observe that each member of set P (A and B) connects to each member of set Q (C, D, and E), no member of set P (A and B) connects to the other, and no member of set Q (C, D, and E) connects to another.

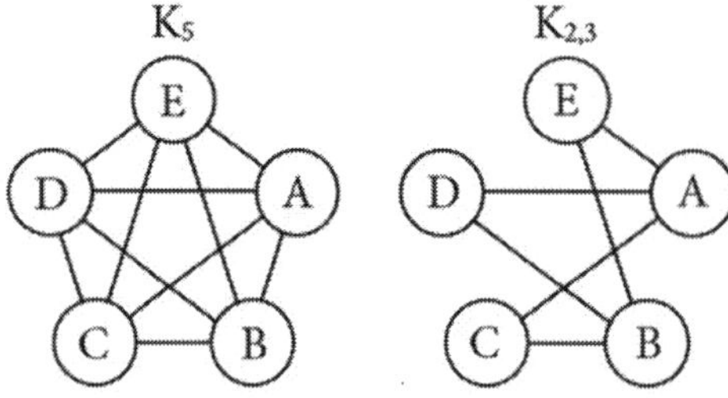

9. K_6 has $E = V(V - 1)/2 = 6(5)/2 = 30/2 = 15$ edges. $K_{3,3}$ has $E = XY = 3(3) = 9$ edges. Subtract these to see that $15 - 9 = 6$ edges need to be removed from K_6 to form $K_{3,3}$. The diagram below shows two ways to remove 6 edges from K_6 to form $K_{3,3}$. In the middle graph, set P includes A, B, and C, while set Q includes D, E, and F. In the right graph, set P includes A, C, and E, while set Q includes B, D, and F. The two ways are really equivalent in the sense that both ways form $K_{3,3}$; the only difference has to do with which vertices belong to the two sets, P and Q.

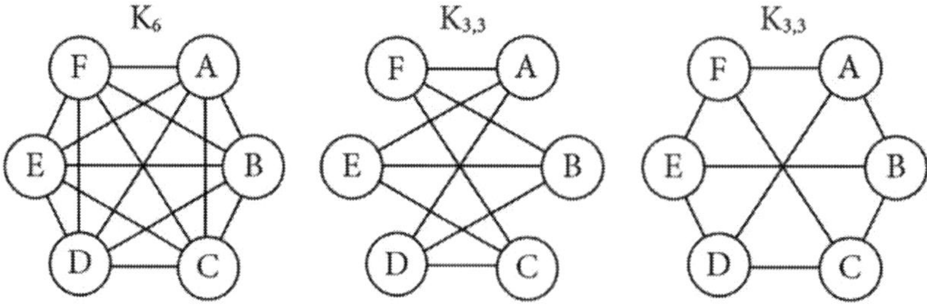

10. K_V has $E = V(V - 1)/2$ edges. $K_{N,N}$ has $E = N(N) = N^2$ edges. Plug in $N = V/2$ to get $E = (V/2)^2 = V^2/4$. (These two values of E are different. You might prefer to call them E_C and E_B instead of using the same symbol E for both.) Subtract $V^2/4$ from $V(V - 1)/2$ to determine how many edges need to be removed from K_V to form $K_{N,N}$. Some algebra is involved, especially the distributive property. Note that $2V(V - 1) = 2V^2 - 2V$. We multiplied the first fraction by 2/2 in order to make a common denominator of 4.

$$\frac{V(V - 1)}{2} - \frac{V^2}{4} = \frac{2V(V - 1)}{4} - \frac{V^2}{4} = \frac{2V^2 - 2V}{4} - \frac{V^2}{4} = \frac{2V^2 - 2V - V^2}{4} = \frac{V^2 - 2V}{4}$$

$$= \frac{V(V - 2)}{4}$$

Let's check this answer using numbers for $V = 10$. In this case, K_{10} has $E = 10(9)/2 = 90/2 = 45$ edges and $K_{5,5}$ has $E = 5(5) = 25$ edges. The difference is $45 - 25 = 20$ edges. This agrees with the formula $V(V - 2)/4 = 10(10 - 2)/4 = 10(8)/4 = 80/4 = 20$ edges. The resulting graph with the edges removed may or may not be $K_{5,5}$; it depends on which edges are removed.

Chapter 6 Maximal Planar Graphs

1. The graph below has $V = 5$ vertices. The degrees add up to $2 + 2 + 2 + 3 + 3 = 12$. Divide the sum of the degrees by 2 to get the number of edges: $E = 12/2 = 6$. A MPG with 5 vertices would have $E = 3V - 6 = 3(5) - 6 = 15 - 6 = 9$ edges. This graph doesn't have enough edges to be a MPG. Based on the exclusion test, at this point "it could be a PG, but not a MPG." It turns out that this graph is a PG, as the redrawn graph below on the right shows. (Recall our definition of PG from the end of the chapter.)

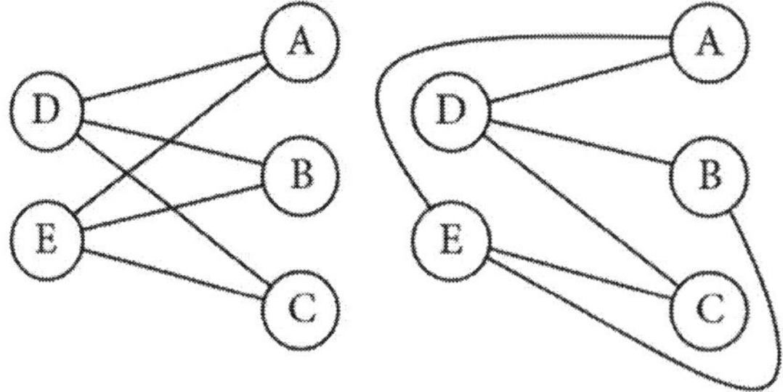

The graph below has $V = 7$ vertices. The degrees add up to $4 + 4 + 4 + 3 + 3 + 3 + 3 = 24$. (This graph is obviously $K_{3,4}$.) Divide the sum of the degrees by 2 to get the number of edges: $E = 24/2 = 12$. A MPG with 7 vertices would have $E = 3V - 6 = 3(7) - 6 = 21 - 6 = 15$ edges. This graph doesn't have enough edges to be a MPG. Based on the exclusion test, at this point "it could be a PG, but not a MPG." It turns out that this graph is nonplanar. The graph below on the right shows that the left graph has $K_{3,3}$ as a subgraph. Here we used Kuratowski's theorem.

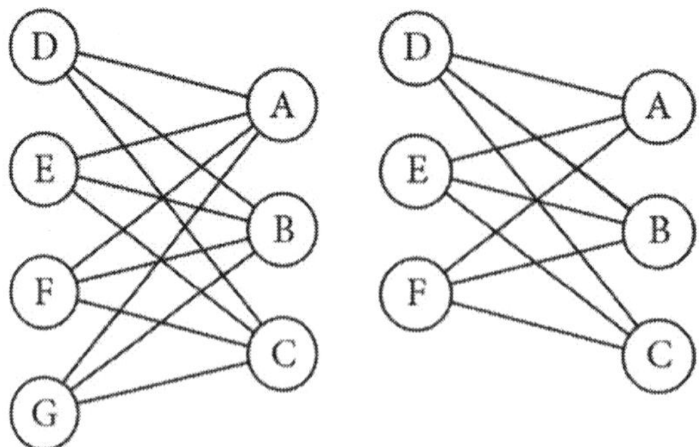

The graph below has V = 6 vertices. The degrees add up to 4 + 4 + 4 + 4 + 4 + 4 = 24. Divide the sum of the degrees by 2 to get the number of edges: E = 24/2 = 12. A MPG with 6 vertices would have E = 3V − 6 = 3(6) − 6 = 18 − 6 = 12 edges. This graph has the right number of edges to be a MPG. Based on the exclusion test, at this point "it could be a MPG." It turns out that this graph is a MPG, as the redrawn graph below on the right shows.

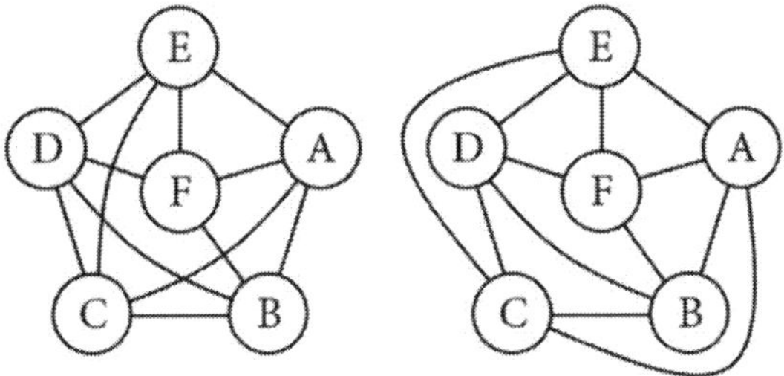

The graph below has V = 8 vertices. The degrees add up to 4 + 4 + 4 + 4 + 4 + 4 + 7 + 7 = 38. Divide the sum of the degrees by 2 to get the number of edges: E = 38/2 = 19. A MPG with 8 vertices would have E = 3V − 6 = 3(8) − 6 = 24 − 6 = 18 edges. This graph has too many edges to be a MPG. The exclusion test shows that "it is definitely nonplanar." The graph below on the right shows that the left graph contains a subgraph that is a subdivision of $K_{3,3}$. Set P is A, D, and H and set Q is F, C, and G. To see $K_{3,3}$, ignore B and E, and look at how the members of sets P and Q connect. Here we used Kuratowski's theorem.

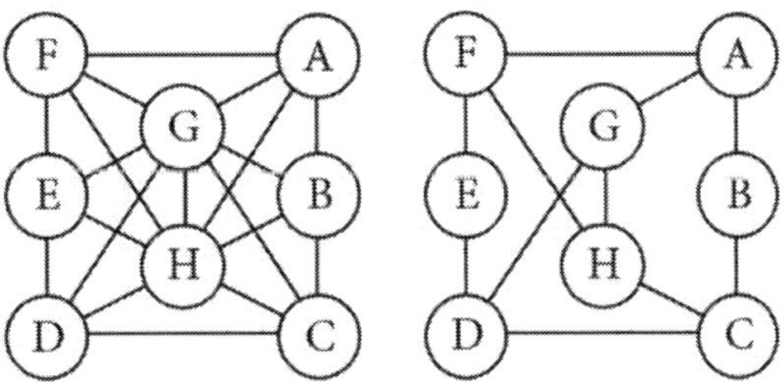

The graph below has V = 8 vertices. The degrees add up to $6 + 3 + 3 + 5 + 5 + 3 + 4 + 5 = 34$. Divide the sum of the degrees by 2 to get the number of edges: $E = 34/2 = 17$. A MPG with 8 vertices would have $E = 3V - 6 = 3(8) - 6 = 24 - 6 = 18$ edges. This graph doesn't have enough edges to be a MPG. Based on the exclusion test, at this point "it could be a PG, but not a MPG." It turns out that this graph is a $\boxed{\text{PG}}$, as the redrawn graph below on the right shows. (You could turn it into a MPG by adding edge BH.)

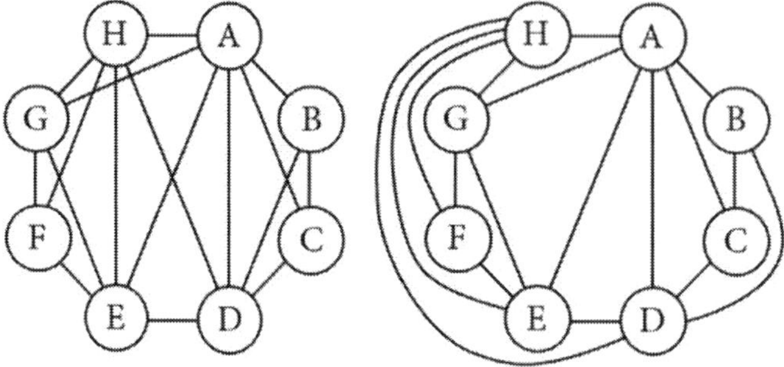

The graph below has V = 8 vertices. The degrees add up to $4 + 5 + 5 + 3 + 6 + 3 + 5 + 5 = 36$. Divide the sum of the degrees by 2 to get the number of edges: $E = 36/2 = 18$. A MPG with 8 vertices would have $E = 3V - 6 = 3(8) - 6 = 24 - 6 = 18$ edges. This graph has the right number of edges to be a MPG. Based on the exclusion test, at this point "it could be a MPG." It turns out that this graph is $\boxed{\text{nonplanar}}$. The graph below on the right shows that the left graph's minors include K_5. When we contracted EF, AF became AE. When we contracted CD, DG became CG. In the last step, it doesn't matter if you contract AB or AH. Here we used Wagner's theorem.

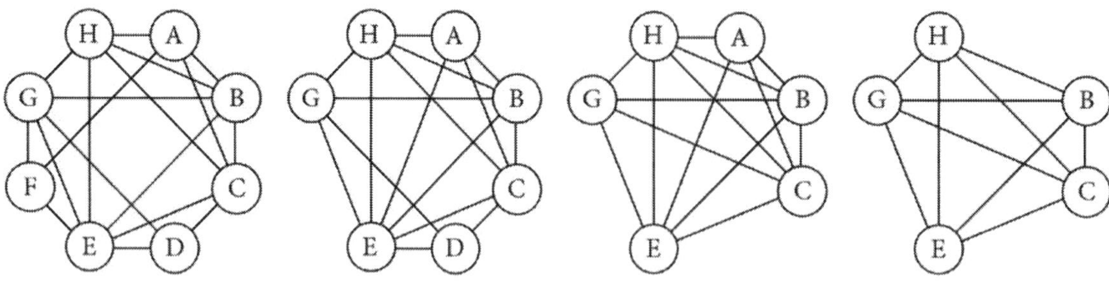

The graph below has V = 10 vertices. The degrees add up to 4 + 3 + 4 + 3 + 4 + 4 + 3 + 4 + 3 + 4 = 36. Divide the sum of the degrees by 2 to get the number of edges: E = 36/2 = 18. A MPG with 10 vertices would have E = 3V − 6 = 3(10) − 6 = 30 − 6 = 24 edges. This graph doesn't have enough edges to be a MPG. Based on the exclusion test, at this point "it could be a PG, but not a MPG." It turns out that this graph is nonplanar. The graph below on the right shows that the left graph's minors include $K_{3,3}$. We first contracted BC and CD. When we contracted AC, AH became CH. Lastly, we contracted CE. Set P is C, G, and I and set Q is F, H, and J. To see $K_{3,3}$, look at how the members of sets P and Q connect. Here we used Wagner's theorem.

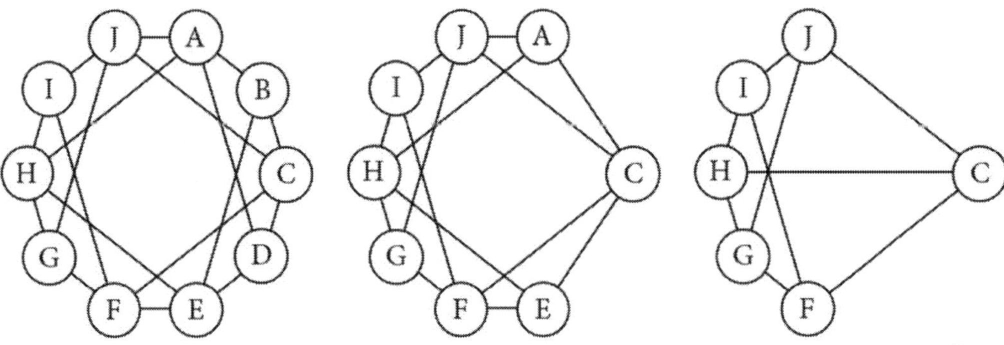

The graph below has V = 10 vertices. The degrees add up to 5 + 4 + 3 + 3 + 5 + 5 + 3 + 3 + 5 + 4 = 40. Divide the sum of the degrees by 2 to get the number of edges: E = 40/2 = 20. A MPG with 10 vertices would have E = 3V – 6 = 3(10) – 6 = 30 – 6 = 24 edges. This graph doesn't have enough edges to be a MPG. Based on the exclusion test, at this point "it could be a PG, but not a MPG." It turns out that this graph $\boxed{\text{nonplanar}}$. The graph below on the right shows that the left graph's minors include K_5. We first contracted HI, then GH, then CD, then DE, and finally AJ. Here we used Wagner's theorem.

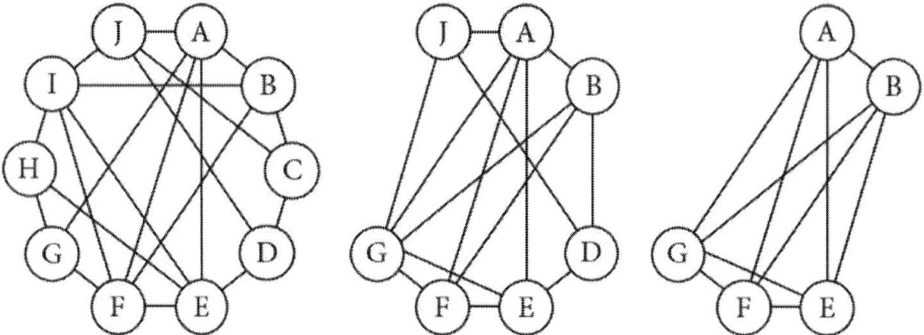

The graph below has V = 9 vertices. The degrees add up to 4 + 4 + 4 + 4 + 4 + 5 + 5 + 5 + 5 = 40. (This graph is obviously $K_{5,4}$.) Divide the sum of the degrees by 2 to get the number of edges: E = 40/2 = 20. A MPG with 9 vertices would have E = 3V – 6 = 3(9) – 6 = 27 – 6 = 21 edges. This graph doesn't have enough edges to be a MPG. Based on the exclusion test, at this point "it could be a PG, but not a MPG." It turns out that this graph is $\boxed{\text{nonplanar}}$. The graph below on the right shows that the left graph has $K_{3,3}$ as a subgraph. Here we used Kuratowski's theorem.

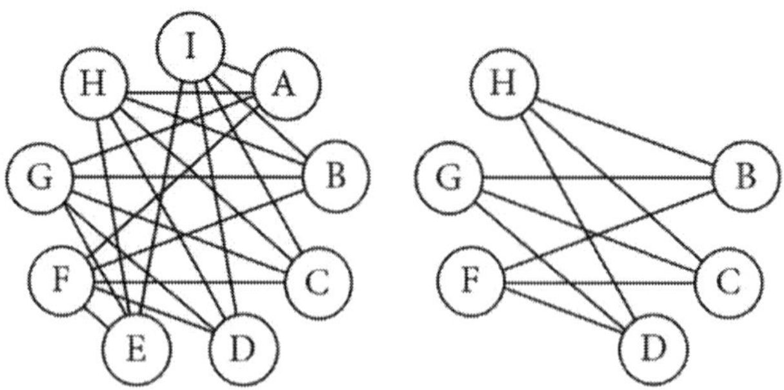

The graph below has V = 5 vertices. The degrees add up to 3 + 4 + 4 + 4 + 3 = 18. Divide the sum of the degrees by 2 to get the number of edges: E = 18/2 = 9. A MPG with 5 vertices would have E = 3V − 6 = 3(5) − 6 = 15 − 6 = 9 edges. This graph has the right number of edges to be a MPG. Based on the exclusion test, at this point "it could be a MPG." It turns out that this graph is a MPG, as the redrawn graph below on the right shows.

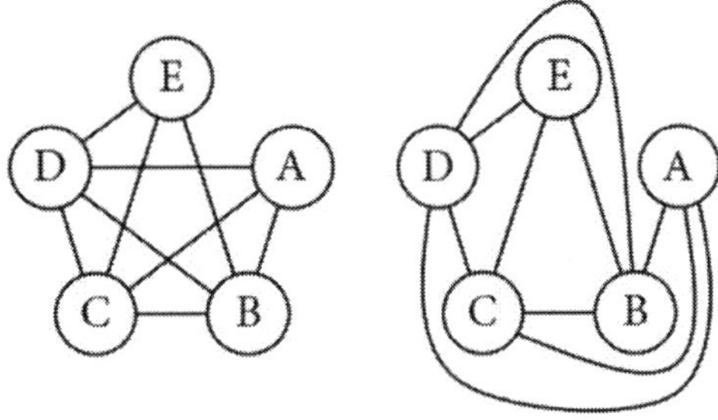

The graph below has V = 9 vertices. The degrees add up to 3 + 4 + 5 + 6 + 8 + 4 + 3 + 6 + 3 = 42. Divide the sum of the degrees by 2 to get the number of edges: E = 42/2 = 21. A MPG with 9 vertices would have E = 3V − 6 = 3(9) − 6 = 27 − 6 = 21 edges. This graph has the right number of edges to be a MPG. Based on the exclusion test, at this point "it could be a MPG." It turns out that this graph is a MPG, as the redrawn graph below on the right shows.

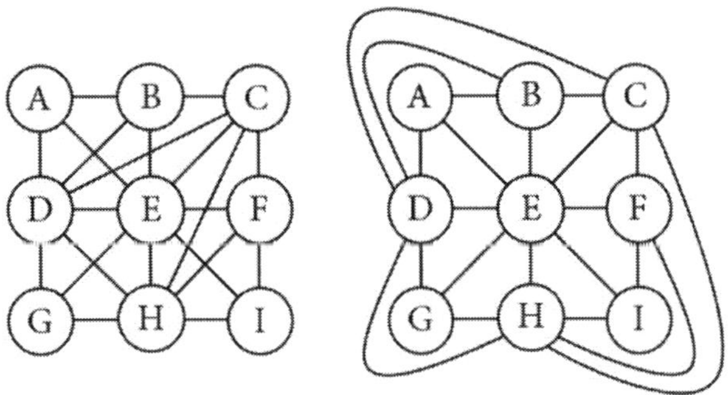

The graph below has V = 10 vertices. The degrees add up to 2 + 2 + 2 + 2 + 2 + 2 + 1 + 2 + 2 + 3 = 20. Divide the sum of the degrees by 2 to get the number of edges: E = 20/2 = 10. A MPG with 10 vertices would have E = 3V − 6 = 3(10) − 6 = 30 − 6 = 24 edges. This graph doesn't have enough edges to be a MPG. Based on the exclusion test, at this point "it could be a PG, but not a MPG." It turns out that this graph is a PG, as the redrawn graph below on the right shows.

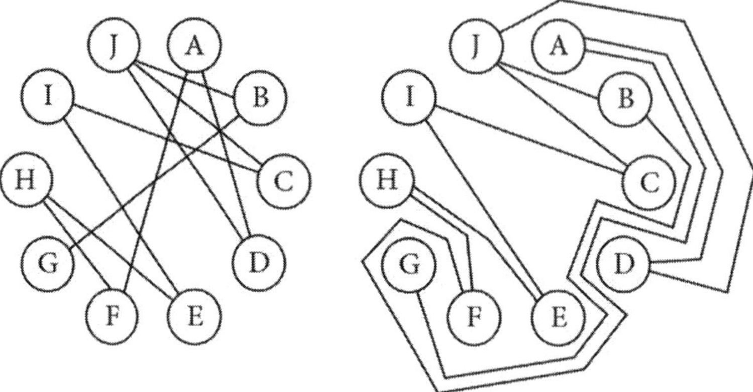

Challenge Problem: The answer key doesn't include answers to challenge problems. These problems are intended to encourage you to think about the ideas.

Chapter 7 Kempe Chains

1. The new graph with the longer section of the G-Y Kempe chains reversed is shown below on the right. The reversed chain is shaded gray. The new coloring satisfies the four-color theorem. Any section of any Kempe chain may be reversed in the new graph and still satisfy the four-color theorem (as long as the entire section is reversed). If the coloring of a graph satisfies the four-color theorem and sections of Kempe chains are reversed one at a time, the new coloring always satisfies the four-color theorem. Each step begins with a coloring that satisfies the four-color theorem and reverses the colors for a single section of Kempe chains. (By single section, we mean separated from other sections of the same Kempe chain.)

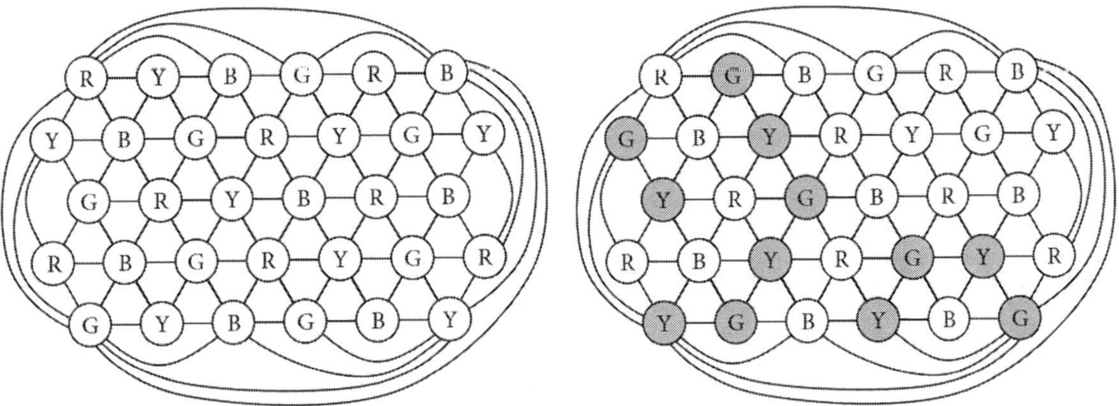

2. The graph given in this problem doesn't in any way impact Kempe's argument for a vertex with degree four. The A-B and C-D Kempe chains drawn are completely irrelevant. There is no reason to use Kempe chains involving neighboring vertices (such as A and B). Kempe's argument involves non-adjacent vertices (such as A and C). If A and C happen to be part of the same section of an A-C Kempe chain, we can use a color reversal to make vertices B and D have the same color; otherwise, we can use a color reversal to make vertices A and C have the same color.

Challenge Problems: The answer key doesn't include answers to challenge problems. These problems are intended to encourage you to think about the ideas.

Chapter 8 A Few Notable Planar Graphs

1. Five structurally different heptahedral MPG's are shown below. There is more than one way to draw each graph. You can distinguish these graphs by the degrees of the vertices:

- top left: two 6's (A and F), three 4's (B, D, and E), two 3's (C and G).

- top right: one 6 (A), three 5's (C, E, and F), three 3's (B, D, and G).

- bottom left: one 6 (A), two 5's (E and F), two 4's (C and D), two 3's (B and G).

- bottom center: three 5's (B, D, and F), three 4's (A, C, and G), one 3 (E).

- bottom right: two 5's (C and D), five 4's (A, B, E, F, and G).

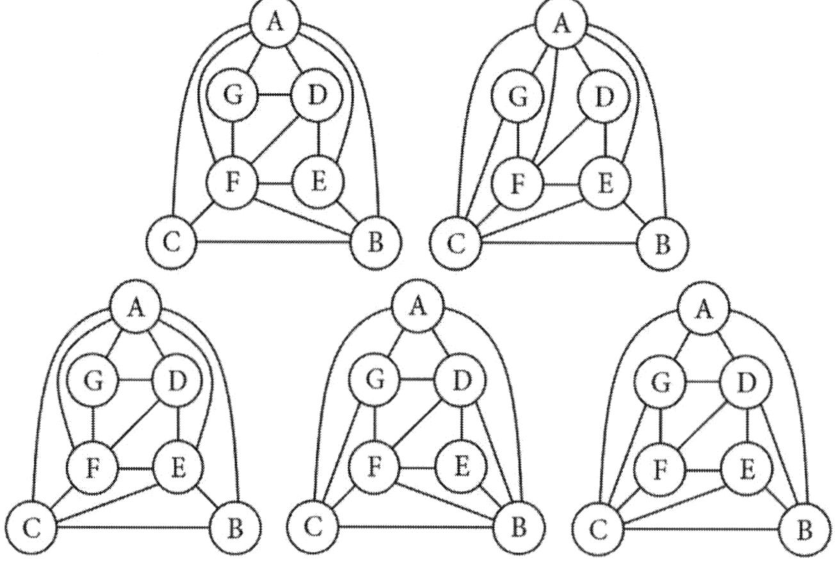

2. Each MPG below has 8 vertices. In the left MPG, no vertex has degree 3. In the right MPG, two vertices have degree 6 (A and C) and two vertices have degree 3 (F and H).

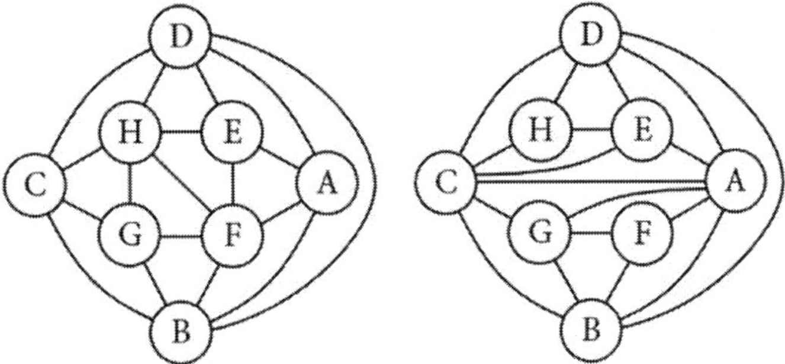

3. The MPG below has 9 vertices (and in order to be a MPG, it has 21 edges). Three of the vertices have degree 6 (G, H, and I).

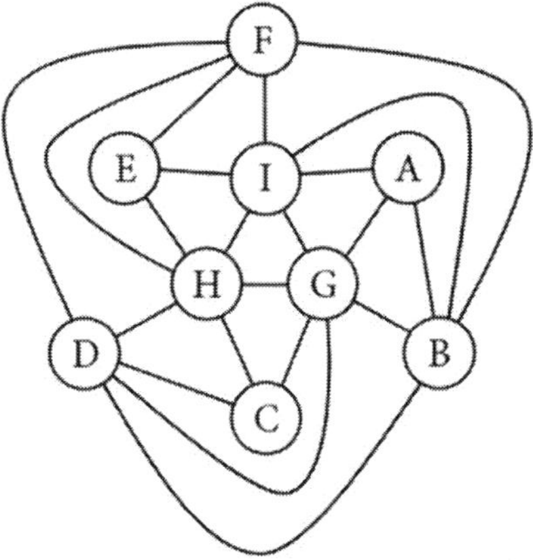

4. The graph below on the left has the colors 2 and 4 reversed in the 2-4 Kempe chain that involves vertex A. The graph below on the right has the colors 2 and 3 reversed in the 2-3 Kempe chain that involves vertex E. Vertex I isn't four-colorable, since, as an unintended consequence, vertex G changed from 3 to 2 when E changed from 2 to 3. Vertex I still connects to vertices with four different colors.

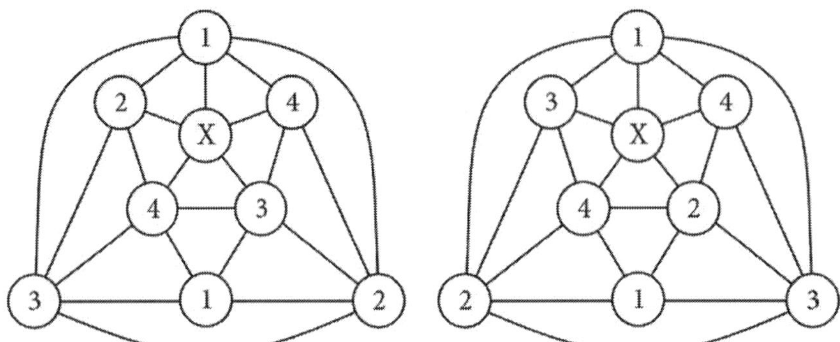

Both Kempe chains are reversed simultaneously in the graph below. This time, vertex I is four-colorable: it may be color 2. However, the graph isn't properly four-colored. As an unintended consequence of reversing two different Kempe chains "at the same time," vertices B and D are each color 2 and are connected by an edge. Since two vertices of the same color aren't allowed to share an edge, this coloring isn't allowed. This highlights the problem of reversing two Kempe chains simultaneously.

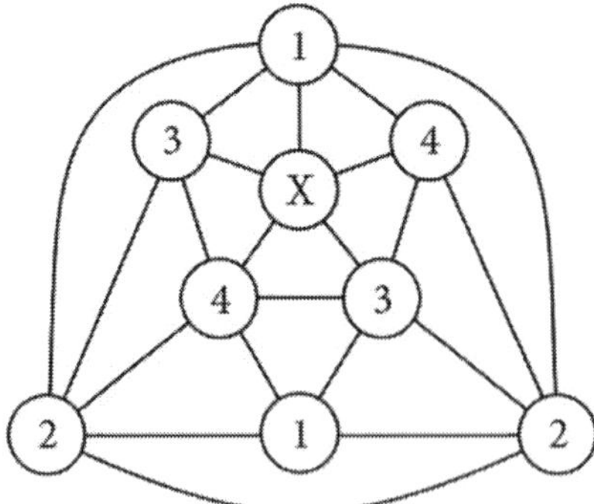

Although the Fritsch MPG illustrates a problem with Kempe's attempted proof of the four-color theorem, the Fritsch MPG can be properly four-colored. It would be a good exercise to four-color the Fritsch MPG.

Chapter 9 Counting Ways

1. One solution for each graph is shown below. Study one color at a time and check that no two vertices of that color share an edge.

- The remaining vertices of the left graph are four-colorable 4 ways. The left graph contains two K_4 subgraphs: GYRB on the left and BGYR on the right. 7 of its 8 vertices participate in K_4 subgraphs.

- The remaining vertices of the middle graph are four-colorable 4 ways. The middle graph contains three K_4 subgraphs: BYGR on the top, RBYG on the right, and BRYG on the left. All 9 of its vertices participate in K_4 subgraphs.

- The remaining vertices of the right graph are four-colorable 5 ways. The right graph doesn't contain any K_4 subgraphs. None of its vertices participate in K_4 subgraphs. Yet this graph is only slightly less restrictive with regard to coloring than the previous two graphs, which shows that there is more to this than the number of K_4 subgraphs and the number of vertices involved in them.

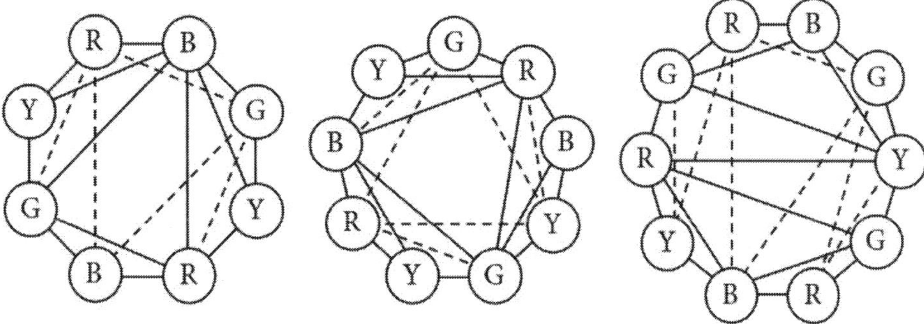

Challenge Problems: The answer key doesn't include answers to challenge problems. These problems are intended to encourage you to think about the ideas.

Chapter 10 Logic Puzzle

1. We first entered ×'s to indicate which vertices share edges and therefore can't be the same color. For the given graph:

- A connects to B, C, D, E, and I.
- B connects to A, C, D, and E.
- C connects to A, B, and D.
- D connects to A, B, C, E, H, and I.
- E connects to A, B, D, F, H, and I.
- F connects to E, G, H, and I.
- G connects to F, H, and I.
- H connects to D, E, F, G, and I.
- I connects to A, D, E, F, G, and H.

	A	B	C	D	E	F	G	H	I	1	2	3	4
A		×	×	×	×				×	✓			
B	×		×	×	×						✓		
C	×	×		×									
D	×	×	×		×			×	×			✓	
E	×	×		×		×		×	×				
F					×		×	×	×				
G						×		×	×				
H				×	×	×	×		×				
I	×			×	×	×	×	×					

Next we determined the colors of the following vertices:

- Since C can't be the same as A, B, or D, it follows that C is color 4.

- Since E can't be the same as A, B, or D, it follows that E is color 4.

- Since I can't be the same as A, D, or E, it follows that I is color 2.

- Since H can't be the same as D, E, or I, it follows that H is color 1.

- Since F can't be the same as E, H, or I, it follows that F is color 3.

- Since G can't be the same as F, H, or I, it follows that G is color 4.

	A	B	C	D	E	F	G	H	I	1	2	3	4
A		×	×	×	×				×	✓			
B	×		×	×	×						✓		
C	×	×		×									✓
D	×	×	×		×			×	×			✓	
E	×	×		×		×		×	×				✓
F					×		×	×	×			✓	
G						×		×	×				✓
H				×	×	×	×		×	✓			
I	×			×	×	×	×	×			✓		

The graph below is colored according to the logic table. Note that every vertex of this MPG participates in one of its many K_4 subgraphs. The coloring of this MPG has a single solution. The dashed lines show that this graph is a MPG. Imagine pushing the dashed lines outside of the polygon. (Recall our definition of PG from the end of Chapter 6.)

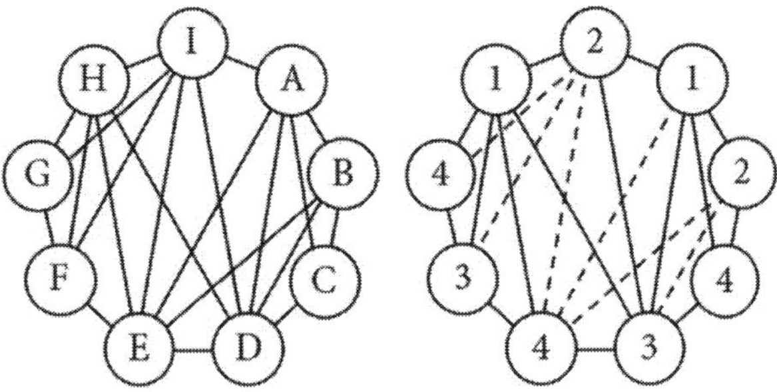

2. We first entered ×'s to indicate which vertices share edges and therefore can't be the same color. For the given graph:

- A connects to B, C, F, and G.

- B connects to A, C, D, E, and G.

- C connects to A, B, D, and F.

- D connects to B, C, E, and F.

- E connects to B, D, F, and G.

- F connects to A, C, D, E, and G.

- G connects to A, B, E, and F.

	A	B	C	D	E	F	G	1	2	3	4
A		×	×			×	×	✓			
B	×		×	×	×		×				
C	×	×		×		×			✓		
D		×	×		×	×					
E		×		×		×	×				
F	×		×	×	×		×			✓	
G	×	×			×	×					

Next we determined that B must be color 3. Why? If you try color 4 for B, this forces G to be color 2 and forces D to be color 1, and then E would connect to colors 1, 2, 3, and 4, so no color would be available for E. This means that B can't be 4. The only other choice for B is color 3.

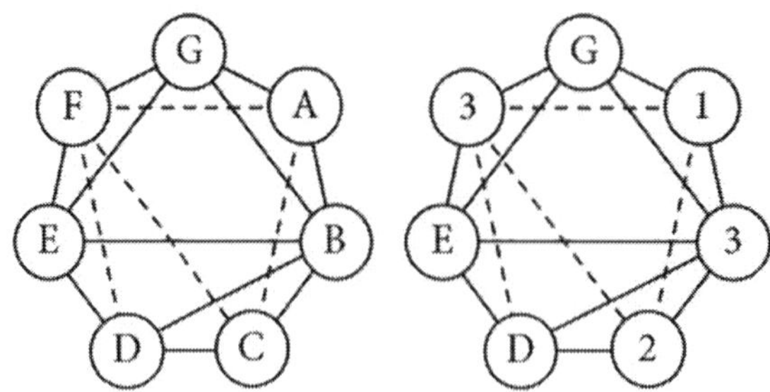

	A	B	C	D	E	F	G	1	2	3	4
A		×	×			×	×	✓			
B	×		×	×	×		×			✓	
C	×	×		×		×			✓		
D		×	×		×	×					
E		×		×		×	×				
F	×		×	×	×		×			✓	
G	×	×			×	×					

There isn't a unique solution to this logic table. There are five possible solutions, as shown below. There are two solutions when G is color 2 and three solutions when G is color 4. This graph doesn't contain any K_4 subgraphs, and the coloring isn't very restrictive. (The dashed lines show that this graph is a MPG. Imagine pushing the dashed lines outside of the polygon.)

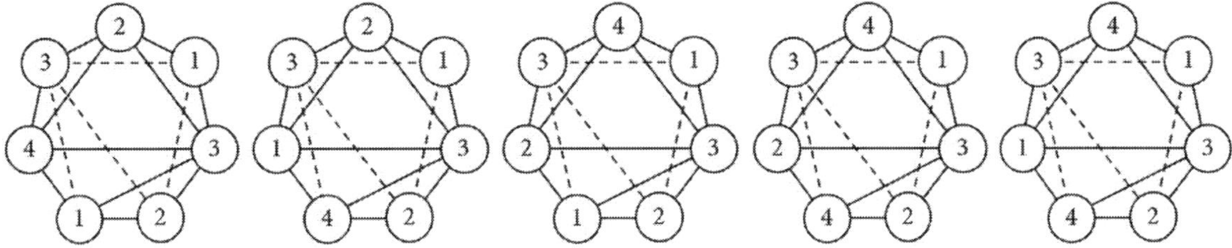

3. We first entered ×'s to indicate which vertices share edges and therefore can't be the same color. For the given graph:

- A connects to B, C, D, F, and G.
- B connects to A, C, and F.
- C connects to A, B, D, F, and G.
- D connects to A, C, E, F, and G.
- E connects to D and F.
- F connects to A, B, C, D, E, and G.
- G connects to A, C, D, and F.

	A	B	C	D	E	F	G	1	2	3	4
A		×	×	×		×	×				
B	×		×			×					
C	×	×		×		×	×	✓			
D	×		×		×	×	×		✓		
E				×		×				✓	
F	×	×	×	×	×		×				
G	×		×	×		×					

This logic table doesn't have a solution. This graph isn't four-colorable. Oh wow, did we just disprove the four-color theorem? No! There is a much simpler explanation. This graph isn't planar. As shown below, this graph contains a K_5 subgraph (which makes the graph non-planar according to Kuratowski's theorem; see Chapter 6).

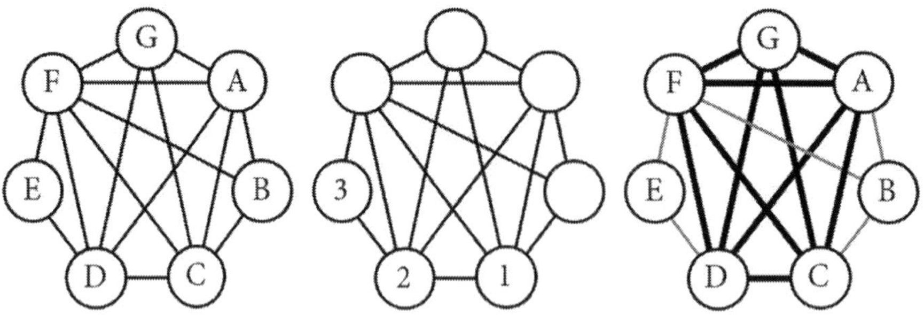

4. We first entered ×'s to indicate which vertices share edges and therefore can't be the same color. For the given graph:

- A connects to B, E, and F.
- B connects to A, C, and F.
- C connects to B, D, and F.
- D connects to C, E, and F.
- E connects to A, D, and F.
- F connects to A, B, C, D, and E.

	A	B	C	D	E	F	1	2	3
A		×			×	×	✓		
B	×		×			×		✓	
C		×		×		×			
D			×		×	×			
E	×			×		×			
F	×	×	×	×	×				

Next we determined the colors of the following vertices:

- Since F can't be the same as A or B, it follows that F is color 3.
- Since E can't be the same as A or F, it follows that E is color 2.
- Since D can't be the same as E or F, it follows that D is color 1.

	A	B	C	D	E	F	1	2	3
A		×			×	×	✓		
B	×		×			×		✓	
C		×		×		×			
D			×		×	×	✓		
E	×			×		×		✓	
F	×	×	×	×	×				✓

Now we run into a contradiction. Since C can't be the same as B, D, or F (which are colors 2, 1, and 3, respectively), no color is available for vertex C. This graph isn't three-colorable. You can see the problem visually on the following page.

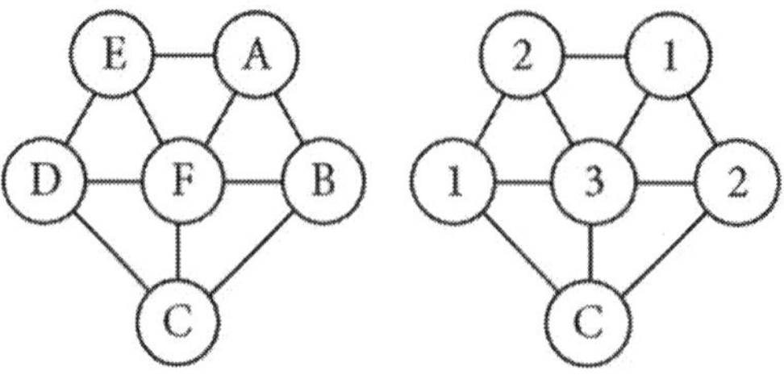

(Note that this graph is a PG, but not a MPG. However, that's irrelevant to the main idea here.) Why did this happen? This PG doesn't contain any K_4 subgraphs, which had been the problem in the last example of this chapter. One way to see that this type of graph isn't three-colorable is to realize that F is a vertex with degree five. In order for F to be three-colorable, A thru E would need to use only two colors, but that's not possible for a cycle (which is a closed loop) consisting of an odd number of vertices. If we add a vertex to make F have degree six (or remove one vertex so that F has degree four), then the graph would be three-colorable. (Consider how this relates to the challenge problem of this chapter and also the challenge problem from Chapter 6, if at all.)

Challenge Problem: The answer key doesn't include answers to challenge problems. These problems are intended to encourage you to think about the ideas.

Chapter 11 Trivial Four-Coloring

1. One possible solution for how to order the vertices and color the MPG is shown below. There are other possible solutions. Check your answers:

- For the numerical order, study one vertex at a time in order. Make sure that no vertex connects to more than three vertices with a smaller number. For example, if vertex 8 connects to vertices 2, 4, 6, and 7, that would be a mistake. (It's okay for a vertex to connect to four or more vertices, provided that no more than three of the vertices have a smaller number.)

- For the coloring, study one color at a time and check that no two vertices of that color share an edge.

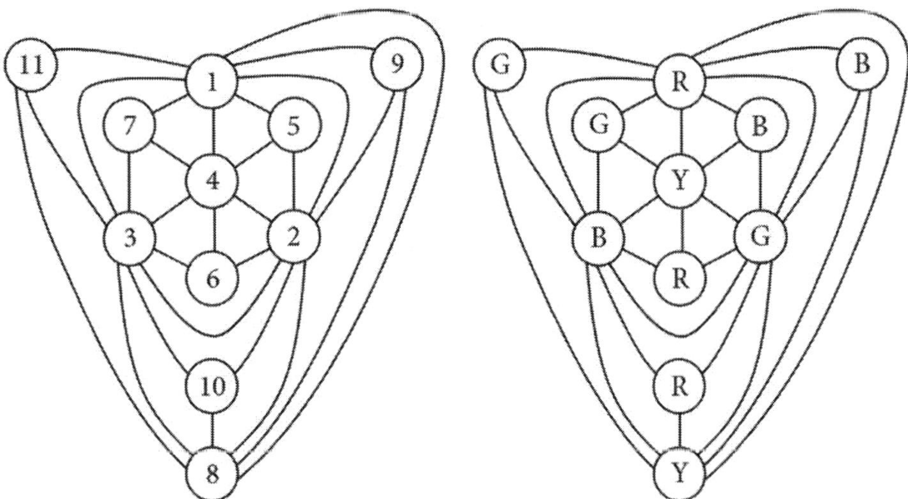

2. One possible solution for how to order the regions and color the map is shown below. There are other possible solutions. Check your answers:

- For the numerical order, study at one region at a time in order. Make sure that no region shares an edge with more than three regions with a smaller number. For example, if region 8 shares an edge with regions 2, 4, 6, and 7, that would be a mistake. (It's okay for a region to share an edge with four or more regions, provided that no more than three of the regions have a smaller number.)

- For the coloring, study one color at a time and check that no two regions of that color share an edge.

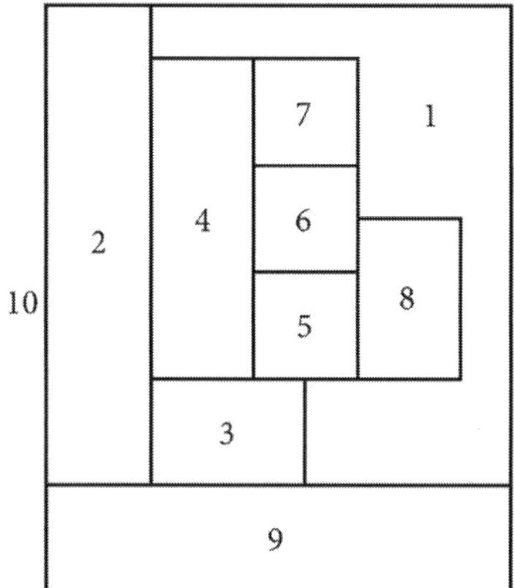 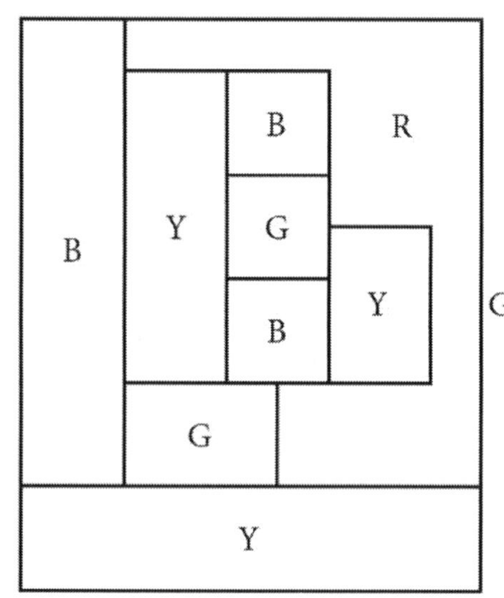

3. The diagrams below show the original graph and the new graph with all of the vertices with degree three removed. The two vertices shaded gray with dashed lines are also removed. These two vertices became degree three after the other 13 were removed. This graph won't completely unravel unless vertices with degree four are also (successively) removed.

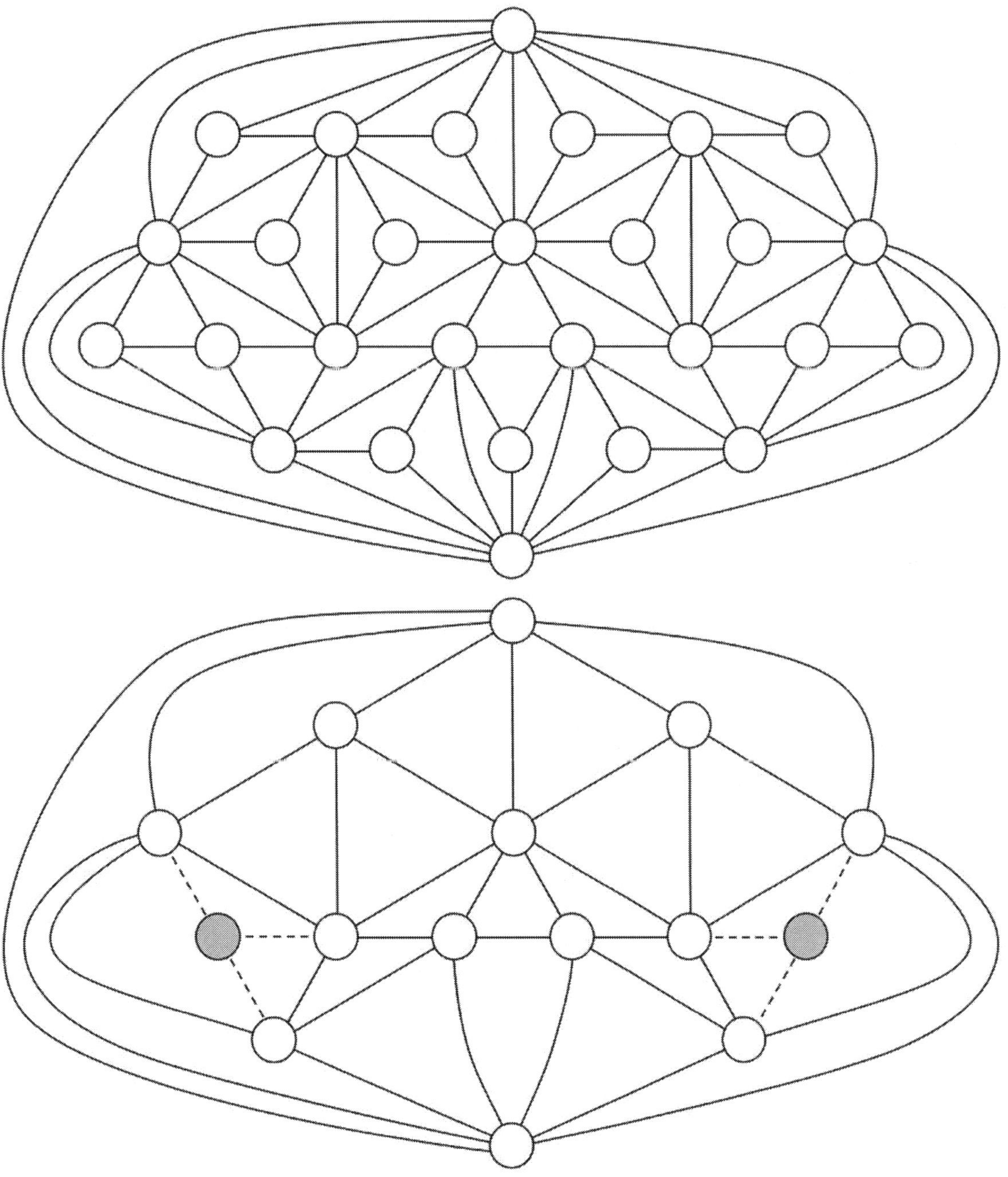

4. One possible solution for how to partially color the MPG for the order given is shown below. There are other possible solutions. Check your answer: Study one color at a time and check that no two vertices of that color share an edge.

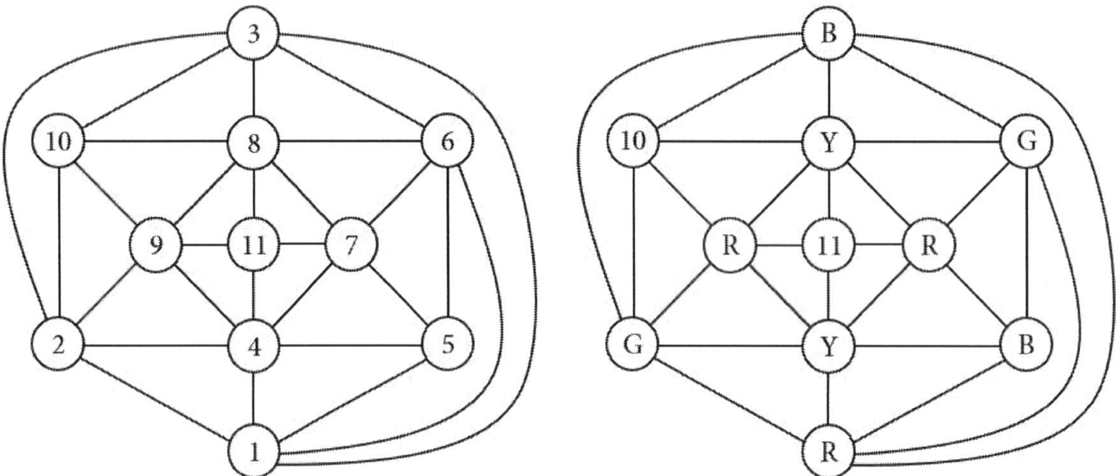

The graph below shows one way to color the two remaining vertices. For vertex 10, we first reversed R and B for one section of a B-R Kempe chain; this section consisted of a single R all by itself (since 10 had not yet been colored), so we reversed that R to a B. (The alternative in our solution would have been to reverse 2 B's and 2 R's on the other side.) After changing the R to a B, we colored vertex 10 R. These two vertices are shaded gray below. Since vertex 11 doesn't connect to four different colors, it turned out to be trivial to color (but that wasn't by construction or planning; it just happened that way).

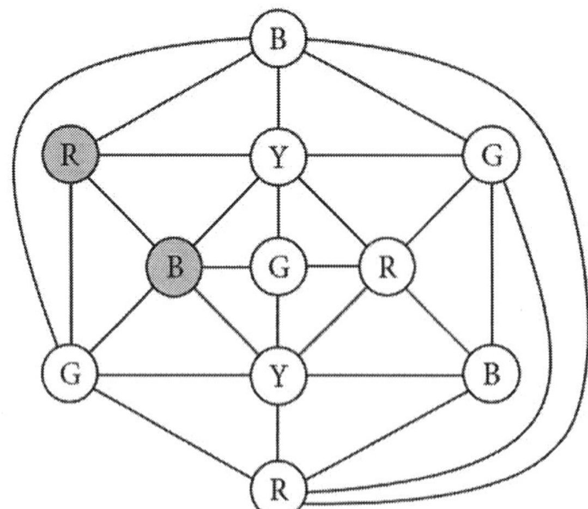

5. We were able to build the icosahedral MPG such that the first 10 (out of 12) vertices can be colored in a trivially four-colorable way. However, the 11th and 12th vertices may require extensive recoloring (if their neighboring vertices happen to be four different colors).

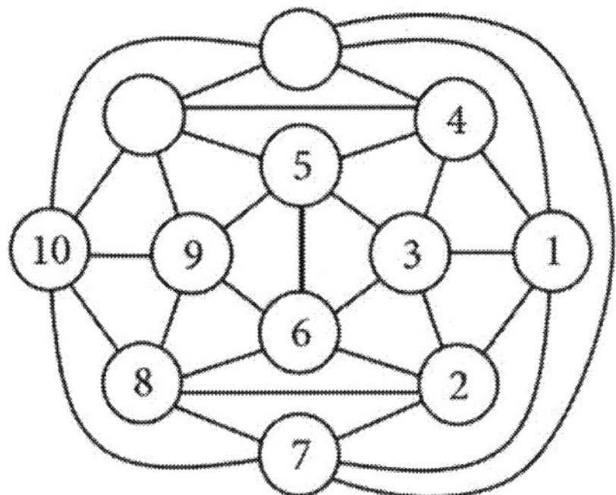

Challenge Problems: The answer key doesn't include answers to challenge problems. These problems are intended to encourage you to think about the ideas.

Chapter 12 Separating Triangles

1. The MPG on the left has separating triangles BGE, DFG, and EFG. Separating triangle EFG, for example, has edges EF, FG, and GE. This triangle isn't a face; it has at least two faces inside of it and at least two faces outside of it.

The MPG on the right contains multiple separating triangles: BCD, CDE, DEF, DFH, DGI, and DHI.

2. Separating triangle CEG can be used to divide the MPG into two smaller graphs as shown below.

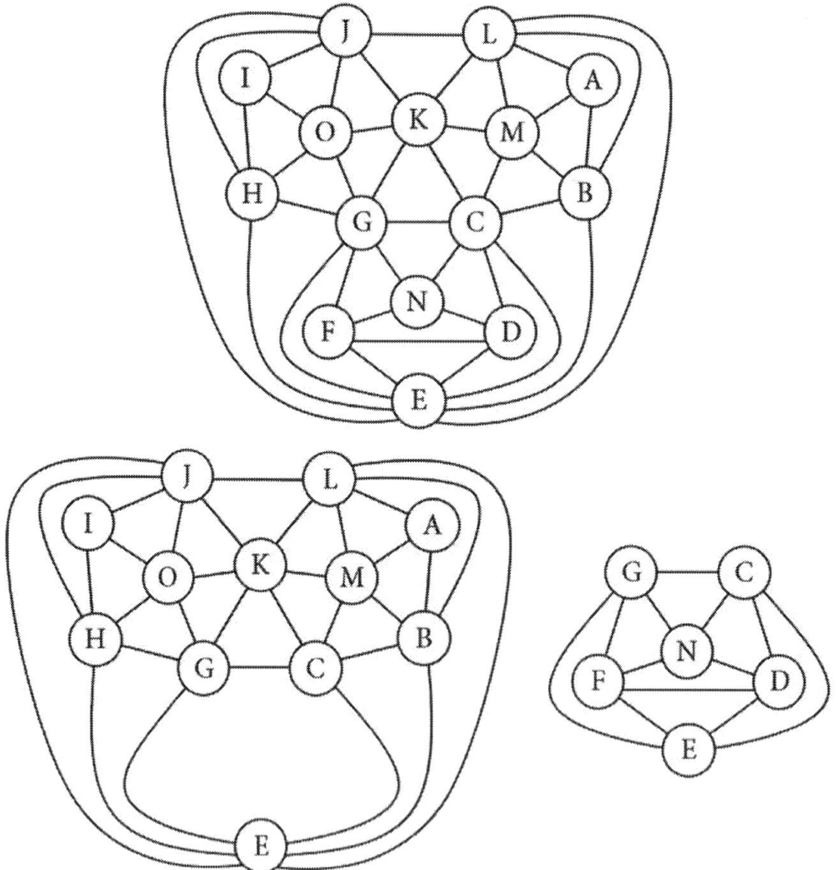

3. One solution for how to color the two smaller graphs and use the same coloring for the complete graph is shown below. Study one color at a time and check that no two vertices of that color share an edge.

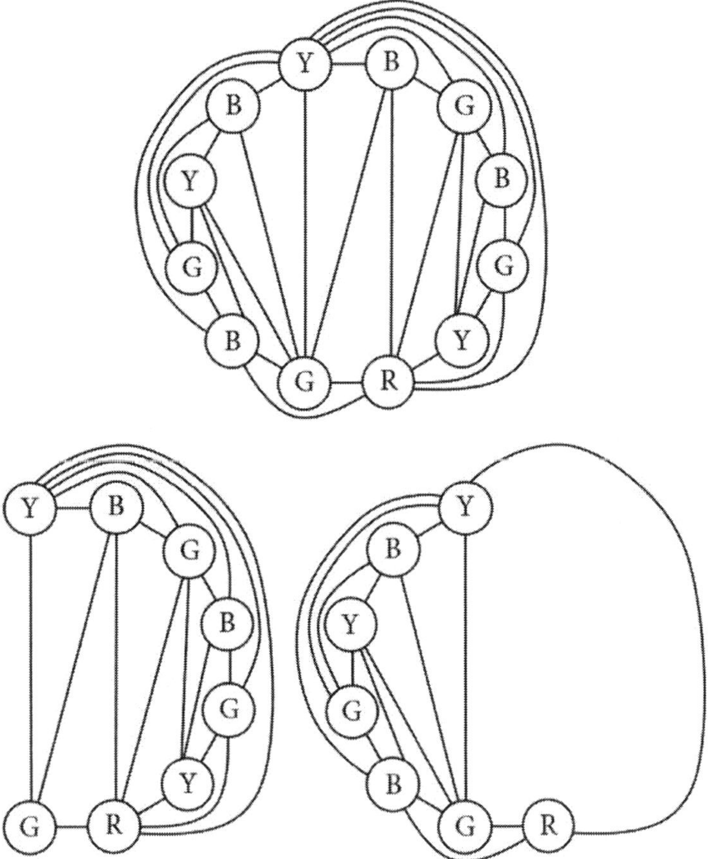

4. One way is to recolor the second graph so that all of its current reds are changed to greens, all of its current greens are changed to blues, all of its current blues are changed to yellows, and all of its current yellows are changed to reds. After this, both graphs will have E green, K blue, and P yellow. This consistent coloring will work for the original MPG. (It would have been simpler to color E green, K blue, and P yellow in each smaller graph to start with.)

5. A MPG with 10 vertices can have as many as 6 ST's, as the graphs below demonstrate.

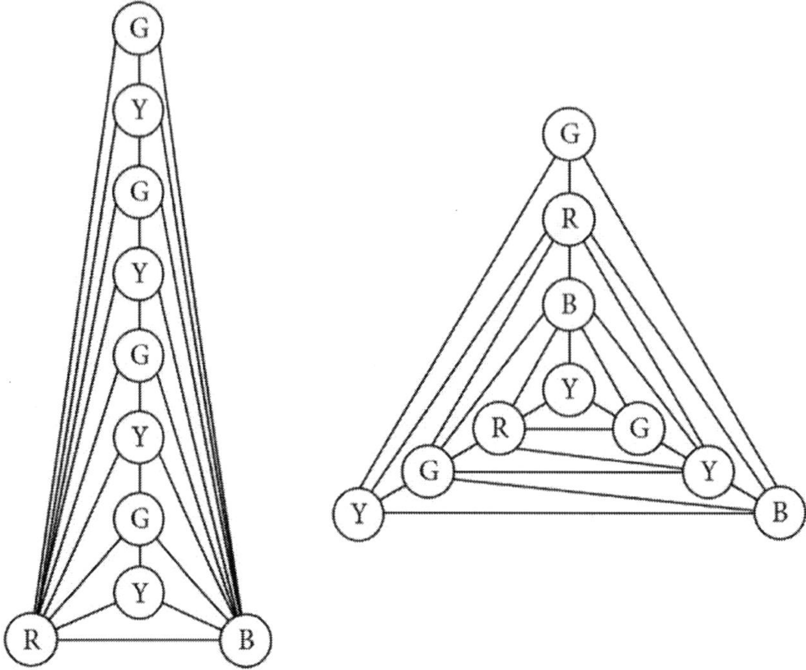

Challenge Problems: The answer key doesn't include answers to challenge problems. These problems are intended to encourage you to think about the ideas.

Chapter 13 Hamiltonian Cycles

1. All four graphs have HC's, as shown below. These are not the only possible HC's for these graphs. Three of the graphs have ST's:

- The top left graph has 3 ST's: ABG, ACG, and BCG.
- The top right graph has 9 ST's: ABM, AFM, BCM, CDM, DEM, EFM, BDM, BFM, and DFM.
- The bottom left graph (the icosahedral graph) doesn't have any ST's.
- The bottom right graph has 5 ST's: AKL, BLM, CMN, DNO, and EKO.

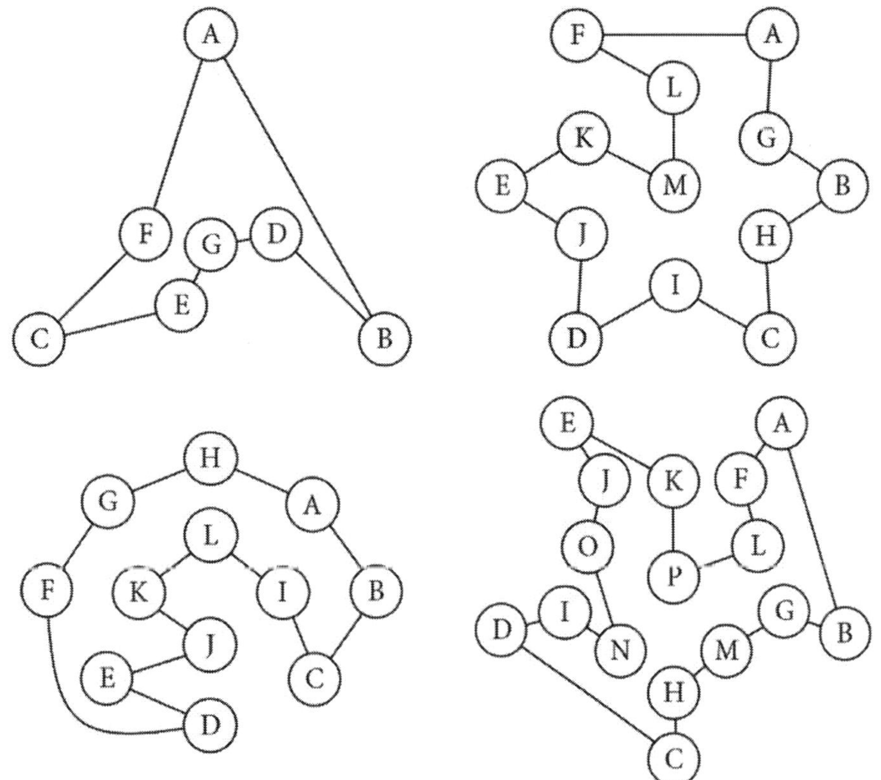

Chapter 14 Polygon Graphs

1. One possible solution is shown below. Note that much different solutions are possible.

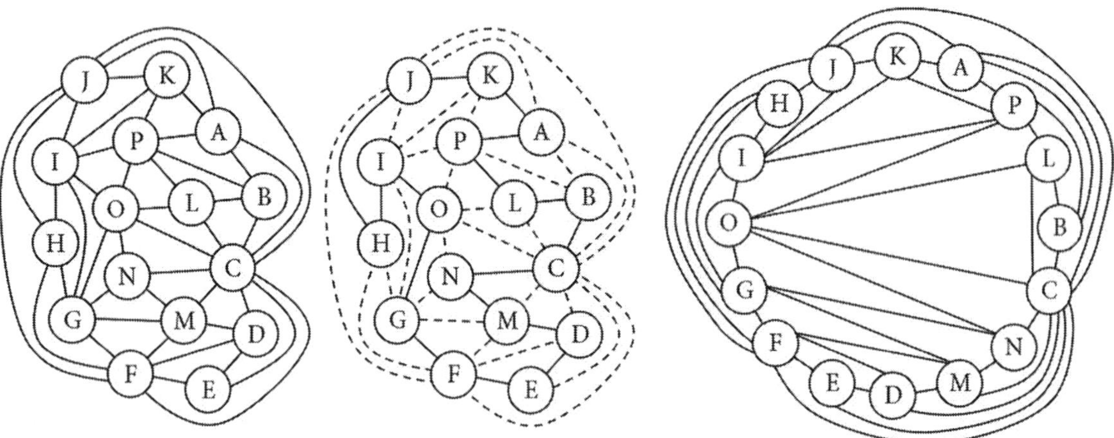

2. The solution is shown below.

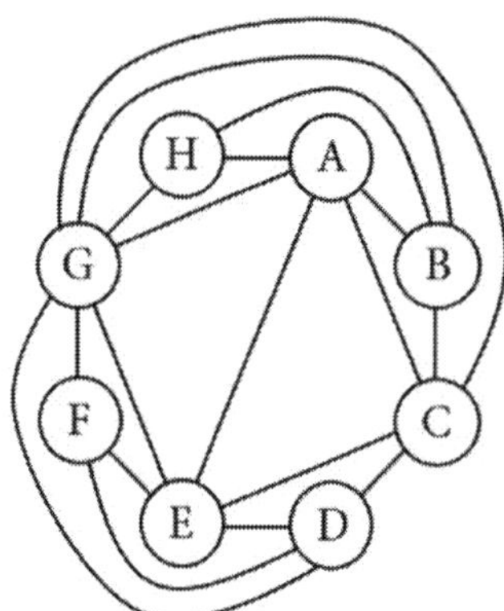

3. It has V – 3 = 24 – 3 = 21 inside edges and V – 3 = 24 – 3 = 21 outside edges. It has E = 3V – 6 = 3(24) – 6 = 72 – 6 = 66 edges in total (21 inside, 21 outside, and 24 making the closed convex polygon). It has V – 2 = 24 – 2 = 22 inside faces and V – 2 = 24 – 2 = 22 outside faces. It has F = 2V – 4 = 2(24) – 4 = 48 – 4 = 44 faces in total (including the infinite area outside).

Challenge Problems: The answer key doesn't include answers to challenge problems. These problems are intended to encourage you to think about the ideas.

Chapter 15 Adding Edges

1. The left graph and middle graph can be four-colored one way by adding a single edge as illustrated below, while the right graph can be four-colored one way by adding two edges. There are other viable solutions besides those shown below, but beware that not any edge can be added to these graphs (some added edges result in a graph that isn't four-colorable). If you found different ways to add edges, the coloring of your graph may be different from the colorings shown below. If the new graph is four-colorable, a valid four-coloring of the new graph will work for the original MPG.

- The left graph didn't contain any K_4 subgraphs originally. The new graph contains 3 K_4 subgraphs that involve 6 of the 8 vertices.

- The middle graph contained 2 K_4 subgraphs originally, which involved 7 of the 8 vertices. The new graph contains 5 K_4 subgraphs that involve all 8 vertices.

- The right graph contained 1 K_4 subgraph originally, which involved 4 of the 8 vertices. The new graph contains 5 K_4 subgraphs that involve all 8 vertices. (It is interesting to compare the left graph with the right graph.)

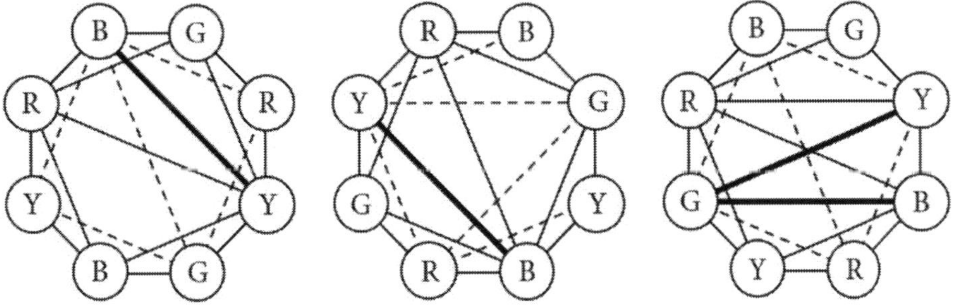

2. We can add 6 edges to the left graph, 4 edges to the middle graph, and 5 edges to the right graph such that the same coloring works for the new graph as for the original graph. We will discuss ultimate four-colorable graphs like these in Chapter 16.

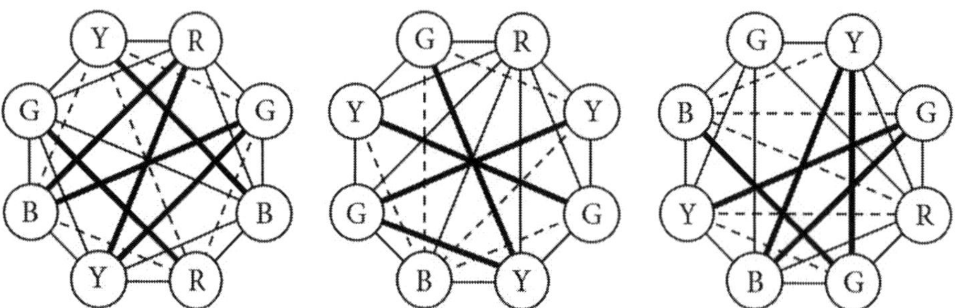

Challenge Problems: The answer key doesn't include answers to challenge problems. These problems are intended to encourage you to think about the ideas.

Chapter 16 Ultimate Four-Coloring

1. The graphs are shown below. Each graph has $V = 10$ vertices.

- For the left graph, $K = 4$, $L = 3$, $M = 2$, and $N = 1$ such that $E = KL + KM + KN + LM + LN + MN = 4(3) + 4(2) + 4(1) + 3(2) + 3(1) + 2(1) = 12 + 8 + 4 + 6 + 3 + 2 = 35$ edges.

- For the right graph, $K = L = 3$ and $M = N = 2$ such that $E = KL + KM + KN + LM + LN + MN = 3(3) + 3(2) + 3(2) + 3(2) + 3(2) + 2(2) = 9 + 6 + 6 + 6 + 6 + 4 = 37$ edges.

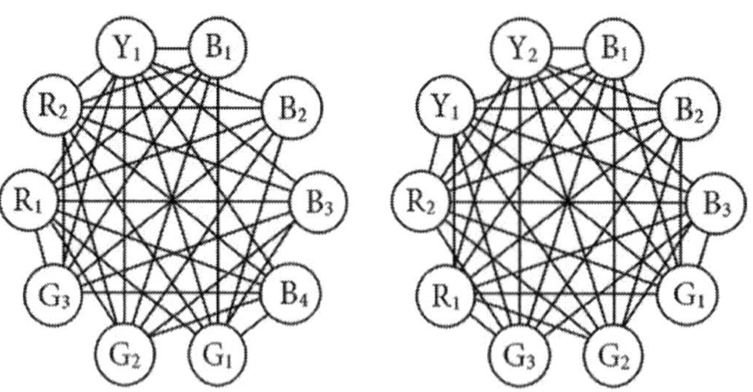

2. The possible values of K, L, M, and N describe the structurally different U4CG's. Note that K, L, M, and N are interchangeable. For example, $K = 4$, $L = 4$, $M = 2$, and $N = 2$ is no different structurally than $K = 4$, $L = 2$, $M = 4$, and $N = 2$. There are 15 possibilities:

- $K = 9, L = 1, M = 1, N = 1$
- $K = 8, L = 2, M = 1, N = 1$
- $K = 7, L = 3, M = 1, N = 1$
- $K = 7, L = 2, M = 2, N = 1$
- $K = 6, L = 4, M = 1, N = 1$
- $K = 6, L = 3, M = 2, N = 1$
- $K = 6, L = 2, M = 2, N = 2$
- $K = 5, L = 5, M = 1, N = 1$
- $K = 5, L = 4, M = 2, N = 1$
- $K = 5, L = 3, M = 3, N = 1$
- $K = 5, L = 3, M = 2, N = 2$
- $K = 4, L = 4, M = 3, N = 1$
- $K = 4, L = 4, M = 2, N = 2$
- $K = 4, L = 3, M = 3, N = 2$
- $K = 3, L = 3, M = 3, N = 3$

3. There are two U4CG's that are MPG's (which are therefore also PG's):

- The U4CG with 4 vertices, for which $K = L = M = N = 1$. This graph is K_4, which is a MPG (Chapter 5) as shown below on the left.

- An U4CG with 5 vertices, for which two vertices will be the same color. This graph is a MPG, as shown below on the right.

Any other U4CG which has 3 colors with a single vertex (like $K = 7$, $L = 1$, $M = 1$, and $N = 1$) has the right number of vertices to be a MPG, but isn't a MPG because it contains $K_{3,3}$ as a subgraph (Chapter 6). With $L = M = N = 1$ (or any other case where 3 of the symbols K thru N equal 1), we get $K = V - 3$ and $E = KL + KM + KN + LM + LN + MN$ $= (V - 3)(1) + (V - 3)(1) + (V - 3)(1) + (1)(1) + (1)(1) + (1)(1) = V - 3 + V - 3 + V - 3$ $+ 1 + 1 + 1 = 3V - 9 + 3 = 3V - 6$, which is the same as the formula for a MPG (Chapter 4). However, if $K > 2$ (meaning that $V > 5$), three of the vertices of the repeated color (corresponding to K in our math) along with the three vertices of a single color (corresponding to L, M, and N) form a $K_{3,3}$ subgraph, showing that these U4CG's are nonplanar when there are 6 or more vertices. (If at least two colors have multiple vertices, such as $K = 2$, $L = 2$, $M = 1$, and $N = 1$, then the U4CG will have too many edges to be a PG; you can show that $E > 3V - 6$ in this case.)

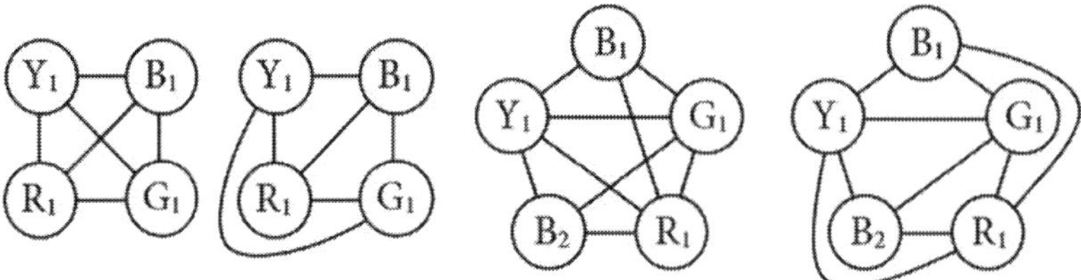

4. A MPG is triangulated (Chapter 3); each face is bounded by three edges that connect to three vertices. Each edge is shared by two triangular faces. Every edge is therefore part of a quadrilateral such as the one shown below. No more than 2 vertices of any of these quadrilaterals can be the same color, which shows that no more than one-half of the vertices of any MPG can possibly be the same color. With a little more work, you should be able to show that even this upper limit (where exactly one-half of the vertices are the same color) can't be attained for a MPG. (The only exception is when $V < 3$, but that wouldn't be a MPG.)

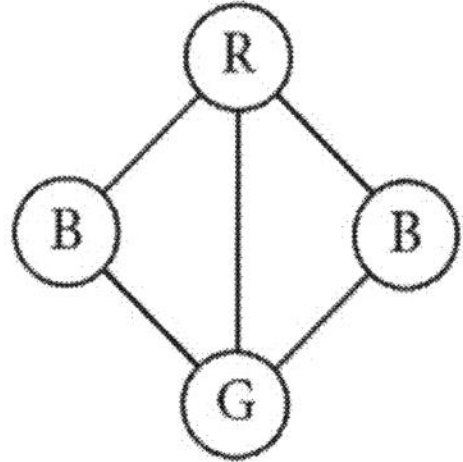

5. Let's call the color with the most vertices color 1. Choose a vertex with color 1 as a starting point. Pick a vertex of a different color for the second vertex. The first vertex must connect to the second vertex for any U4CG since they have different colors. Choose a vertex with color 1 for the third vertex, a vertex that isn't color 1 for the fourth vertex, etc. until running out of color 1 vertices. Since color 1 has fewer than one-half of the vertices, we won't have two color 1 vertices as neighbors. Since color 1 is the dominant color, it can't have fewer than one-fourth of the vertices, which means we will be able to spread out colors 2, 3, and 4 if we choose them wisely so that no two colors neighbor one another once we run out of vertices of color 1. We can thus make a complete cycle with all of the vertices where no two vertices of the same color are chosen one after the other. Since any two vertices with different colors are connected in a U4CG, this makes a HC. If we arrange this HC in the shape of a convex polygon, the result will resemble the example below.

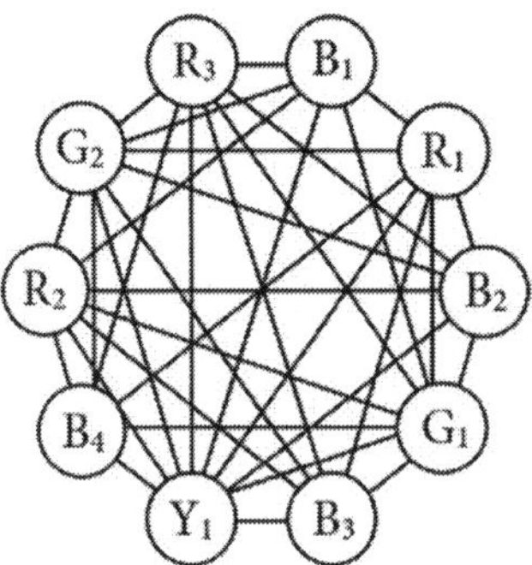

6. One possible solution is shown below. The HC is shown as thick lines on the left, and the polygon form of the graph is shown to the right. The two graphs below are isomorphic.

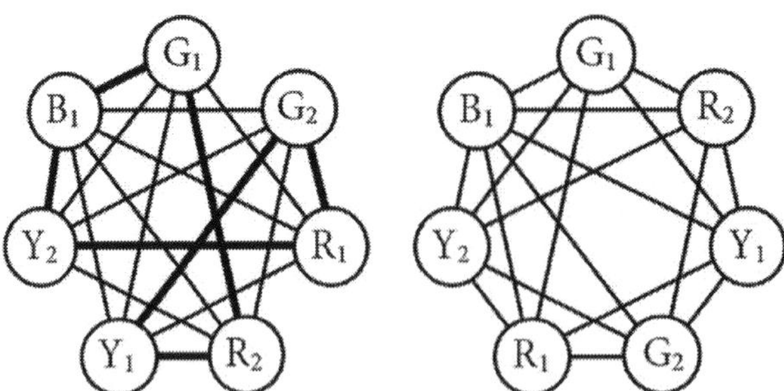

7. $K + L + M + N = (V - 1)/3 + (V - 1)/3 + (V - 1)/3 + 1 = V/3 - 1/3 + V/3 - 1/3 + V/3 - 1/3 + 1 = 3V/3 - 3/3 + 1 = V - 1 + 1 = V$. For example, suppose that $V = 10$. In that case, $K = L = M = (10 - 1)/3 = 9/3 = 3$, such that $K + L + M + N = 3 + 3 + 3 + 1 = 10 = V$.

$$E = KL + KM + KN + LM + LN + MN$$

$$E = \frac{(V-1)}{3}\frac{(V-1)}{3} + \frac{(V-1)}{3}\frac{(V-1)}{3} + \frac{(V-1)}{3}(1) + \frac{(V-1)}{3}\frac{(V-1)}{3}$$
$$+ \frac{(V-1)}{3}(1) + \frac{(V-1)}{3}(1)$$

$$E = \frac{(V-1)^2}{9} + \frac{(V-1)^2}{9} + \frac{(V-1)}{3} + \frac{(V-1)^2}{9} + \frac{(V-1)}{3} + \frac{(V-1)}{3}$$

$$E = \frac{3(V-1)^2}{9} + \frac{3(V-1)}{3} = \frac{(V-1)^2}{3} + \frac{3(V-1)}{3} = \frac{V^2 - 2V + 1}{3} + \frac{3V - 3}{3}$$

$$E = \frac{V^2 - 2V + 1 + 3V - 3}{3} = \frac{V^2 + V - 2}{3} = \frac{(V+2)(V-1)}{3}$$

8. $K + L + M + N = V/6 + V/6 + V/3 + V/3 = 2V/6 + 2V/3 = V/3 + 2V/3 = 3V/3 = V$.

$$E = KL + KM + KN + LM + LN + MN$$

$$E = \frac{V}{6}\frac{V}{6} + \frac{V}{6}\frac{V}{3} + \frac{V}{6}\frac{V}{3} + \frac{V}{6}\frac{V}{3} + \frac{V}{6}\frac{V}{3} + \frac{V}{3}\frac{V}{3}$$

$$E = \frac{V^2}{36} + \frac{V^2}{18} + \frac{V^2}{18} + \frac{V^2}{18} + \frac{V^2}{18} + \frac{V^2}{9}$$

$$E = \frac{V^2}{36} + \frac{4V^2}{18} + \frac{V^2}{9} = \frac{V^2}{36} + \frac{8V^2}{36} + \frac{4V^2}{36}$$

$$E = \frac{V^2 + 8V^2 + 4V^2}{36} = \frac{13V^2}{36}$$

9. The maximum possible degree of a single vertex in this MPG is 17. In this extreme case, all 23 other vertices must have degree 5.

- $E = 3V - 6 = 3(24) - 6 = 72 - 6 = 66$ is the total number of edges (Chapter 4).
- $2E = 2(66) = 132$ is the total number of degrees. (Recall from Chapter 4 that $DV = 2E$ is the sum of the degrees for a MPG.)
- If all of the vertices but one are degree 5, these vertices would add up to $23(5) = 115$ degrees.
- Subtract: $132 - 115 = 17$. The last vertex can have a degree equal to 17.

Each vertex of this U4CG has a degree equal to 18. For an equal distribution of colors, there will be 6 blues, 6 greens, 6 reds, and 6 yellows. Each vertex connects to 18 other vertices. An equal number of colors would work for this MPG since $18 > 17$.

10. The maximum possible degree of a single vertex in this MPG is 93. In this extreme case, all 99 other vertices must have degree 5.

- $E = 3V - 6 = 3(100) - 6 = 300 - 6 = 294$ is the total number of edges (Chapter 4).
- $2E = 2(294) = 588$ is the total number of degrees. (Recall from Chapter 4 that $DV = 2E$ is the sum of the degrees for a MPG.)
- If all of the vertices but one are degree 5, these vertices would add up to $99(5) = 495$ degrees.
- Subtract: $588 - 495 = 93$. The last vertex can have a degree equal to 93.

Each vertex of this U4CG has a degree equal to 75. For an equal distribution of colors, there will be 25 blues, 25 greens, 25 reds, and 25 yellows. Each vertex connects to 75 other vertices. This shows that an equal number of colors wouldn't work for this MPG. Challenge yourself to figure which combinations of colors could work.

Challenge Problem: The answer key doesn't include answers to challenge problems. These problems are intended to encourage you to think about the ideas.

Chapter 17 Removing Edges

1. There are 4 sets of vertices: B_1, B_2, and B_3; G_1, G_2, and G_3; R_1 and R_2; and Y_1. Every blue connects to every non-blue, but doesn't connect to another blue, and similarly for the other colors.

- $K = 3$, $L = 3$, $M = 2$, and $N = 1$.

- $V = K + L + M + N = 3 + 3 + 2 + 1 = 9$.

- $E = KL + KM + KN + LM + LN + MN = 3(3) + 3(2) + 3(1) + 3(2) + 3(1) + 2(1) = 9 + 6 + 3 + 6 + 3 + 2 = 29$.

- An MPG with $V = 9$ vertices would have $E = 3V - 6 = 3(9) - 6 = 27 - 6 = 21$ edges.

- Since $29 - 21 = 8$, we need to remove 8 edges from the U4CG to form a MPG.

- One possible solution is shown below.

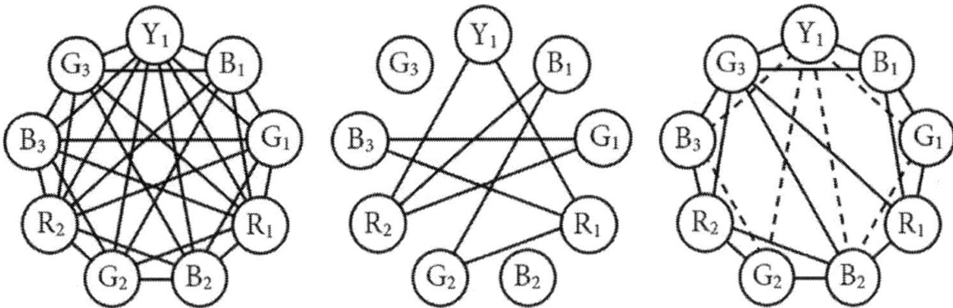

2. $V = 8$ and $E = 3V - 6 = 3(8) - 6 = 24 - 6 = 18$.

- One possible solution is shown below.

- In our solution, $K = 3$, $L = 2$, $M = 2$, and $N = 1$.

- In our solution, $E = KL + KM + KN + LM + LN + MN = 3(2) + 3(2) + 3(1) + 2(2) + 2(1) + 2(1) = 6 + 6 + 3 + 4 + 2 + 2 = 23$ edges.

- Since $23 - 18 = 5$, we need to add 5 edges to the MPG to form an U4CG.

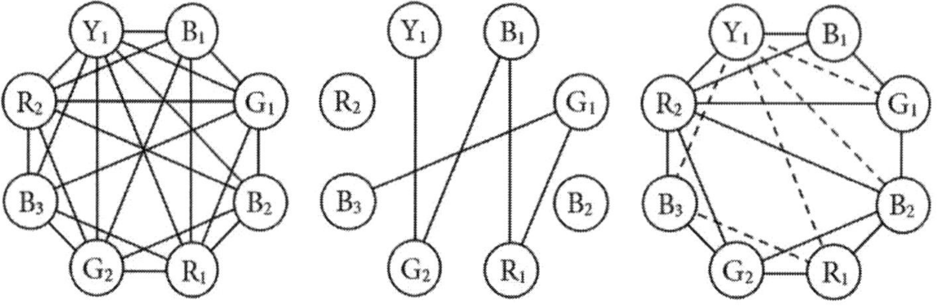

3. $V = 7$ and $E = V(V - 1)/2 = 7(6)/2 = 42/2 = 21$.

- A MPG with $V = 7$ vertices would have $E = 3V - 6 = 3(7) - 6 = 21 - 6 = 15$ edges.

- Since $21 - 15 = 6$, we need to remove 6 edges from K_7 to form a MPG.

- One possible solution is shown below.

- There are 4 active removed edges (solid) and 2 passive removed edges (dotted).

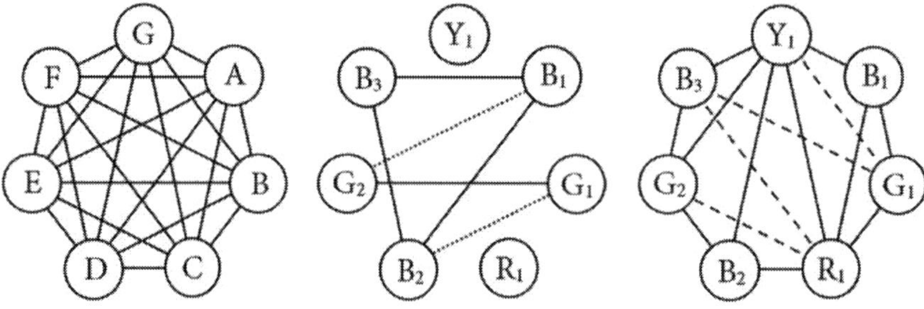

4. $V = 10$ and $E = V(V - 1)/2 = 10(9)/2 = 90/2 = 45$.

- A MPG with $V = 10$ vertices would have $E = 3V - 6 = 3(10) - 6 = 30 - 6 = 24$ edges.

- Since $45 - 24 = 21$, we need to remove 21 edges from K_{10} to form a MPG.

- One possible solution is shown below.

- In our solution, the removed complete subgraphs are K_4, K_3, K_2, and K_1.

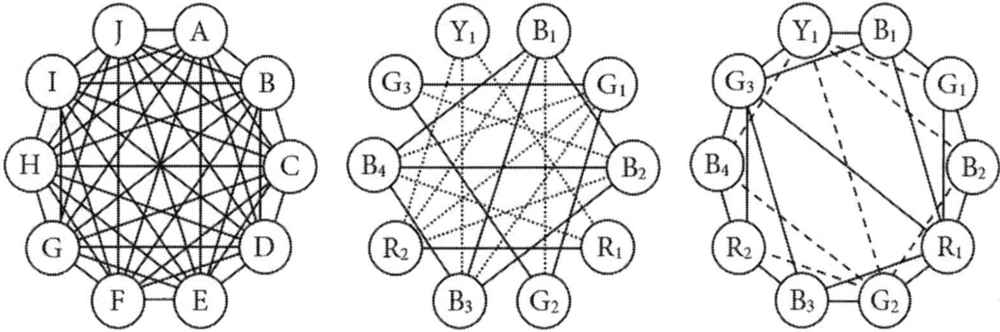

5. A MPG with $V = 11$ vertices would have $E = 3V - 6 = 3(11) - 6 = 33 - 6 = 27$ edges. The possible U4CG's with $V = 11$, where no single color has more than one-half of the vertices, are (with K thru N ordered from largest to smallest; there is no reason to interchange any of the values since, for example, 5, 3, 2, 1 is structurally equivalent to 3, 5, 1, 2):

- $K = 5, L = 4, M = 1, N = 1$. In this case, $E = KL + KM + KN + LM + LN + MN = 5(4) + 5(1) + 5(1) + 4(1) + 4(1) + 1(1) = 20 + 5 + 5 + 4 + 4 + 1 = 39$ edges. Since $39 - 27 = 12$, we need to remove 12 edges from this U4CG to form a MPG.

- $K = 5, L = 3, M = 2, N = 1$. In this case, $E = KL + KM + KN + LM + LN + MN = 5(3) + 5(2) + 5(1) + 3(2) + 3(1) + 2(1) = 15 + 10 + 5 + 6 + 3 + 2 = 41$ edges. Since $41 - 27 = 14$, we need to remove 14 edges from this U4CG to form a MPG.

- $K = 5, L = 2, M = 2, N = 2$. In this case, $E = KL + KM + KN + LM + LN + MN = 5(2) + 5(2) + 5(2) + 2(2) + 2(2) + 2(2) = 10 + 10 + 10 + 4 + 4 + 4 = 42$ edges. Since $42 - 27 = 15$, we need to remove 15 edges from this U4CG to form a MPG.

- $K = 4, L = 4, M = 2, N = 1$. In this case, $E = KL + KM + KN + LM + LN + MN = 4(4) + 4(2) + 4(1) + 4(2) + 4(1) + 2(1) = 16 + 8 + 4 + 8 + 4 + 2 = 42$ edges. Since $42 - 27 = 15$, we need to remove 15 edges from this U4CG to form a MPG.

- $K = 4$, $L = 3$, $M = 3$, $N = 1$. In this case, $E = KL + KM + KN + LM + LN + MN = 4(3) + 4(3) + 4(1) + 3(3) + 3(1) + 3(1) = 12 + 12 + 4 + 9 + 3 + 3 = 43$ edges. Since $43 - 27 = 16$, we need to remove 16 edges from this U4CG to form a MPG.

- $K = 4$, $L = 3$, $M = 2$, $N = 2$. In this case, $E = KL + KM + KN + LM + LN + MN = 4(3) + 4(2) + 4(2) + 3(2) + 3(2) + 2(2) = 12 + 8 + 8 + 6 + 6 + 4 = 44$ edges. Since $44 - 27 = 17$, we need to remove 17 edges from this U4CG to form a MPG.

- $K = 3$, $L = 3$, $M = 3$, $N = 2$. In this case, $E = KL + KM + KN + LM + LN + MN = 3(3) + 3(3) + 3(2) + 3(3) + 3(2) + 3(2) = 9 + 9 + 6 + 9 + 6 + 6 = 45$ edges. Since $45 - 27 = 18$, we need to remove 18 edges from this U4CG to form a MPG.

6. For a complete graph, $E = V(V - 1)/2$. For a MPG, $E = 3V - 6$. If we subtract these, we get $(V - 3)(V - 4)/2$ as shown in the chapter.

- For $V = 4$, we get $1(0)/2 = 0$. No edges need to be removed from K_4.
- For $V = 5$, we get $2(1)/2 = 2/2 = 1$. We need to remove 1 edge from K_5.
- For $V = 6$, we get $3(2)/2 = 6/2 = 3$. We need to remove 3 edges from K_6.
- For $V = 7$, we get $4(3)/2 = 12/2 = 6$. We need to remove 6 edges from K_7.
- For $V = 8$, we get $5(4)/2 = 20/2 = 10$. We need to remove 10 edges from K_8.
- For $V = 9$, we get $6(5)/2 = 30/2 = 15$. We need to remove 15 edges from K_9.
- For $V = 10$, we get $7(6)/2 = 42/2 = 21$. We need to remove 21 edges from K_{10}.
- For $V = 11$, we get $8(7)/2 = 56/2 = 28$. We need to remove 28 edges from K_{11}.
- For $V = 12$, we get $9(8)/2 = 72/2 = 36$. We need to remove 36 edges from K_{12}.

The pattern is 0, 1, 3, 6, 10, 15, 21, 28, 36, etc. Look at the differences. Add 1 to 0 to get 1, add 2 to get 3, add 3 to get 6, add 4 to get 10, add 5 to get 15, add 6 to get 21, add 7 to get 28, add 8 to get 36, etc. Each time, add one more than the previous time.

7. We need to remove $(V - 3)(V - 4)/2 = 4(3)/2 = 12/2 = 6$ edges from K_7 to form a MPG.

- K_3, K_2, K_1, and K_1. This requires removing $3 + 2 = 5$ edges. (No edges are removed for a K_1.) A passive edge also needs to be removed in this case.
- K_2, K_2, K_2, and K_1. This requires removing $2 + 2 + 2 = 6$ edges.

Since the question specified sets of 4 complete subgraphs, our solution doesn't include any possibilities with just 3 complete subgraphs (like K_3, K_3, and K_1). Also, we didn't include K_4, K_1, K_1, and K_1 because more than one-half of the vertices would be the same color, which is a problem if trying to form a MPG, which we noted in Chapter 16 (which still applies, even though in this problem we are removing edges from a complete graph, not an U4CG). It would be instructive to draw the case K_4, K_1, K_1, and K_1 and show that the resulting graph isn't a MPG (since it contains a $K_{3,3}$ subgraph.)

8. We need to remove $(V - 3)(V - 4)/2 = 9(8)/2 = 72/2 = 36$ edges from K_{12} to form a MPG.

- K_5, K_5, K_1, and K_1.
- K_5, K_4, K_2, and K_1.
- K_5, K_3, K_3, and K_1.
- K_5, K_3, K_2, and K_2.
- K_4, K_4, K_3, and K_1.
- K_4, K_4, K_2, and K_2.
- K_4, K_3, K_3, and K_2.
- K_3, K_3, K_3, and K_3.

Since the question specified sets of 4 complete subgraphs, our solution doesn't include any solutions with just 3 complete subgraphs (like K_4, K_4, and K_4). Also, we didn't include cases where half of the vertices would be the same color, for the same reason that we discussed at the end of the solution to Problem 7.

9. No, it isn't possible for a MPG with 7 vertices to be three-colorable.

- Removing K_3, K_2, and K_2 from K_7 won't work because that requires removing $3 + 2 + 2 = 7$ edges. Since K_7 has E $7(6)/2 = 42/2 = 21$ edges and a MPG with V = 7 has E $= 3V - 6 = 3(7) - 6 = 21 - 6 = 15$ edges, we need to remove just $21 - 15 = 6$ edges to form a MPG. Removing 7 would be too many.

- The alternative, removing K_3, K_3, and K_1 also won't work. In this case, the number of edges isn't a problem, since $3 + 3 = 6$. (Note that no edges are removed for a K_1.) In this case, the problem is that after removing the two K_3's, the new graph will contain a $K_{3,3}$ subgraph, which shows that it isn't a MPG. See if you can show this.

10. K_9 has V = 9 vertices and E $= V(V - 1)/2 = 9(8)/2 = 72/2 = 36$ edges. There are structurally different U4CG's with V = 9 vertices, and the different cases have different numbers of edges. The possibilities include (where no color has half of the vertices):

- K = 4, L = 3, M = 1, and N = 1. In this case, E = KL + KM + KN + LM + LN + MN $= 4(3) + 4(1) + 4(1) + 3(1) + 3(1) + 1(1) = 12 + 4 + 4 + 3 + 3 + 1 = 27$ edges. Since $36 - 27 = 9$, we need to remove 9 edges from K_9 to form this U4CG.

- K = 4, L = 2, M = 2, and N = 1. In this case, E = KL + KM + KN + LM + LN + MN $= 4(2) + 4(2) + 4(1) + 2(2) + 2(1) + 2(1) = 8 + 8 + 4 + 4 + 2 + 2 = 28$ edges. Since $36 - 28 = 8$, we need to remove 8 edges from K_9 to form this U4CG.

- K = 3, L = 3, M = 2, and N = 1. In this case, E = KL + KM + KN + LM + LN + MN $= 3(3) + 3(2) + 3(1) + 3(2) + 3(1) + 2(1) = 9 + 6 + 3 + 6 + 3 + 2 = 29$ edges. Since $36 - 29 = 7$, we need to remove 7 edges from K_9 to form this U4CG.

- K = 3, L = 2, M = 2, and N = 2. In this case, E = KL + KM + KN + LM + LN + MN $= 3(2) + 3(2) + 3(2) + 2(2) + 2(2) + 2(2) = 6 + 6 + 6 + 4 + 4 + 4 = 30$ edges. Since $36 - 30 = 6$, we need to remove 6 edges from K_9 to form this U4CG.

11. We will do this in reverse. Since $K + L + M + N = V$, it follows that $V - K = L + M + N$, $V - L = K + M + N$, $V - M = K + L + N$, and $V - N = K + L + M$.

$$E = \frac{K(V - K) + L(V - L) + M(V - M) + N(V - N)}{2}$$

$$E = \frac{K(L + M + N) + L(K + M + N) + M(K + L + N) + N(K + L + M)}{2}$$

$$E = \frac{KL + KM + KN + KL + LM + LN + KM + LM + MN + KN + LN + MN}{2}$$

$$E = \frac{2KL + 2KM + 2KN + 2LM + 2LN + 2MN}{2}$$

$$E = KL + KM + KN + LM + LN + MN$$

For $K = 7$, $L = 5$, $M = 3$, and $N = 1$, one formula gives (since $V = K + L + M + N = 7 + 5 + 3 + 1 = 12 + 4 = 16$)

$$E = \frac{K(V - K) + L(V - L) + M(V - M) + N(V - N)}{2}$$

$$E = \frac{7(16 - 7) + 5(16 - 5) + 3(16 - 3) + 1(16 - 1)}{2}$$

$$E = \frac{7(9) + 5(11) + 3(13) + 1(15)}{2} = \frac{63 + 55 + 39 + 15}{2} = \frac{118 + 54}{2} = \frac{172}{2} = 86$$

and the other formula gives

$$E = KL + KM + KN + LM + LN + MN = 7(5) + 7(3) + 7(1) + 5(3) + 5(1) + 3(1)$$

$$E = 35 + 21 + 7 + 15 + 5 + 3 = 56 + 22 + 8 = 86$$

Both formulas give $E = 86$ for this case.

The formula with a denominator of 2 states that there are K blue vertices that each connect to $V - K$ non-blue vertices, L green vertices that each connect to $V - L$ non-green vertices, M red vertices that each connect to $V - M$ non-red vertices, and N yellow vertices that each connect to $V - N$ non-yellow vertices.

Challenge Problems: The answer key doesn't include answers to challenge problems. These problems are intended to encourage you to think about the ideas.

Chapter 18 Vertex Splitting

1. The middle diagram below shows VS1; the central vertex is pushed out into the right face. Regardless of which vertex is pushed and into which face it is pushed (including the infinite area outside), the new graph will have three vertices with degree four and two with degree three in way that is structurally equivalent to that shown below. The right diagram below shows VS2; the central vertex is pushed upward along an edge toward the top vertex. Regardless of which vertex is pushed and along which edge it is pushed, the new graph will still be structurally equivalent. (The middle and right graphs below are structurally equivalent, too.)

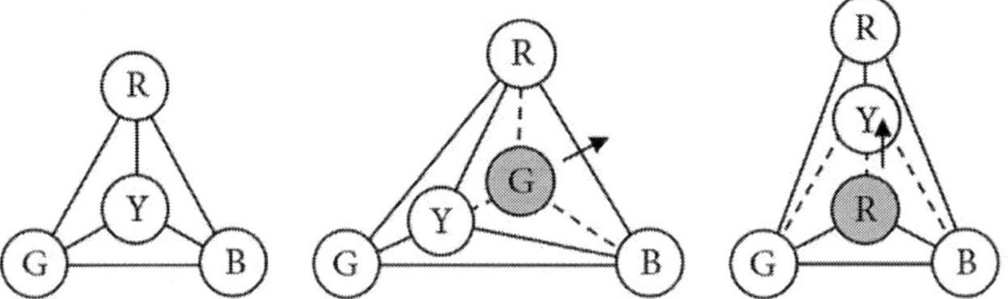

2. One possible solution is to push the green vertex along the edge toward the blue vertex with VS2, as shown below. Observe that all of the vertices are degree four now.

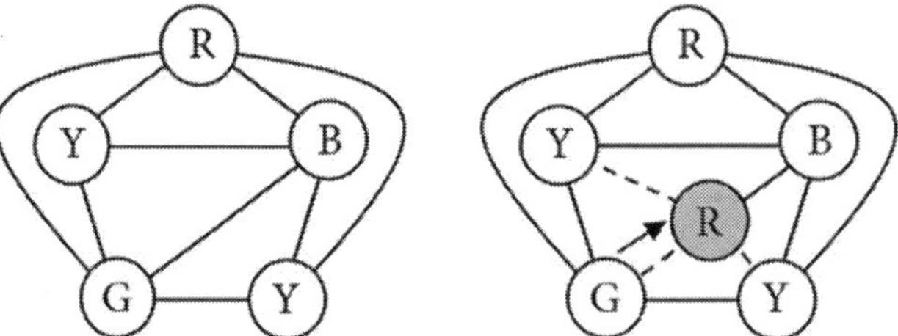

3. One possible solution is shown below. We reversed what had been a separated G-Y Kempe chain so that X could be colored G.

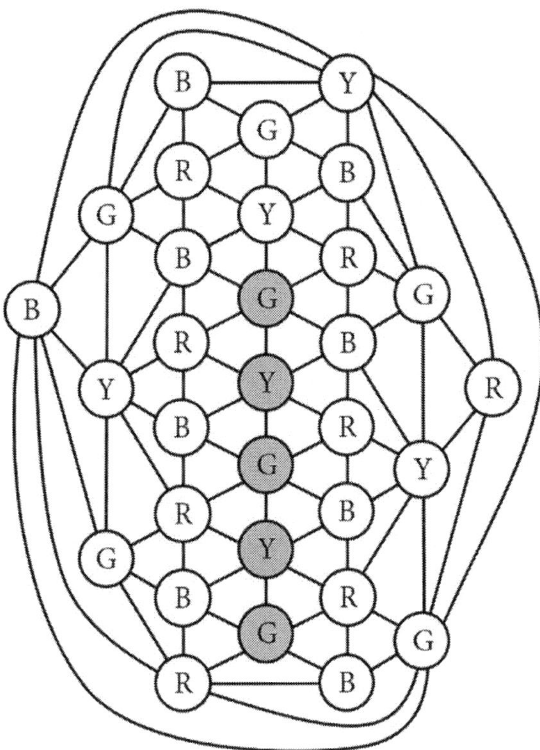

4. One possible solution is shown below. We applied VS2 to the top left and bottom right vertices. Each vertex left behind is degree four, making each split a case of VS2.

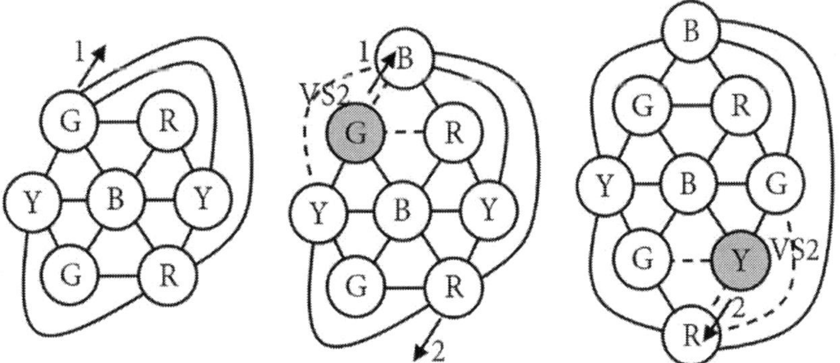

5. The hard to way to do this is to start with K_4 and figure out how to split one vertex at a time to arrive at the icosahedral MPG. Since every vertex in the finished MPG has degree five, it might seem intuitive to try to keep the degrees of the vertices equitable while splitting the vertices, making the MPG of Problem 2 along the way. However, it may be easier if you make the degrees unevenly distributed (as we did in the solution below). The easy way to do this is to start with the icosahedral MPG and contract edges to form minors until arriving at K_4, and then figure out how to split vertices to draw effectively the same process in reverse. As you study our solution, think through the splitting carefully, including the three new edges, which can get a little tricky. It helps to identify the two split vertices and the two diamond vertices to correctly identify the three new edges.

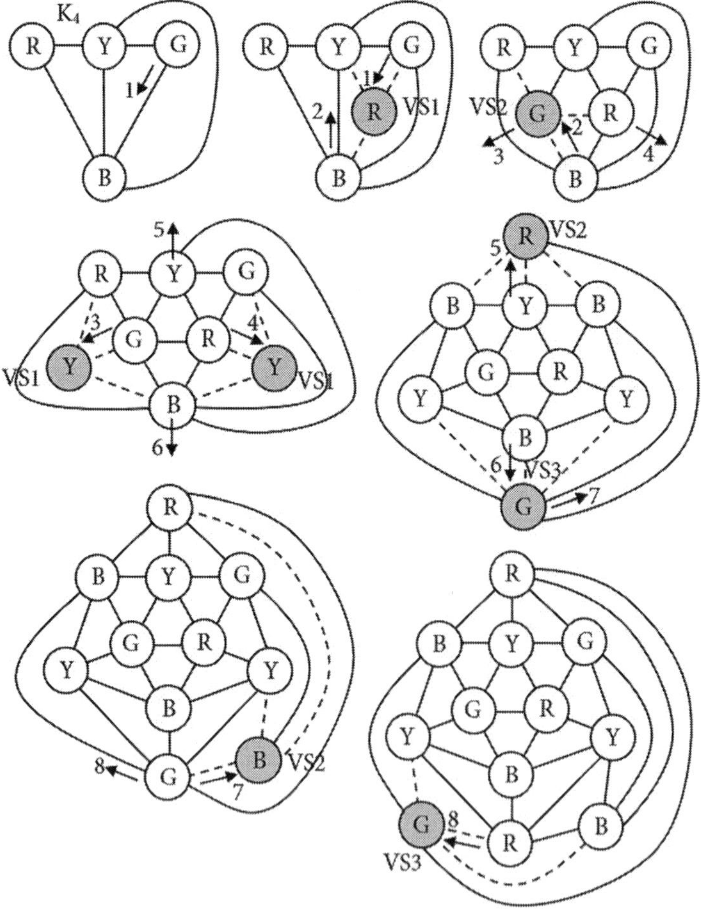

6. The picture produces a PG, but not a MPG. If you add another edge, this will be VS1, VS2, or VS3. The idea behind triangulation (which is the process used to make MPG's) is that we only need to work with MPG's: if we can prove the four-color theorem for MPG's, it will follow that all PG's are four-colorable (Chapter 3).

Challenge Problems: The answer key doesn't include answers to challenge problems. These problems are intended to encourage you to think about the ideas.

Chapter 19 Quadrilateral Switching

1. One way to transform the given MPG into 5 structurally different MPG's is shown below. The numbers indicate the degrees of the vertices. (This method of labeling MPG's may not always specify a unique MPG, as we'll see in Chapter 28.)

- The given MPG appears at the top left.
- Turn AE to CF to transform the top left MPG into the top center MPG.
- Turn CE to DF to transform the top center MPG into the top right MPG.
- Turn CG to BD to transform the top right MPG into the bottom left MPG.
- Turn FG to AE to transform the bottom left MPG into the bottom center MPG.

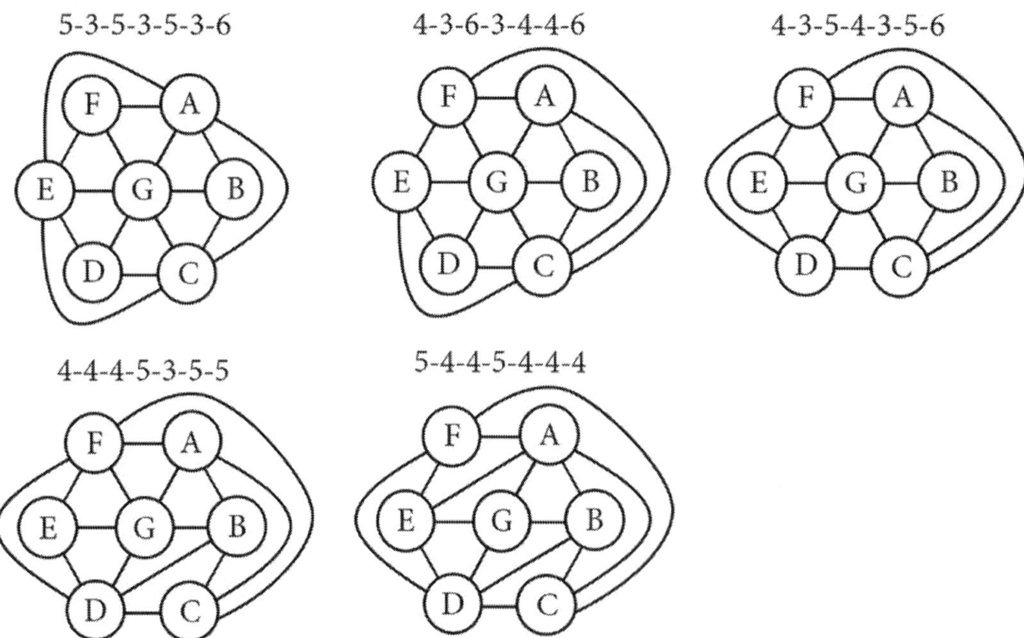

2. One way to color the given MPG and transform it into 4 structurally different MPG's by turning color-preserving quadrilateral switches is shown below. The MPG shown at the top right requires turning color-changing quadrilateral switches. **Note that if you color the given MPG differently, your solution will be different from ours.**

- The given MPG appears at the top left. We show one way to four-color it.
- Turn 2-4 to 1-3 to transform the top left MPG into the top center MPG.
- Turn 1-4 to 2-3 to transform the top left MPG into the bottom left MPG.
- Turn 2-4 to 1-3 to transform the top left MPG into the bottom center MPG. This involves a different 4 than the similar switch from earlier.
- The MPG at the top right requires turning a color-changing quadrilateral switch. Turn 2-4 to 3-3 to transform the bottom center MPG into the top right MPG. The shaded vertices show one way to recolor the graph.

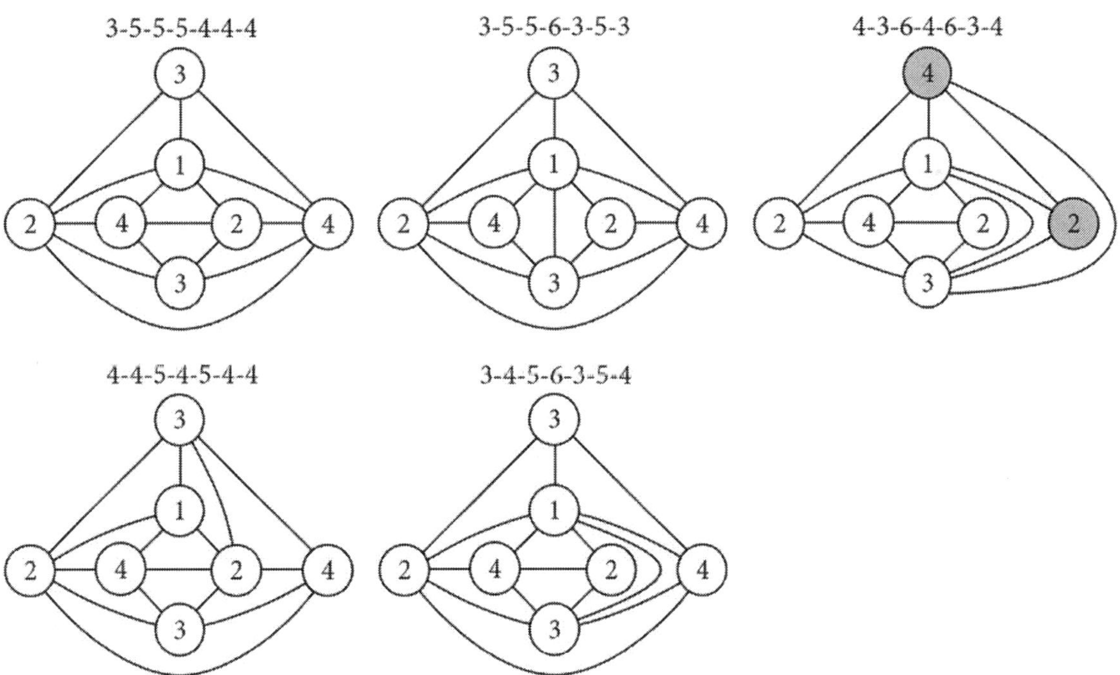

3. One possible solution is shown below.

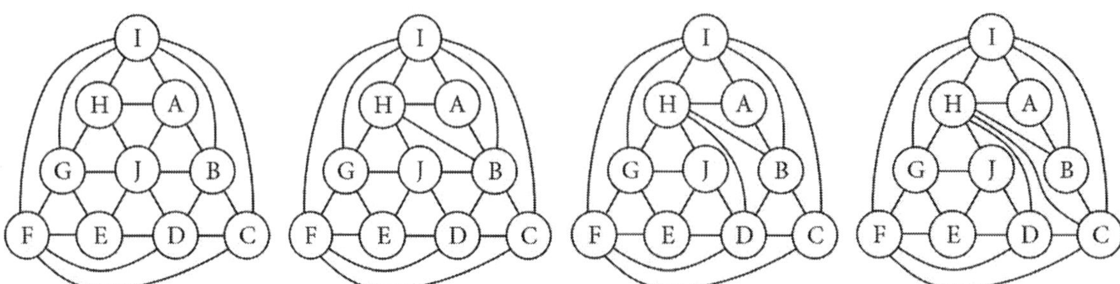

4. Only the vertices on the corners of the quadrilateral switch have their degree numbers affected. If AC is turned to BD in the quadrilateral switch below, the degrees of A and C will each decrease by 1 and the degrees of B and D will each increase by 1.

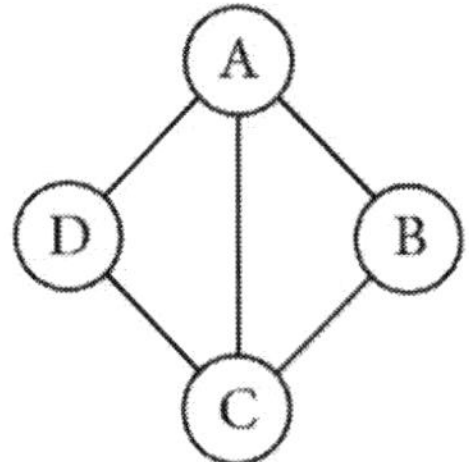

5. If a MPG with degrees 8-8-4-4-4-4-4-3-3 (not necessarily in that order) is transformed into a MPG with degrees 5-5-5-5-5-5-4-4-4, using the answer to Problem 4, this would require turning a minimum of 3 quadrilateral switches. Why 3? Each degree 8 vertex needs to be reduced by 3 degrees. In the best possible scenario, both 8's can be reduced by 1 degree with a single switch. (However, if you draw the MPG's, you might discover that this best-case scenario isn't possible, and thus revise your answer.)

Challenge Problems: The answer key doesn't include answers to challenge problems. These problems are intended to encourage you to think about the ideas.

Chapter 20 Kirchhoff's Rules

1. One possible way to four-color the graph is shown below. To check your solution, study one color at a time and check that no two vertices of that color share an edge. Your numbers for the color differences will be different if your coloring is different, but the sums will still be zero.

- G-N-F-G: $(4 - 3) + (1 - 4) + (3 - 1) = 1 + (-3) + 2 = 3 - 3 = 0$.
- K-A-C-G-K: $(2 - 4) + (1 - 2) + (3 - 1) + (4 - 3) = -2 + (-1) + 2 + 1 = 3 - 3 = 0$.
- K-O-G-N-C-M-K: $(1 - 4) + (3 - 1) + (4 - 3) + (1 - 4) + (3 - 1) + (4 - 3) = -3 + 2 + 1 + (-3) + 2 + 1 = 6 - 6 = 0$.
- A-B-C-N-G-O-K-L-A: $(4 - 2) + (1 - 4) + (4 - 1) + (3 - 4) + (1 - 3) + (4 - 1) + (1 - 4) + (2 - 1) = 2 + (-3) + 3 + (-1) + (-2) + 3 + (-3) + 1 = 9 - 9 = 0$.
- J-K-L-A-B-C-D-E-F-G-H-I-J: $(4 - 3) + (1 - 4) + (2 - 1) + (4 - 2) + (1 - 4) + (3 - 1) + (2 - 2) + (1 - 3) + (3 - 1) + (4 - 3) + (2 - 4) + (3 - 2) = 1 + (-3) + 1 + 2 + (-3) + 2 + 0 + (-2) + 2 + 1 + (-2) + 1 = 10 - 10 = 0$.

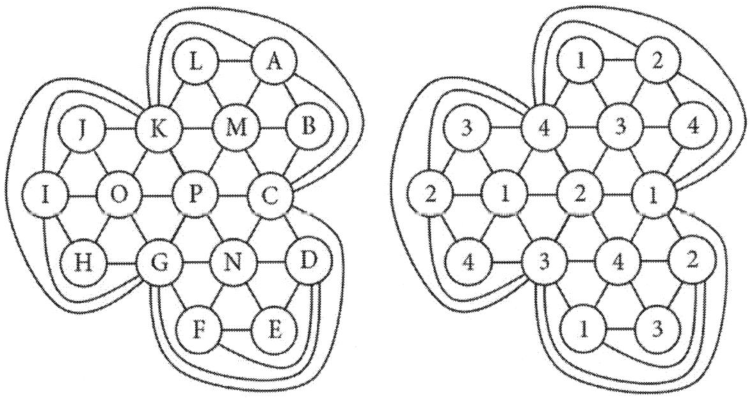

Challenge Problem: The answer key doesn't include answers to challenge problems. These problems are intended to encourage you to think about the ideas.

Chapter 21 Building Blocks

1. One possible solution is shown below.

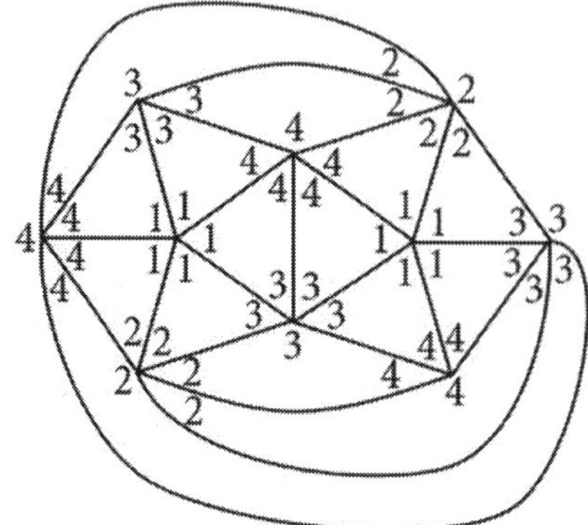

2. The edges of the graph below correspond to the coloring of the graph above.

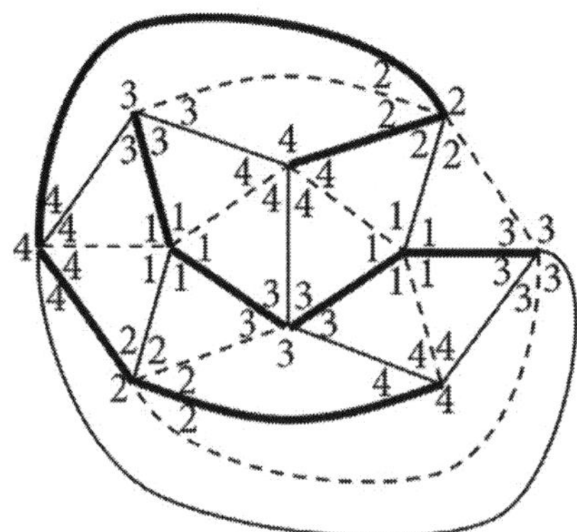

Challenge Problem: The answer key doesn't include answers to challenge problems. These problems are intended to encourage you to think about the ideas.

Chapter 22 Four-Coloring by Pairing Faces

1. One possible solution is shown below. In our solution, the B-G Kempe chain includes one closed loop with four highlighted edges. The R-Y Kempe chain forms two separated sections: one long section has seven vertices and the other section has a lone R. In contrast, the B-G Kempe chain consists of a single section. Compare these two Kempe chains with the alternative pairs of Kempe chains (B-R and G-Y or B-Y and G-R); the three ways to divide the MPG into Kempe chains sometimes have different structures. Note that your solution may appear quite different from ours.

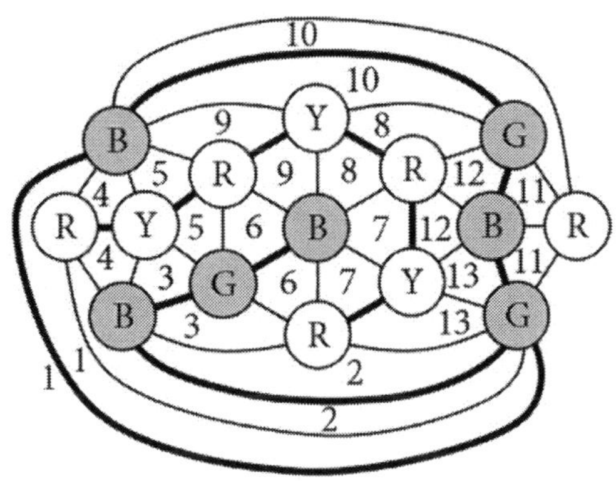

2. One possible solution is shown below. In our solution, the B-G Kempe chain includes two closed loops with an even number of highlighted edges: one has 6, the other has 4. The R-Y Kempe chain forms three separated sections: one section with a single R (bottom right), one section with one R and one Y (top left), and the rest in a long section. In contrast, the B-G Kempe chain consists of a single section. Compare these two Kempe chains with the alternative pairs of Kempe chains (B-R and G-Y or B-Y and G-R); the three ways to divide the MPG into Kempe chains sometimes have different structures. Note that your solution may appear quite different from ours.

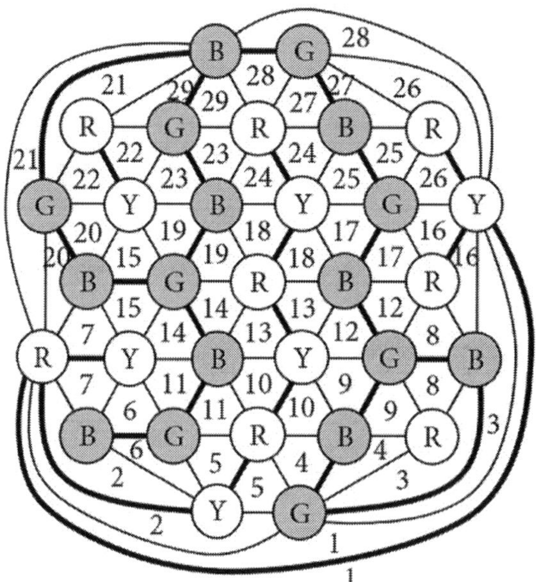

Challenge Problems: The answer key doesn't include answers to challenge problems. These problems are intended to encourage you to think about the ideas.

Chapter 23 The Three-Edges Theorem

1. One possible solution is shown below. Note that your solution may appear quite different from ours.

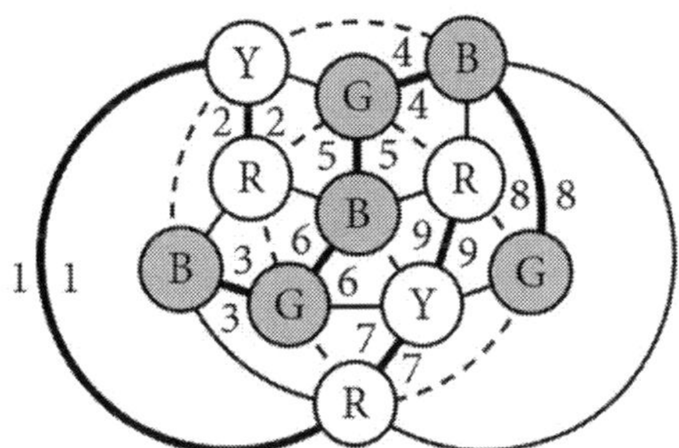

2. One possible solution is shown below. Note that your solution may appear quite different from ours. Our B-G Kempe chain includes a closed loop with 6 edges and our R-Y Kempe chain includes a closed loop with 4 edges. Our solution has one lone separated G and one lone separated R.

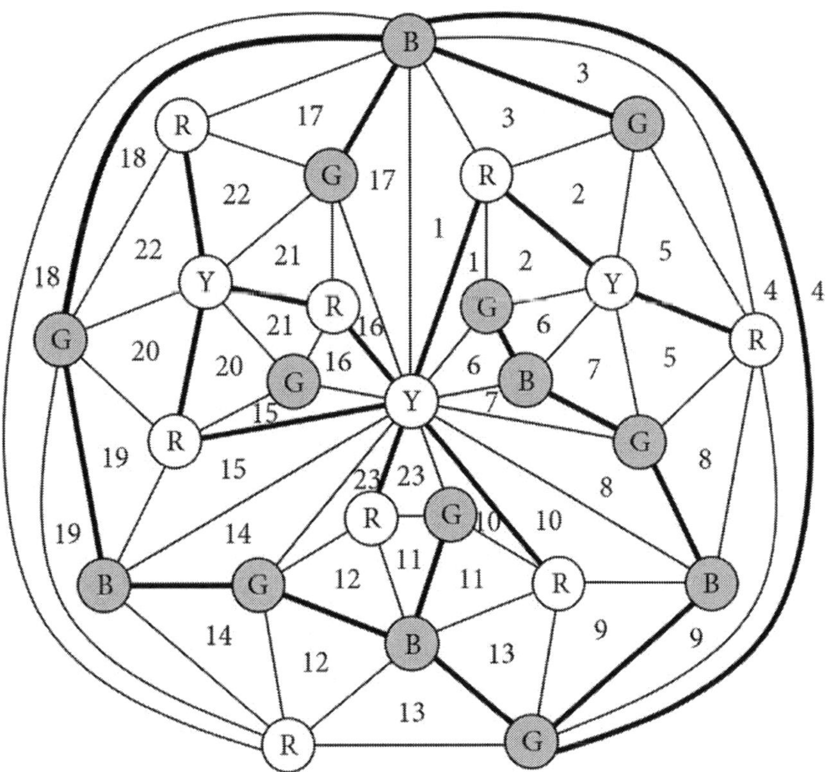

3. This 2 originates in Euler's formula (Chapter 4): $V + F = E + 2$. Note that a triangle has $V = 3$ vertices, $F = 2$ faces (including the infinite area outside, as usual), and $E = 3$ edges. For a triangle, $V + F = E + 2$ (such that $3 + 2 = 3 + 2$). If you split a vertex (Chapter 18) to form a new MPG, you get 1 more vertex, 3 more edges, and 2 more faces. When you add 1 to V, 2 to F, and 3 to E, the formula $V + F = E + 2$ is preserved for the new MPG. Therefore, one way to derive Euler's formula for a MPG is to begin with the triangle graph (Chapter 8), for which the number of vertices plus faces is greater than the number of edges by 2, and use the technique of vertex splitting. That 2 carries over to Euler's formula, $V + F = E + 2$, and to the formulas $E = 3(V - 2)$ and $F = 2(V - 2)$ for a MPG.

4. Plug $V = 32$ into $E = 3(V - 2) = 3(30) = 90$ and $F = 2(V - 2) = 2(30) = 60$ to get the total number of edges and faces. Divide E by 3 to see that there are 30 solid edges, 30 thick solid edges, and 30 dashed edges.

5. Add these numbers together to determine that the total number of edges is $E = 17 + 17 + 17 = 51$. Plug $E = 51$ into the formula $E = 3(V - 2)$ to get $51 = 3(V - 2)$. Divide both sides by 3 to get $17 = V - 2$. Add 2 to both sides to see that there are $V = 17 + 2 = 19$ vertices. Use the formula $F = 2(V - 2) = 2(17) = 34$ to see that there are 34 faces.

6. The subgraph has $16 + 24 = 40$ vertices (16 along its border and 24 inside). The subgraph also has 101 edges (16 along its border and 85 inside). One way to determine this is to first answer the questions regarding the new MPG (see below). Subtract 16 from 186 to get 170. Divide 170 by two to get 85. There are 85 edges inside of the subgraph plus 16 edges along the border of the subgraph, which makes 101 edges. The subgraph has 62 faces: divide the number of faces for the new MPG (which is 124; see below) by two.

The new MPG has $16 + 24 + 24 = 64$ vertices (16 along the border, 24 inside, and 24 outside). The new MPG has $E = 3(V - 2) = 3(62) = 186$ edges (16 along the border, 85 inside, and 85 outside). The new MPG has $F = 2(V - 2) = 2(62) = 124$ faces (62 inside and 62 outside).

Divide the total number of edges (186) by three to determine that there are 62 solid edges, 62 thick solid edges, and 62 dashed edges. All 16 of the edges along the border of the subgraph are dashed edges. Subtract 16 from 62 to see that there are 46 dashed edges left. This means that there are 31 solid edges, 31 thick solid edges, and 23 dashed edges inside of the subgraph and 31 solid edges, 31 thick solid edges, and 23 dashed edges outside of the subgraph (divide the numbers 62 and 46 each by two to determine this).

7. The subgraph has 11 + 14 = 25 vertices (11 along its border and 14 inside). The subgraph also has 61 edges (11 along its border and 50 inside). One way to determine this is to first answer the questions regarding the new MPG (see below). Subtract 11 from 111 to get 100. Divide 100 by two to get 50. There are 50 edges inside of the subgraph plus 11 edges along the border of the subgraph, which makes 61 edges. The subgraph has 37 faces: divide the number of faces for the new MPG (which is 74; see below) by two.

The new MPG has 11 + 14 + 14 = 39 vertices (11 along the border, 14 inside, and 14 outside). The new MPG has E = 3(V − 2) = 3(37) = 111 edges (11 along the border, 50 inside, and 50 outside). The new MPG has F = 2(V − 2) = 2(37) = 74 faces (37 inside and 37 outside).

Divide the total number of edges (111) by three to determine that there are 37 solid edges, 37 thick solid edges, and 37 dashed edges. All 11 edges along the border of the subgraph can't be dashed edges because then we would need to divide 37 by two to determine the number of solid edges inside and outside of the subgraph (and similarly for the thick solid edges).

The obvious reason that there can't be a closed loop of 11 dashed edges if the three-edges theorem is to be satisfied is that a Kempe chain (such as B-G) is two-colored, but a closed loop with an odd number of edges isn't two-colorable.

8. Our solution below uses thick solid edges for the closed loop. In each case, we must either have two solid edges or two dashed edges in the same face inside of the loop.

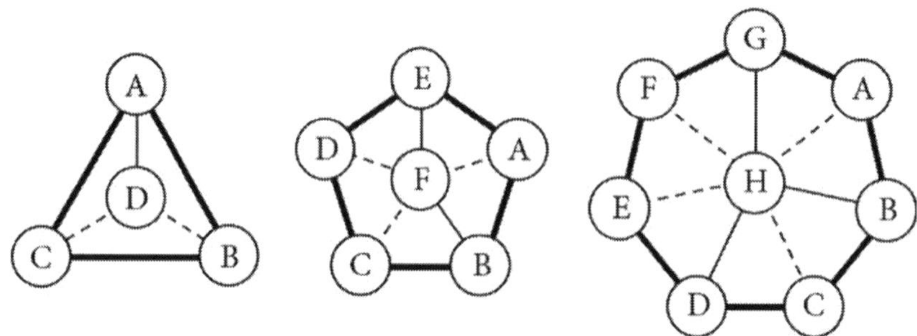

9. The new MPG formed by duplicating the inside vertices and edges and placing them outside is shown below on the left. The coloring for the original MPG shown below on the right also works for the inside and outside of the new MPG on the left.

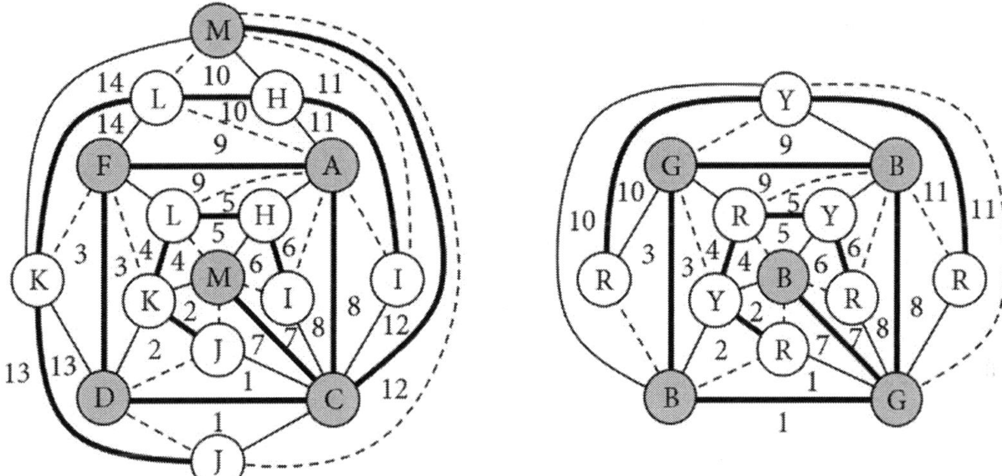

10. The new MPG formed by duplicating the inside vertices and edges and placing them outside is shown below on the left. We must either have two solid edges or two dashed edges in the same face inside of the loop, like faces AGL and AKL below. No, this doesn't violate the three-edges theorem. You can't make every edge of AHIEF a thick solid edge (and still satisfy the three-edges theorem), but you can satisfy the three-edges theorem by making two of the edges of AHIEF **not** thick solid edges, as shown below on the right.

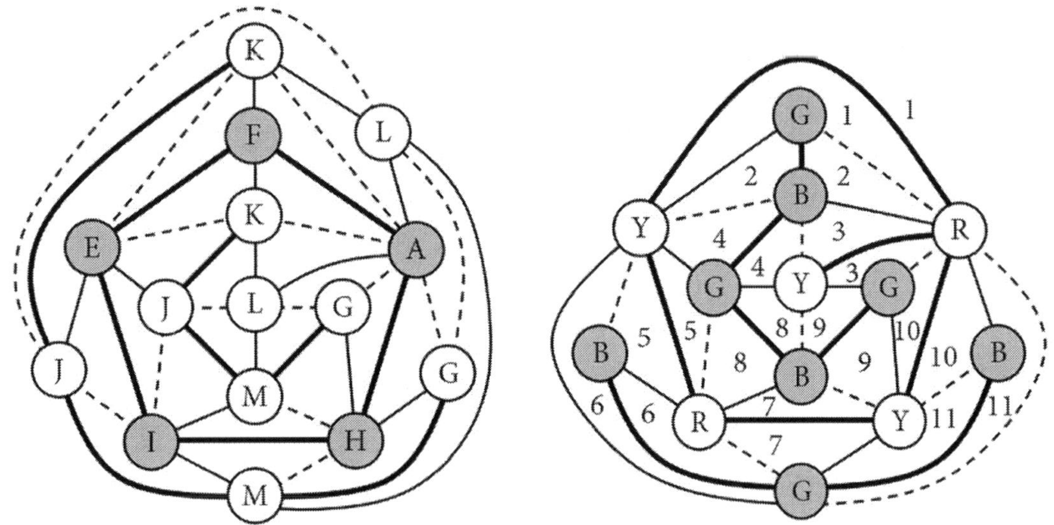

11. A MPG is triangulated, meaning that every face has three edges. A PG that isn't a MPG has at least one face with more than three edges. For any face that has four or more edges, there is no way that every edge of the face can be a different type of edge since there are only three types of edges used in the three-edges theorem.

12. One possible solution is shown below. Compare the structures of the different edge types.

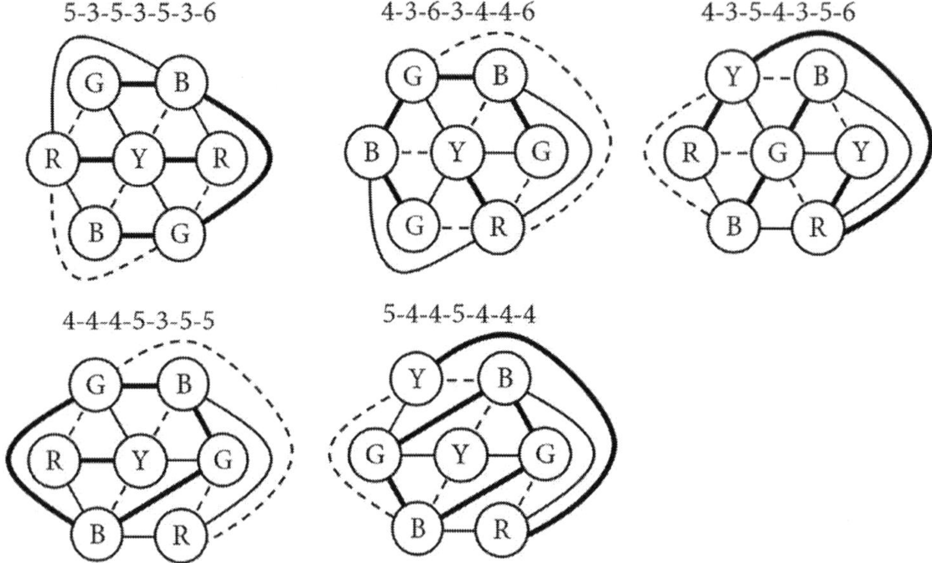

Challenge Problems: The answer key doesn't include answers to challenge problems. These problems are intended to encourage you to think about the ideas.

Chapter 24 A Recoloring Technique

1. One possible solution is shown below. We swapped dashed and solid edges.

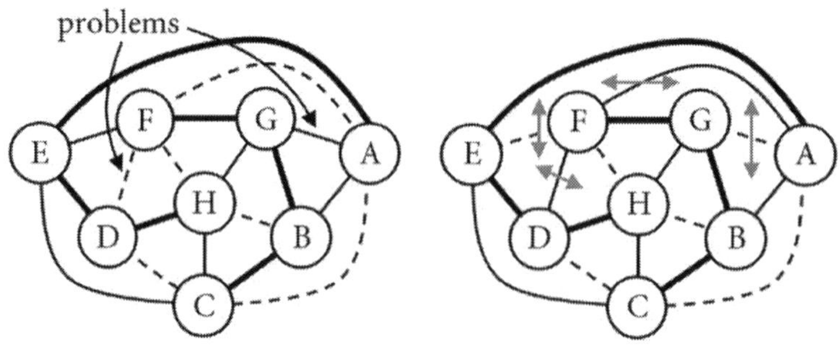

2. One possible solution is shown below. Your solution may look considerably different. We first broke open the square of thick solid edges (ABCD).

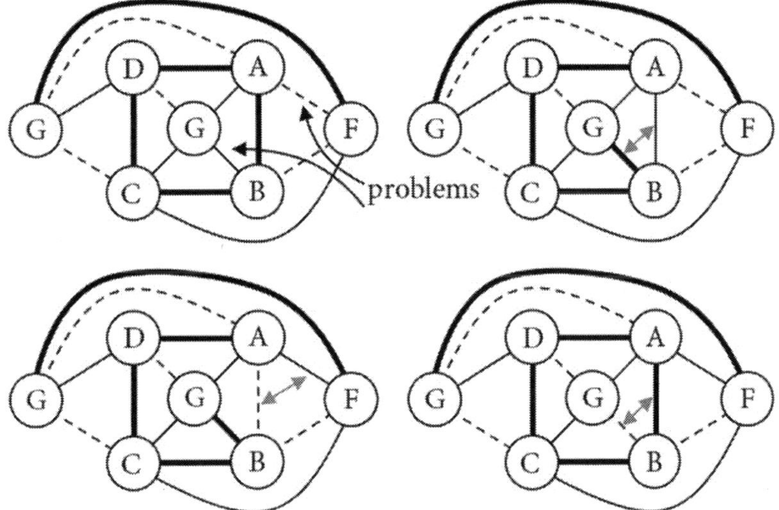

3. One possible solution is shown below. Your solution may look considerably different. We made Y-Y a thick solid edge so that there would be 7 edges of each kind. Our second swap broke open the closed loop of thick solid edges. Be careful not to make a swap that creates a new closed loop separating the problem faces.

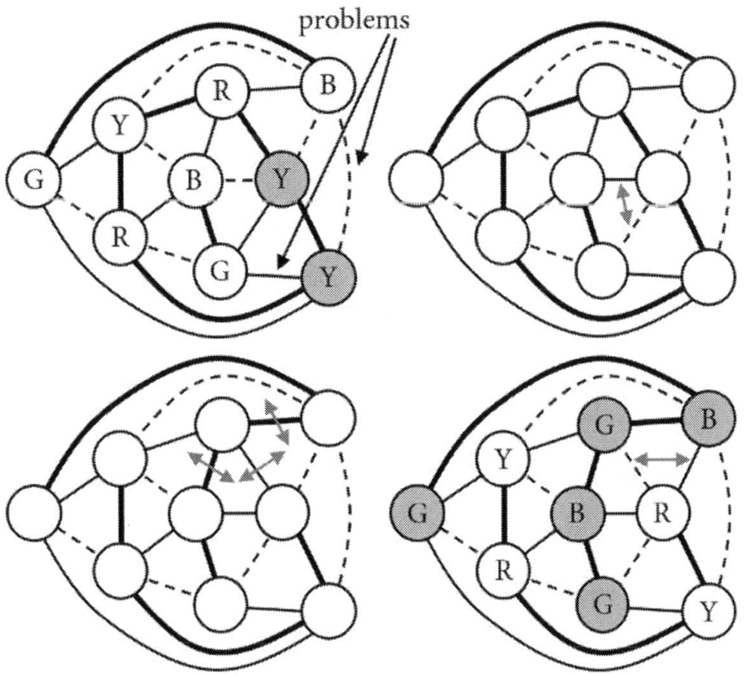

4. One possible solution is shown below. Your solution may look considerably different. We made Y-Y a solid edge so that there would be 10 edges of each kind. Our second swap broke open the closed loop of thick solid edges. Be careful not to make a swap that creates a new closed loop separating the problem faces.

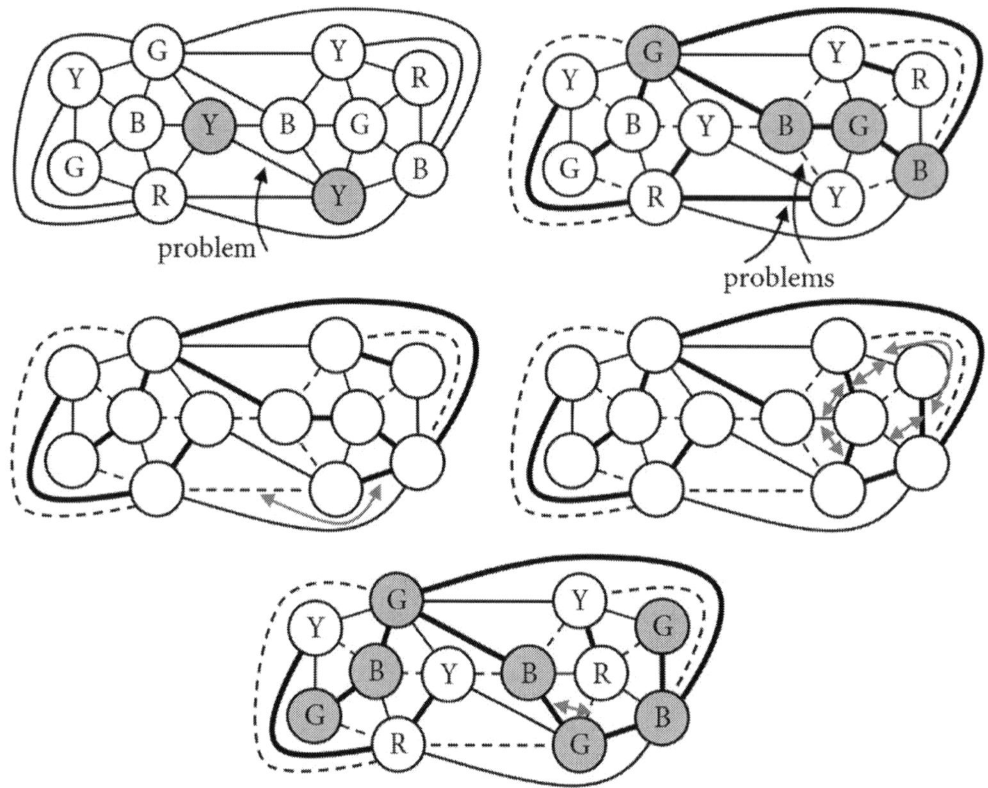

5. One possible solution is shown below. We assigned edges to the gray vertex so that there would be 17 edges of each kind. Then we swapped dashed and solid edges.

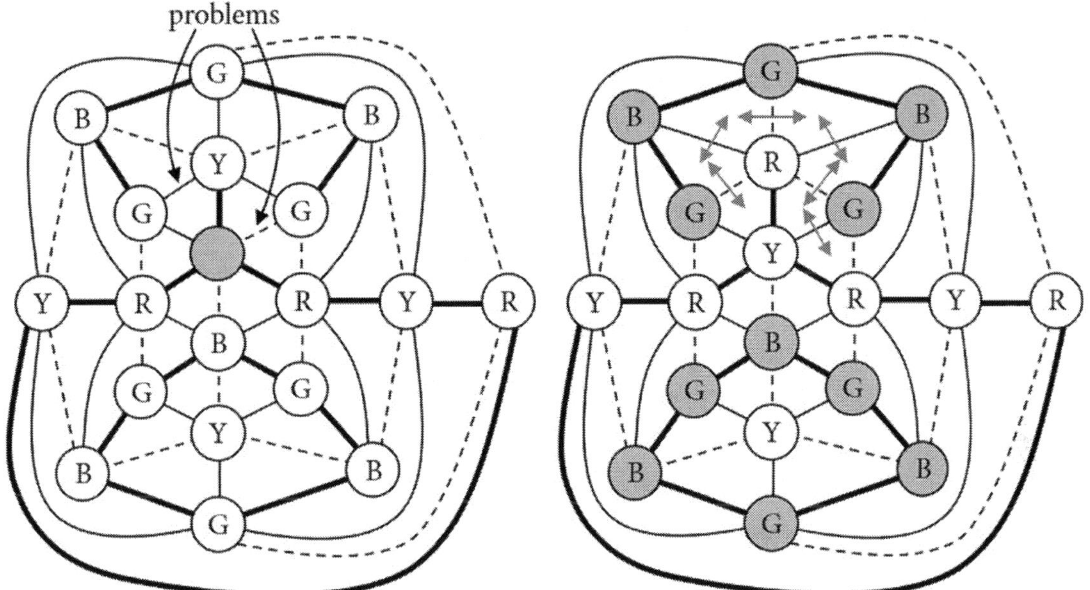

6. One possible solution is shown below. We swapped thick solid and solid edges.

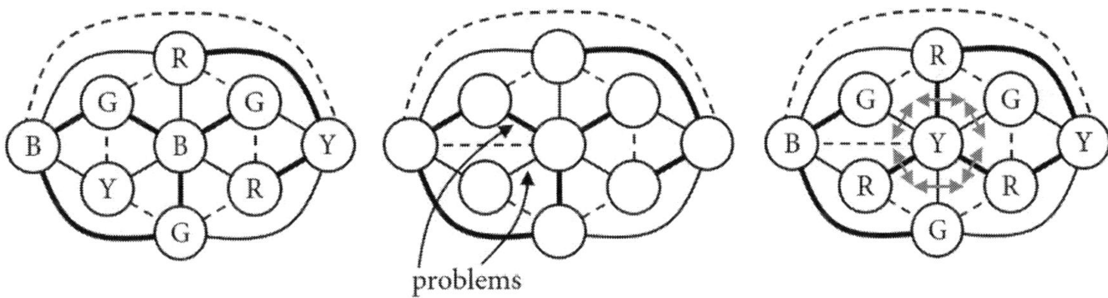

7. A simple solution is shown below.

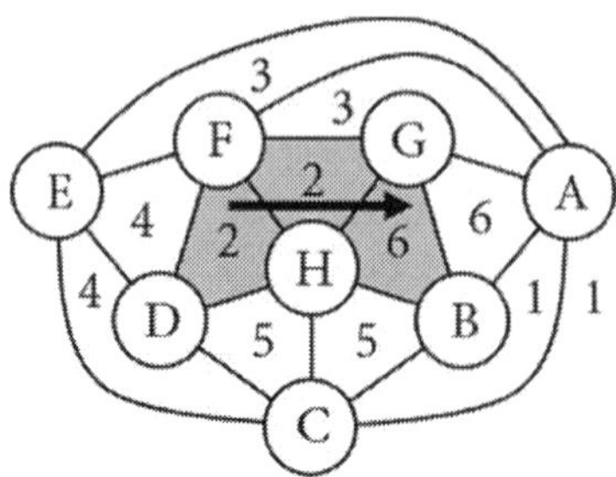

8. A simple solution is shown below.

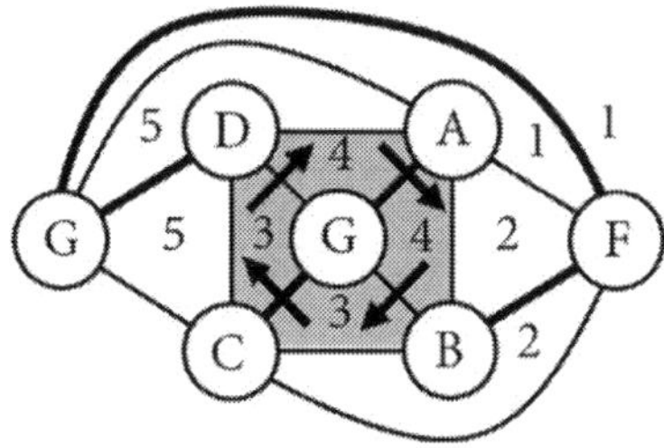

9. One possible solution is shown below. We first used the B-G and R-Y chains as the diagonals of quadrilaterals to form pairs of faces. We also treated Y-Y as the diagonal of a quadrilateral in order to make the 7th pair. (There must be 7 pairs since there are $V = 9$ vertices and $V - 2 = 9 - 2 = 7$ pairs of faces.) The left graph has a closed loop with an odd number of diagonal edges. The arrows in the right figure show how we broke open the loop without creating any island faces or new odd closed loops of diagonal edges.

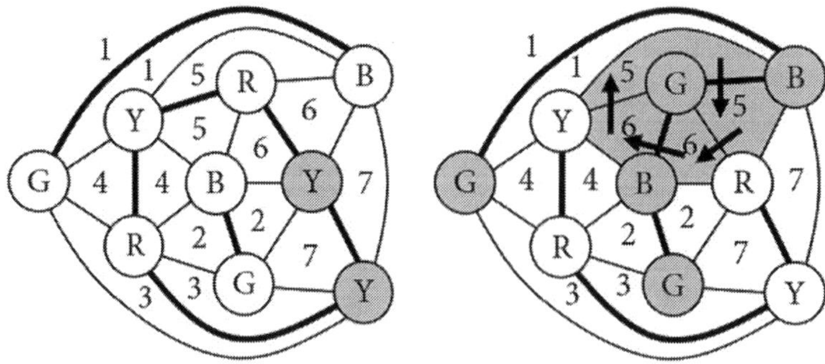

10. One possible solution is shown below. We first used the B-G and R-Y chains as the diagonals of quadrilaterals to form pairs of faces. The left graph has two island faces. The right figure shows how we moved the 10 in order to pair all of the faces. However, a closed loop with an odd number of diagonal edges was formed in the process; this loop is shaded gray in the top right figure. The bottom figure shows how we broke open the closed loop.

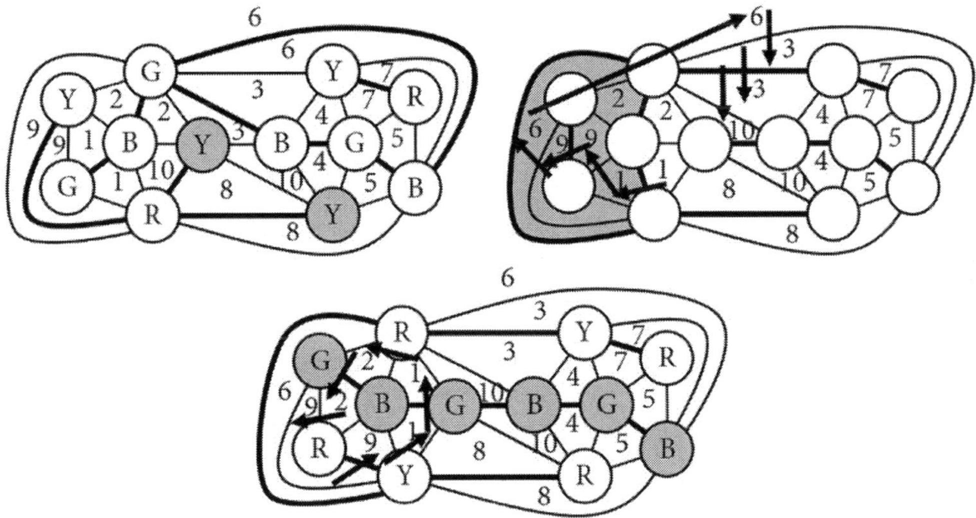

Challenge Problems: The answer key doesn't include answers to challenge problems. These problems are intended to encourage you to think about the ideas.

Chapter 25 Kempe's Problem Revisited

1. One possible solution is shown below. Your solution may look considerably different. We assigned edges to vertex X so that there would be 15 edges of each kind. Beware that it's very easy in this problem to accidentally make closed loops that separate the problem faces. Our diagram on the top left has a closed loop of 6 solid edges, which we broke open in the first step. Each of the top diagrams below has a closed loop of dashed edges; we broke both of these open in (or before) the third step.

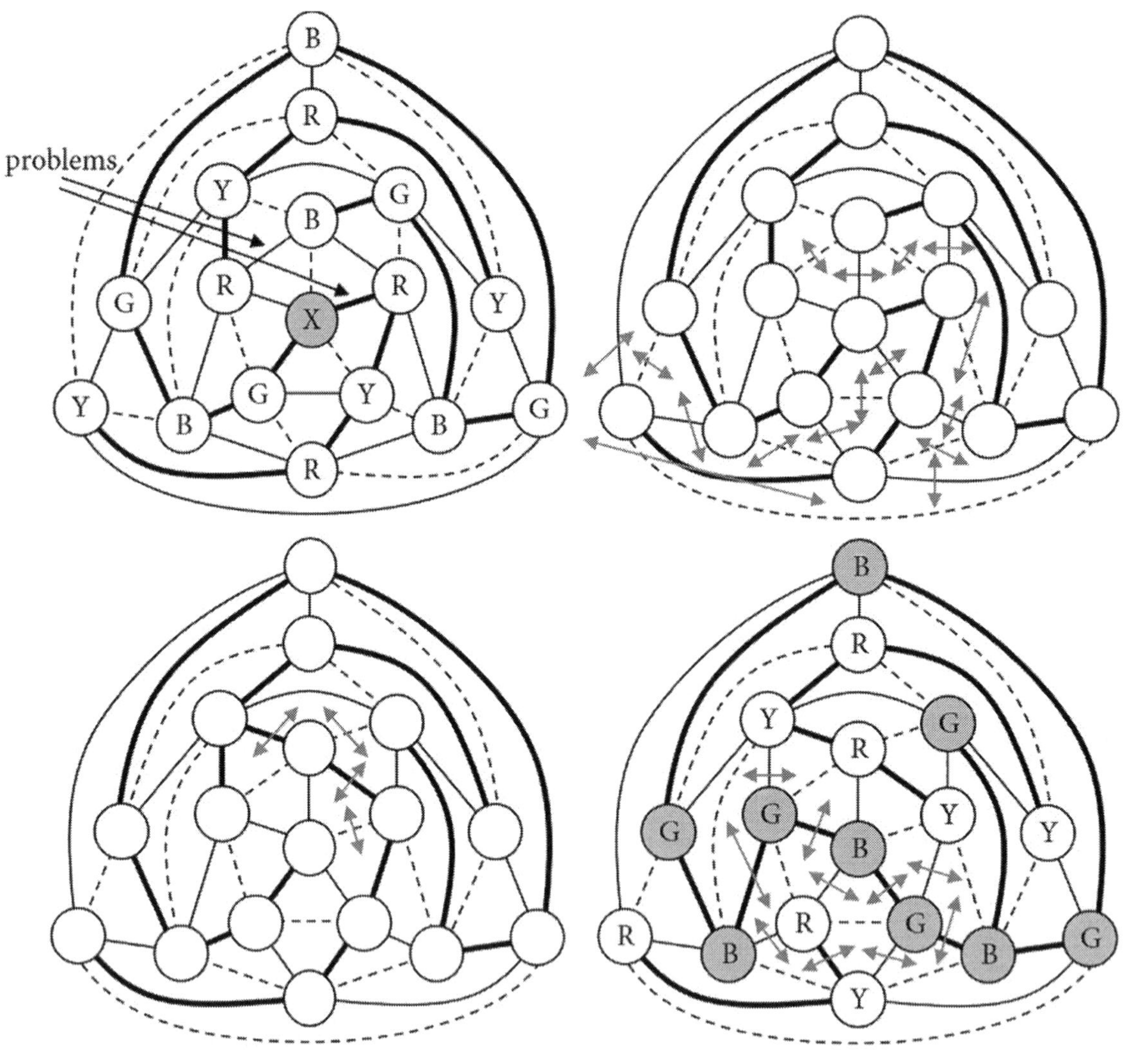

2. One possible solution is shown below. Your solution may look considerably different. We interpreted G-X and B-X as diagonal edges in order to pair the 14th and 15th faces. The left figure below has a closed loop of 9 diagonal edges. In our solution on the right, the closed loop has 8 diagonal edges, which isn't a problem since 8 is even. Beware that it's very easy in this problem to accidentally make a different closed loop with an odd number of diagonal edges in the process of breaking open another one; it's also very easy to accidentally create island faces if you're not careful with the face-sliding technique.

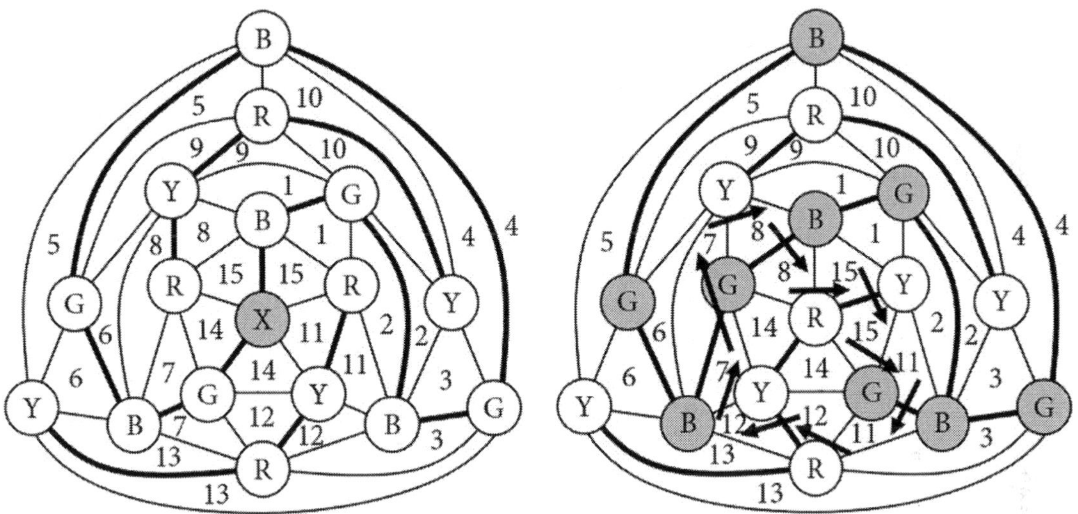

Challenge Problems: The answer key doesn't include answers to challenge problems. These problems are intended to encourage you to think about the ideas.

Chapter 26 Degrees of Separation

1. There is only one correct solution for numbering the degrees of separation from outside to inside (shown below on the left). There are multiple possible solutions for pairing faces and four-coloring the MPG; one example is shown below on the right. The thick solid edges form a closed loop in our solution, which is okay because it has an even number of edges. Note the lone Y near the top (this is a R-Y chain with just one vertex; we could have colored it R instead).

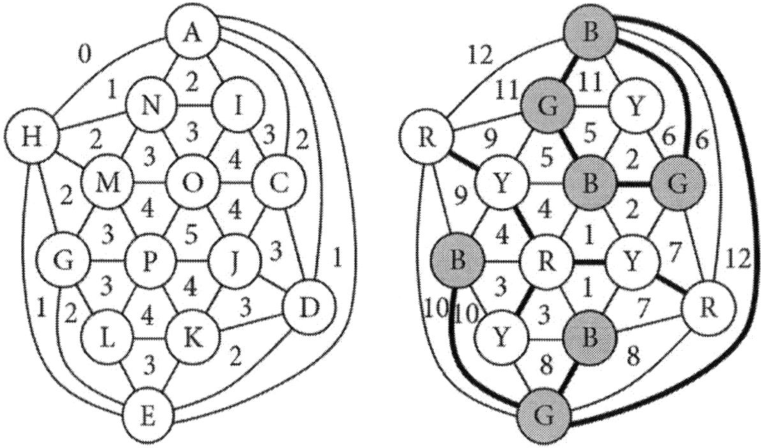

2. There is only one correct solution for numbering the degrees of separation from outside to inside (shown below on the left). There are multiple possible solutions for pairing faces and four-coloring the MPG; one example is shown below on the right. The thick solid edges form a figure eight in our solution, which is okay because each closed loop of thick solid edges (there are three: AIJG, AEJG, and AEJI) has an even number of edges. Note the two lone Y's (these are separated R-Y chains with just one vertex each; we could have colored either one, or both, R instead of Y).

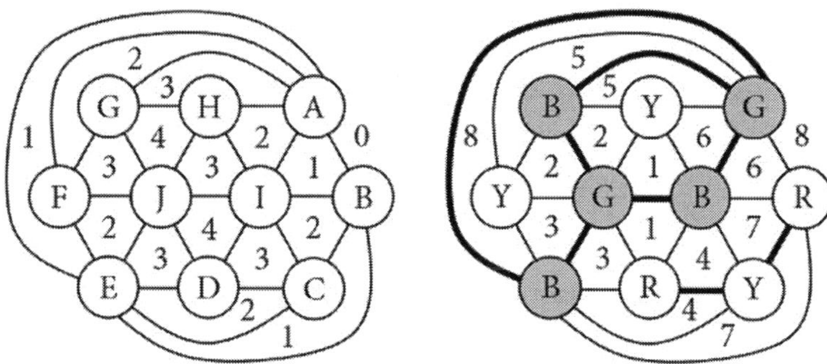

Chapter 27 A Handwaving "Proof" of the 4CT

1. The top left graph is "none of the above" since ACDFG forms the K_5 subgraph shown below.

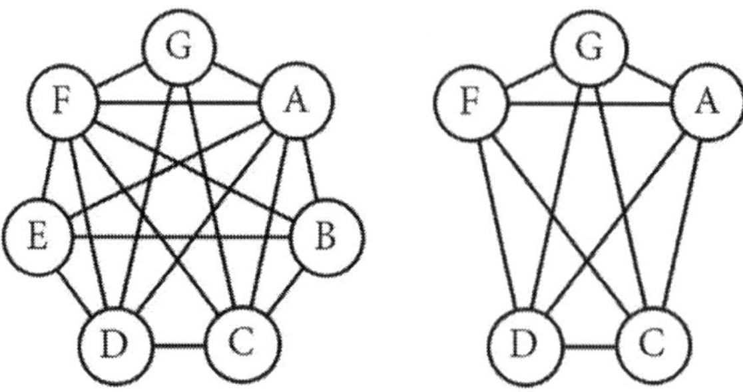

The top right graph is a K_5-less graph, but it isn't a maximal K_5-less graph and it isn't a complete tetrapartite graph. It doesn't contain a K_5 subgraph; there isn't a set of five vertices where every vertex connects to all four of the other vertices. This K_5-less graph isn't maximal since it is possible to add G_1R_2 in the graph below without making a K_5 subgraph. It isn't a complete tetrapartite graph because it isn't a maximal K_5-less graph. (However, this graph could be formed by removing edges from a complete tetrapartite graph. It is a tetrapartite graph, just not a *complete* one.)

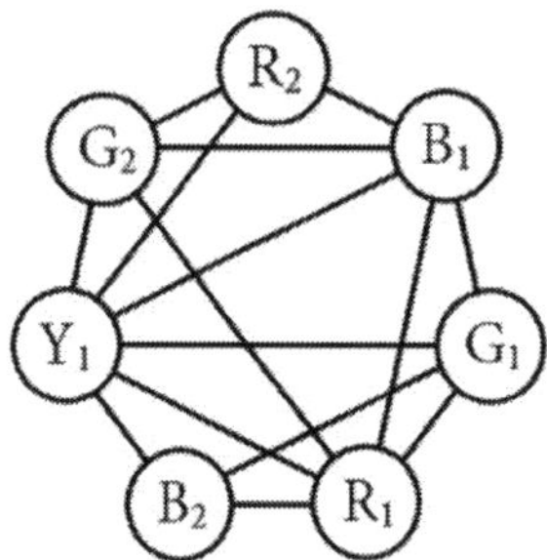

The bottom left graph is "all of the above." It is a maximal K_5-less graph since it doesn't contain a K_5 subgraph and because adding *any* single edge (without making a double edge) will make a K_5 subgraph. Since it is four-colorable, as shown below, in addition to being a maximal K_5-less graph, it is a complete tetrapartite graph.

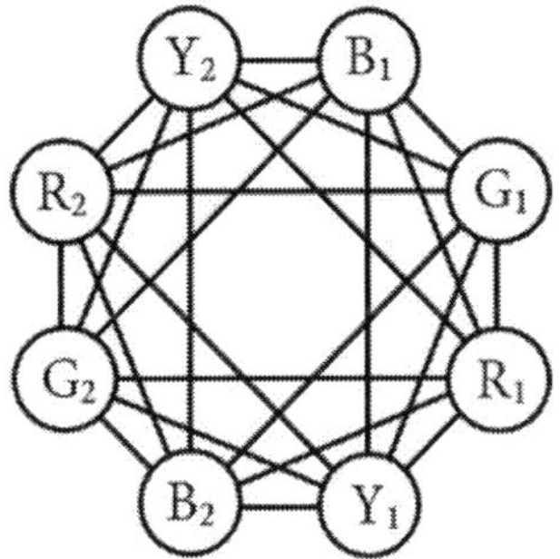

The bottom right graph is a maximal K_5-less graph (and is therefore also a K_5-less graph). It doesn't contain a K_5 subgraph; there isn't a set of five vertices where every vertex connects to all four of the other vertices. To see this in the graph below, note that G_2 and R_1 don't connect to X_1 and R_2 doesn't connect to G_1, so that no G or R is part of a K_5 subgraph. This K_5-less graph is maximal in the sense that adding a single edge (without creating a double edge) will make a K_5 subgraph. For example, adding edge G_1R_2 would make the K_5 subgraph $B_1G_1R_2X_1Y_1$, adding edge G_1G_2 would make the K_5 subgraph $B_1G_1G_2R_1Y_1$, and adding edge R_1R_2 would make the K_5 subgraph $B_1G_2R_1R_2Y_1$. It isn't a complete tetrapartite graph because it isn't four-colorable. In the graph below, B_1, G_1, and R_1 must be three different colors since they all connect to one another. This forces the colors Y_1, B_2, G_2, R_2, B_3, and B_4. The last vertex connects to four different colors, so a fifth color is needed for it (labeled X_1).

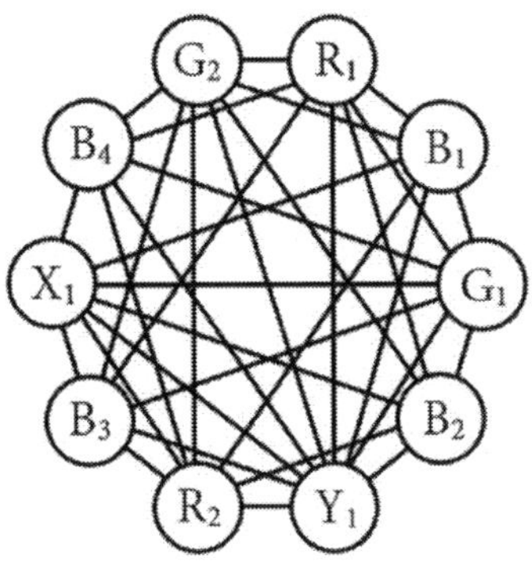

2. One possible solution is shown below. There are other possible solutions. Check that your solution meets the following criteria:

- The graph must have 8 vertices.

- No set of five vertices has every vertex connected to all four of the other vertices. For example, in the graph below, neither G_2 nor R_1 is part of a K_5 subgraph because neither connects to the only color labeled X, and neither G_1 nor R_2 is part of a K_5 subgraph because while they connect to X they don't connect to each other. Green and red are thus not part of any K_5 subgraph (and neither are the other colors).

- If you add any one of the missing edges, it must form a K_5 subgraph. In the graph below, adding B_1B_2 would make $B_1B_2G_1X_1Y_1$ a K_5 subgraph, adding G_1G_2 would make $B_1G_1G_2R_1Y_1$ a K_5 subgraph, adding G_1R_2 would make $B_1G_1R_2X_1Y_1$ a K_5 subgraph, adding R_1R_2 would make $B_1G_2R_1R_2Y_1$ a K_5 subgraph, adding R_1X_1 would make $B_1G_1R_1X_1Y_1$ a K_5 subgraph, and adding G_2X_1 would make $B_1G_2R_2X_1Y_1$ a K_5 subgraph.

- Check that the graph isn't four-colorable. Set three colors for one face and check that a fifth color is required. For example, if X_1 doesn't connect to at least one vertex of every color, your graph is four-colorable; check the coloring carefully.

- Rest your eyes, refresh your brain, and check again. When trying to draw a maximal K_5-less graph that isn't a complete tetrapartite graph, it's really easy to inadvertently draw a graph that (A) contains a K_5 subgraph, (B) is actually four-colorable, or (C) makes it possible to add an edge without making a K_5 subgraph (or a double edge). **<u>Check your solution thoroughly and carefully</u>**.

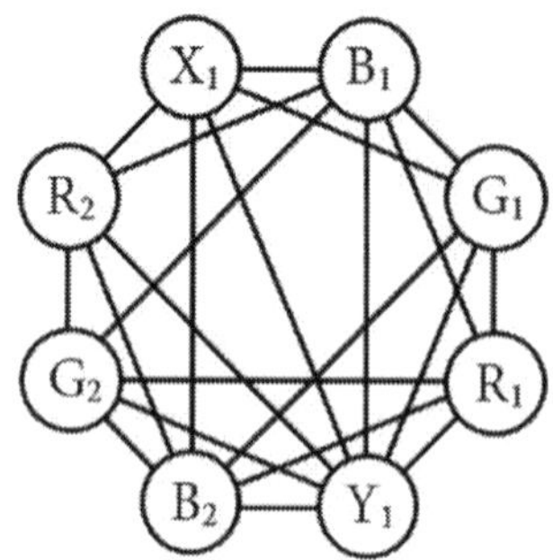

3. The left graph below doesn't have a K_5 subgraph, but does have a subgraph that is a subdivision of K_5, and it is also four-colorable. The middle graph was formed by removing edges from the left graph in order to make it easier to see that the left graph has a subgraph that is a subdivision of K_5. The right graph is a complete tetrapartite graph formed by adding edges to the left graph.

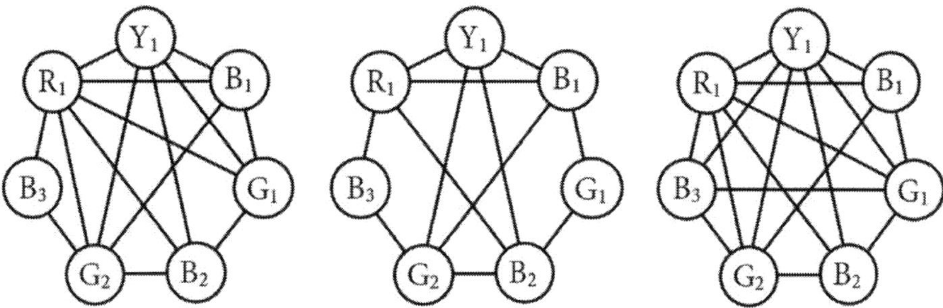

The graph below doesn't have a K_5 subgraph, but does have a subgraph that is a subdivision of K_5, yet it isn't four-colorable. The right graph was formed by removing edges from the left graph in order to make it easier to see that the left graph has a subgraph that is a subdivision of K_5. To see that the left graph isn't four-colorable, first note that $B_2 G_1 R_1 Y_1$ form a K_4 subgraph so these must be four different colors. All of the other colors other than X_1 follow from this. Then X_1 connects to four different colors, requiring X_1 to be a fifth color.

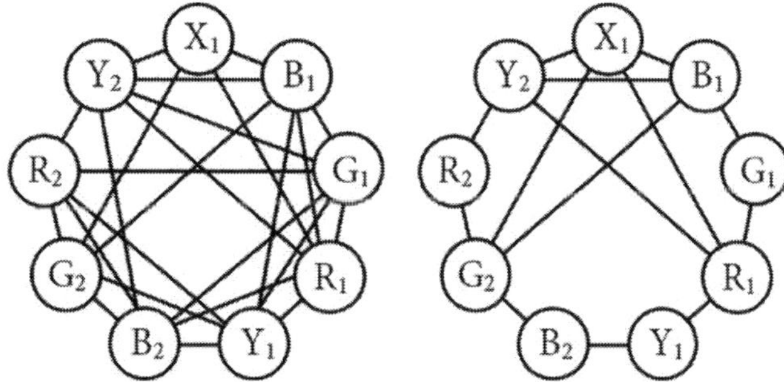

This exercise shows that a graph which has a subgraph that is a subdivision of K_5 may or may not be four-colorable, whereas a graph which has a K_5 subgraph definitely isn't four-colorable.

4. The middle graph below is a maximal K_5-less graph that isn't a complete tetrapartite graph which was formed by adding edges to the given graph (shown to the left). To see that the middle graph is K_5-less, note that G_1 and R_2 aren't part of any K_5 subgraphs since there is no G_1R_2 edge and G_2 and R_1 aren't part of any K_5 subgraphs since there are no R_1X_1 or G_2X_1 edges, which means that no greens or reds are part of K_5 subgraphs. To see that it is maximal, note that adding B_1B_2, B_1B_3, B_1B_4, B_2B_3, B_2B_4, or B_3B_4 would make $BBG_1R_1Y_1$ a K_5 subgraph, adding G_1G_2 would make $B_1G_1G_2R_1Y_1$ a K_5 subgraph, adding G_1R_2 would make $B_1G_1R_2X_1Y_1$ a K_5 subgraph, adding G_2X_1 would make $B_1G_2R_2X_1Y_1$ a K_5 subgraph, adding R_1R_2 would make $B_1G_2R_1R_2Y_1$ a K_5 subgraph, and adding R_1X_1 would make $B_1G_1R_1X_1Y_1$ a K_5 subgraph. To see that it isn't a complete tetrapartite graph, note that it isn't four-colorable. Once the colors B_1, G_1, and R_1 are set, all of the other colors besides X_1 follow, and then X_1 must be a fifth color since it connects to all four colors.

The right graph is a complete tetrapartite graph which was formed by adding edges to the given graph.

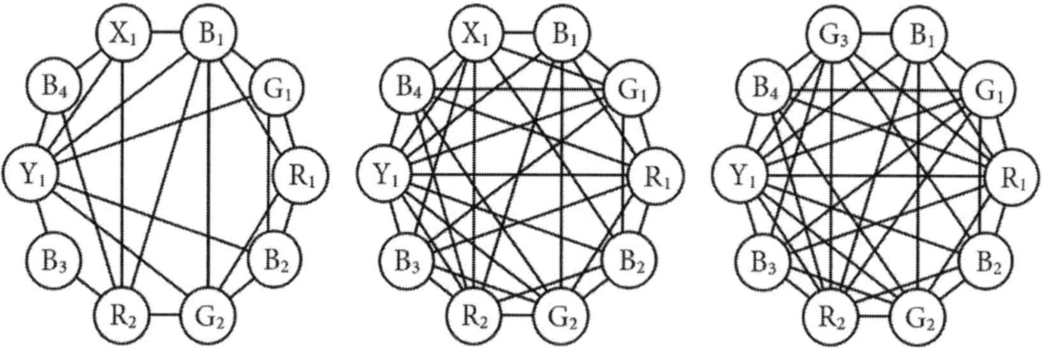

5. Note that $K = 8$, $L = 6$, $M = 5$, $N = 3$. There are $KLMN = (8)(6)(5)(3) = (48)15) = 720$ K_4 subgraphs. Every blue vertex participates in $LMN = (6)(5)(3) = 90$ K_4 subgraphs, every green vertex participates in $KMN = (8)(5)(3) = 120$ K_4 subgraphs, every red vertex participates in $KLN = (8)(6)(3) = 144$ K_4 subgraphs, and every yellow vertex participates in $KLM = (8)(6)(5) = 240$ K_4 subgraphs.

6. One possible solution is given for each graph. Note that these are not the only solutions. The edges that need to be removed from the complete tetrapartite graphs are shown as thick black edges.

- Our solution for the top left graph has $K = 1$, $L = 2$, $M = 2$, and $N = 1$. Just one edge needs to be removed.

- Our solution for the top right graph has $K = 2$, $L = 1$, $M = 2$, and $N = 2$. Three edges need to be removed.

- Our solution for the bottom left graph has $K = 1$, $L = 1$, $M = 5$, and $N = 5$. We need to remove 16 edges.

- Our solution for the bottom right graph (the icosahedral graph) has $K = L = M = N = 3$. We need to remove 24 edges.

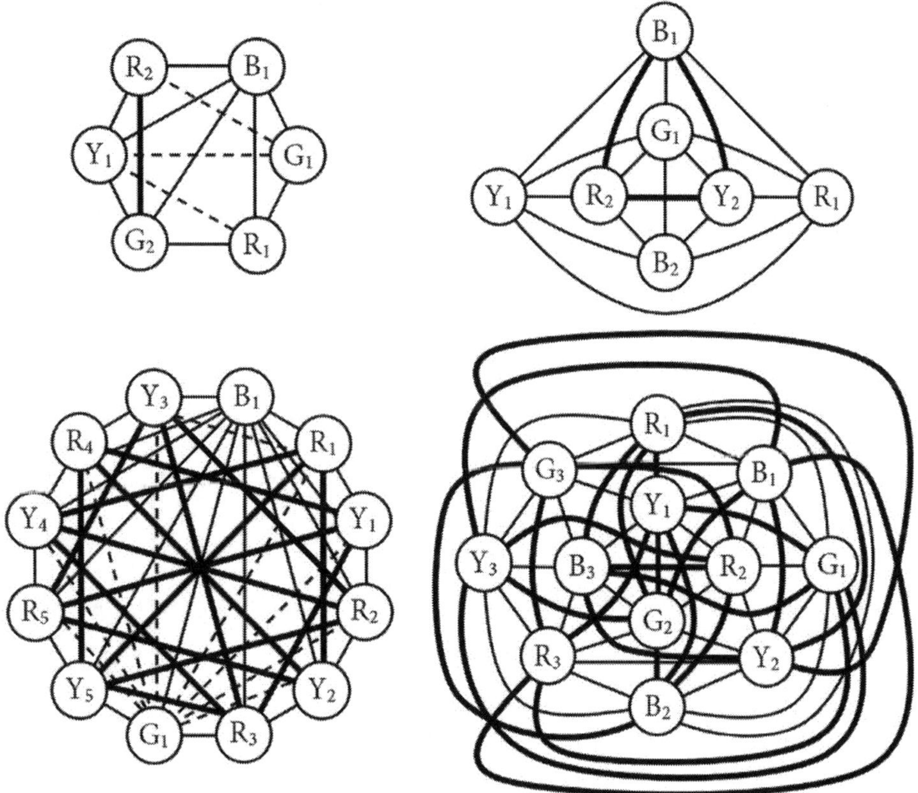

Challenge Problems: The answer key doesn't include answers to challenge problems. These problems are intended to encourage you to think about the ideas.

Chapter 28 Random Notes

1. The serious "problem" is that this graph is disconnected. Below, the same graph has been separated into two different graphs.

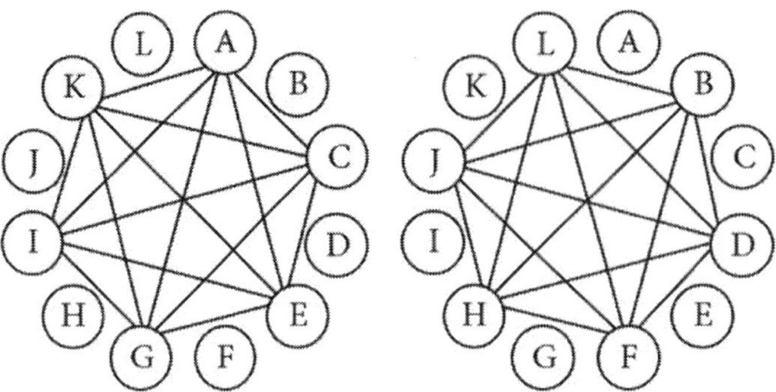

The sum of the degrees of the given "graph" is 12(5) = 60. Divide this by two to determine the number of edges: 60/2 = 30 edges. A MPG with V = 12 vertices would have E = 3V – 6 = 3(12) – 6 = 36 – 6 = 30 edges. The given "graph" does have the right number of edges to be a MPG, but since the graph is disconnected (it's really two separate graphs), it isn't a MPG. The two separated graphs aren't MPG's either. For each separated graph, the sum of the degrees is 5(6) = 30, which means that each separated graph has 30/2 = 15 edges. A MPG with V = 6 vertices would have E = 3V – 6 = 3(6) – 6 = 18 – 6 = 12 edges. Each separated graph has 3 too many edges to be a MPG. The separated graphs are nonplanar (Chapter 6).

2. Vertices A and C in the quadrilateral shown have three edges. Every vertex in a MPG (with at least 4 vertices) connects to the diagonal of a quadrilateral (just like A and C do for the quadrilateral shown). Therefore, every vertex has at least three edges (like A and C).

3. Every edge in any MPG is the diagonal of some quadrilateral such as the one shown for Problem 2. Let vertices A and C represent any two vertices that share an edge in any MPG. If the MPG has at least 5 vertices, we'll see that A and C can't both remain degree 3 (and thus neither can any other pair of vertices connected by an edge) once the graph is triangulated.

Once the graph is triangulated, only one edge may join B and D; we can't join B and D both above and below the quadrilateral or we would make a double edge. We can form triangle ABD or BCD, but not both. Either above or below (or both) the quadrilateral, we will need to connect B or D to some other vertex, and in order to make every face a triangle, we will have to connect another vertex to A (if above) or C (if below), which prevents A and C from both remaining degree 3.

4. The vertex that has degree $V - 1$ connects to every other vertex in the graph. For example, if the vertex with degree $V - 1$ is blue, this means that no other vertex can be blue (provided that the entire graph will be four-colored and that the four-color theorem is true). Therefore, these $V - 1$ vertices must be three-colorable.

Let one of the vertices with degree $V - 1$ be red. No other vertex can be red because it connects to every other vertex. Let the other vertex with degree $V - 1$ be blue. No other vertex can be blue because it connects to every other vertex. The remaining $V - 2$ vertices must be green and yellow (two-colorable). See the solution to Problem 5 from Chapter 12.

5. We can't remove more than V − 4 edges from a single vertex if we don't want the vertex to have a degree less than 3. Try it with numbers. Suppose there are V = 18 vertices. In the complete graph, every vertex originally had degree V − 1 = 18 − 1 = 17 because it connects to all 17 of the other vertices. In this case, V − 4 = 18 − 4 = 14. If we remove 14 edges from a single vertex, that vertex will now be degree 17 − 14 = 3. This isn't less than 3.

We can't remove more than V − 6 edges from a single vertex if we don't want the vertex to have a degree less than 5. Try it with numbers. Suppose there are V = 18 vertices. In the complete graph, every vertex originally had degree V − 1 = 18 − 1 = 17 because it connects to all 17 of the other vertices. In this case, V − 6 = 18 − 6 = 12. If we remove 12 edges from a single vertex, that vertex will now be degree 17 − 12 = 5. This isn't less than 5.

Challenge Problems: The answer key doesn't include answers to challenge problems. These problems are intended to encourage you to think about the ideas.

Index

Was This Book Helpful?

A great deal of effort and thought was put into this book, such as:

- Numerous diagrams to help illustrate the concepts.
- Exploring a wide variety of fascinating ideas relating to the four-color theorem.
- Including relevant topics from graph theory without assuming any background.
- Breaking down the examples and solutions to help make the ideas clear.
- Careful selection of examples and problems for their instructional value.
- Full solutions to the problems at the back of the book.

If you appreciate the effort that went into making this book possible, there is a simple way that you could show it:

<u>Please take a moment to post an honest review.</u>

For example, you can review this book at Amazon.com or Goodreads.com.

Even a short review can be helpful and will be much appreciated. If you're not sure what to write, following are a few ideas, though it's best to describe what's important to you.

- Did you learn anything about basic graph theory?
- Did you learn anything about the four-color theorem?
- Were the solutions at the back of the book helpful?
- Would you recommend this book to others? If so, why?

Do you believe that you found a mistake? Please email the author, Chris McMullen, at greekphysics@yahoo.com to ask about it. One of two things will happen:

- You might discover that it wasn't a mistake after all and learn why.
- You might be right, in which case the author will be grateful and future readers will benefit from the correction. Everyone is human.

About the Author

Dr. Chris McMullen has over 20 years of experience teaching university physics in California, Oklahoma, Pennsylvania, and Louisiana. Dr. McMullen is also an author of math and science workbooks. Whether in the classroom or as a writer, Dr. McMullen loves sharing knowledge and the art of motivating and engaging students.

The author earned his Ph.D. in phenomenological high-energy physics (particle physics) from Oklahoma State University in 2002. Originally from California, Chris McMullen earned his Master's degree from California State University, Northridge, where his thesis was in the field of electron spin resonance.

As a physics teacher, Dr. McMullen observed that many students lack fluency in fundamental math skills. In an effort to help students of all ages and levels master basic math skills, he published a series of math workbooks on arithmetic, fractions, long division, algebra, geometry, trigonometry, and calculus entitled *Improve Your Math Fluency*. Dr. McMullen has also published a variety of science books, including astronomy, chemistry, and physics workbooks.

Author, Chris McMullen, Ph.D.

A Fourth Dimension of Space

The Visual Guide to Extra Dimensions includes numerous diagrams (many unique) of the fourth dimension of space. Learn about hypercubes, polytopes, hypersurfaces, and more. Volume 1 focuses on geometry, while Volume 2 gets into the physics of extra dimensions.

Full-Color Illustrations of the Fourth Dimension offer some cool and unique glimpses into the nature of the fourth dimension. Volume 2 helps you visualize what it would be like to live in a universe with four dimensions of space.

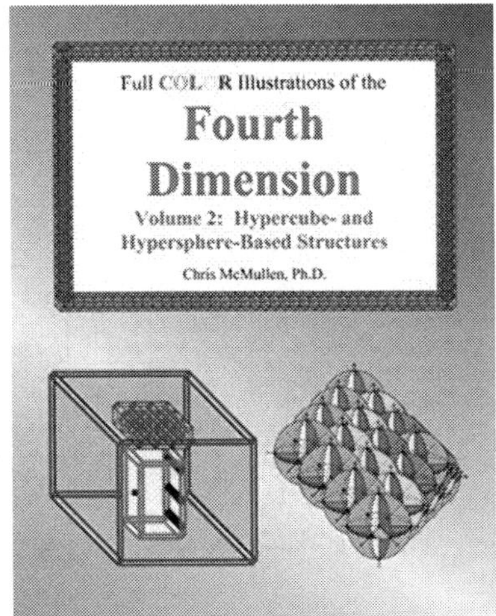

Enjoy Puzzles?

300+ Mathematical Pattern Puzzles introduces several different types of mathematical puzzles, with a different topic covered in each chapter.

Pyramid Math Puzzle Challenge is more challenging. You don't know which way the pattern will go, and the underlying ideas vary considerably from puzzle to puzzle.

Got Math?

Want to refresh essential calculus or trigonometry skills, put your calculus skills to the challenge, or test your general math knowledge?

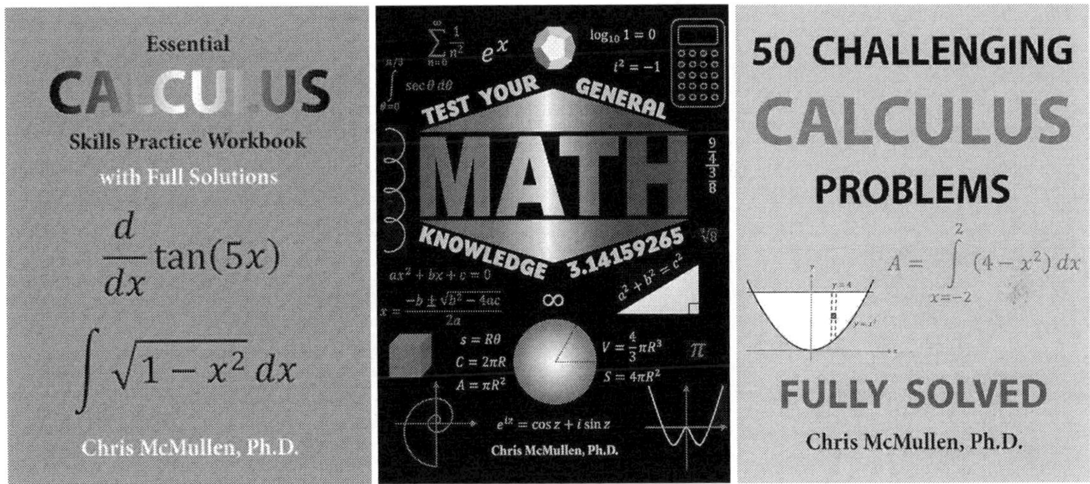

The *Improve Your Math Fluency* series includes algebra, geometry, calculus, trigonometry, long division, fractions, arithmetic, and more. Search for math books by Chris McMullen.

Science

Learn or review essential physics concepts and problem-solving skills (calculus-based and trig-based levels available). Search for physics books by Chris McMullen (also includes Electricity & Magnetism, Modern Physics, and Thermal Physics).

Review chemistry or astronomy concepts, or practice balancing chemical reactions (which make for good math puzzles).

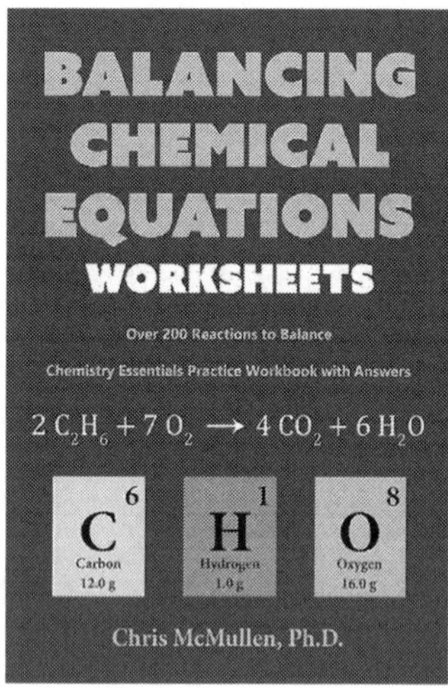

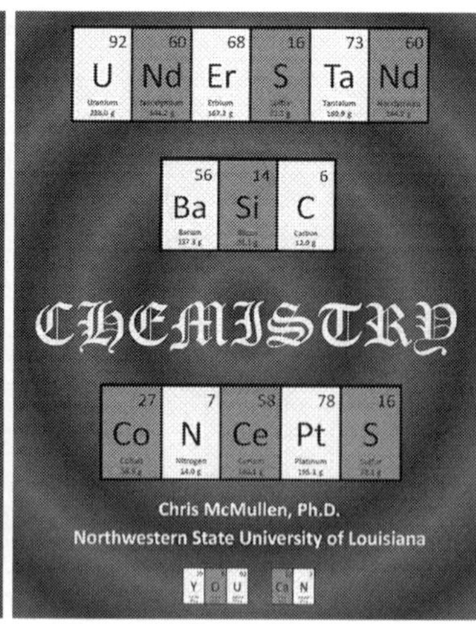

Anagrams for Math Lovers

If you enjoy math and you enjoy anagrams, VErBAl ReAcTiONS is a rather unique type of puzzle that may appeal to you. These puzzles look like chemical reactions, but they are really just word jumbles with a twist. Each word was formed by using elements from the periodic table. For example, the word BRaIn consists of three elements (B for boron, Ra for radium, and In for Indium). If a "chemical word" has two or more of the same element, we use this as a coefficient to set up the VErBAl ReAcTiON. For example, consider the puzzle below.

$$C + 3S + Ta + 2Ti \rightarrow \underline{\ }\,\underline{\ }\,\underline{\ }\,\underline{\ }\,\underline{\ }\,\underline{\ }\,\underline{\ }\,\underline{\ }$$

This puzzle is telling you that the answer has 1 C, 3 S's, 1 Ta, and 2 Ti's. We can rearrange C, S, S, S, Ta, Ti, and Ti to form the chemical word STaTiSTiCS. These puzzles are available in Easy, Medium, and Hard levels.

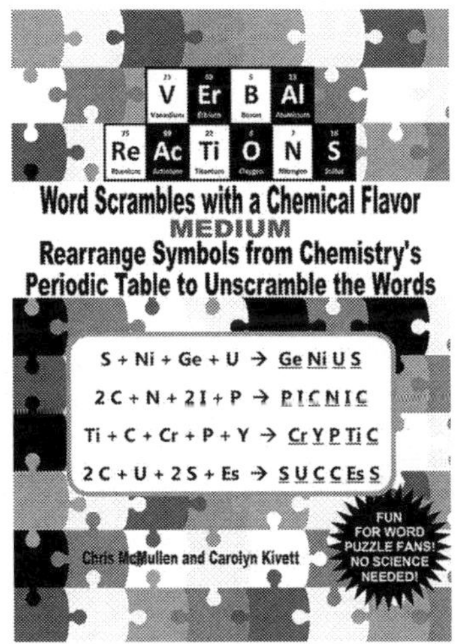

Made in the USA
San Bernardino, CA
08 June 2020

72830275R00237